GEOLOGY OF NATIONAL PARKS

GEOLOGY OF NATIONAL PARKS

ANN G. HARRIS
YOUNGSTOWN STATE UNIVERSITY

KENDALL/HUNT PUBLISHING COMPANY
DUBUQUE, IOWA

With the exception of the maps appearing on pages 28 and 32 the map reproductions contained herein are maps of the national parks from *A Guide to the National Parks—Their Landscape and Geology* by William H. Matthews III. Copyright © 1968, 1973 by William H. Matthews III. Reprinted by permission of Doubleday & Company, Inc.

Cover design: Scratchboard drawing by Cheri Mohn is from Glacier-Waterton National Peace Park.

Library of Congress Catalog Card Number: 74-25041

ISBN 0—8403—1092—7

Printed in the United States of America

CONTENTS

Kendall/Hunt Pub Co.

4/95

7.95

PREFACE

This book came about because of the questions that students asked. Many of them, after being exposed to the beginning courses, would go on vacations or trips and see the geologic features they learned about. They came back filled with questions, many of which I was unable to answer, about the National Parks.

Because of the high interest, the Department decided to offer a course on Geology of National Parks. At first it was offered on an experimental basis under the Continuing Education Department. But later it became a credit course under the regular curriculum because of the tremendous response.

The individual chapters are divided into three sections that are more or less independent of one another. The first section contains the local history and is not really related to the geology but builds up the background of the region. I found it helped to make the area a reality to the students.

The second—geologic features—explains many (but by no means all) of the features that can be seen in the parks. How they formed and why they are there. The instructor can, if he wishes, put the emphasis on this section and treat it more like a Physical Geology course. The glossary in the back defines the terms that appear in boldface print so if the chapters are discussed out of sequence the students have a source to learn the meaning of unfamiliar terminology.

The last section—geologic history—takes a person step-by-step through the sequence of events that lead to the formation of the park. Emphasis could be placed on this section if a more historical approach is desired. The sequence of events is outlined in the bold-faced captions.

I have found in this format the students have an easier time to relate to the subject matter, especially if they have been lucky enough to have visited some of these parks. However, the students who have never been exposed to geology before this course have no trouble comprehending the subject matter.

Showing slides of the individual parks at the end of each lecture enhances the understanding of the subject matter. These can be obtained by copying pictures out of books or borrowing slides from the students and having them duplicated. Also the use of transparencies with an overhead projector to fill in many of the gaps. I have learned that this helps the students comprehension when they can see what is being discussed.

Some of the geologic columns are only tentative because very little work has been done in the area, or it is so complex that the sequence hasn't been completely worked out. The complexity of some of the parks is the reason that some have more than one column.

The geologic timetable is as up-to-date as possible. It uses the new United States Geological Survey method of dating the Precambrian, but the old classification has been also included since most of the available literature still refers to it.

This book would have not been produced without the help of many people. I would like to thank the various park superintendents and park naturalists for their suggestions and literature they sent. Also the National Park Service for the photographs and permission to use their maps in this book.

Special thanks also go to fellow geologists Mr. Rowland Tabor, Dr. Richard Stewart, Mr. Philip F. Van Cleave and Mr. Don Esterbrook for their help in updating the geology in many of the parks. A special bouquet goes to Hildegard Schnuttgen of the Youngstown State University library for obtaining many of the references I needed.

I am deeply in debt to Donna Restivo, Jill Ansevin, Amy Horvat, Marilyn Lawson and Melcina Mooney for all the typing. My final indebtedness is to my husband Earl who put up with all the inconveniences and took over much of the care of our daughters so I could devote most of my time to writing this book.

AGH

Youngstown, Ohio
August 1974

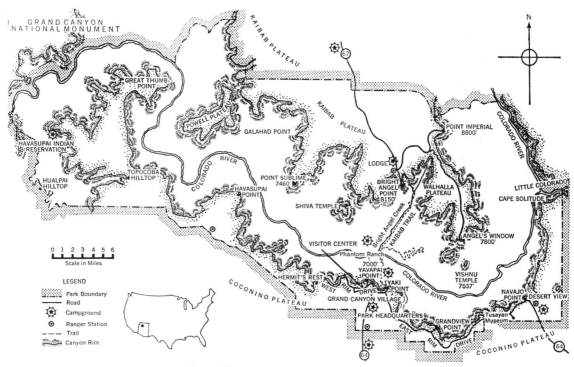

Grand Canyon National Park, Arizona.

Geologic Column of Grand Canyon National Park

Erathem	System	Formation
Paleozoic	Permian	Kaibab Formation
		Toroweap Formation
		Coconino Formation
		Hermit Shale
	Pennsylvanian	Supai Formation
	Mississippian	Redwall Limestone
	Cambrian	Muav Formation*
		Bright Angel Shale
		Tapeats Sandstone
Precambrian	Precambrian Y and Z	Chuar Group†
		Nankoweap Formation†
		Rama Formation†
		Dox Formation†
		Shinumo Sandstone†
		Hakati Shale†
		Bass Limestone†
	Precambrian W	Zoraster Granite
		Brahma Schist
		Vishnu Schnist

*Includes discontinuous layers of Devonian Temple Butte Formation.
†Member of Unkar Group.

Modified after Maxon et. al.

PART I
FORMED BY STREAM EROSION AND WEATHERING

CHAPTER 1
GRAND CANYON NATIONAL PARK

Location: N.W. Arizona
Area: 673,575 acres; 1,052 square miles
Established: 1919

LOCAL HISTORY

Around 3,000 years ago some Indian hunters took twigs and shaped them in forms which resembled animals, perhaps in the hope that the Great Spirit would help them in their hunt. This is the first record of man in the park area. These artifacts have been found in some of the caves located in Grand Canyon National Park.

Following the hunters at a later time came the culture called the Basket Makers. They were followed at a later time by the Pueblo culture who built the first apartment houses. The five hundred plus Indian sites in the park indicate they lived here around 600 years ago.

Three tribes which still call this region their home are the Navajo and Hopi east of the park and the Havasupai in a valley in the western section of the park which is irrigated by Havasu Creek. The tribal name of the Havasupai means "people of the blue-green water."

In 1540, Don Lopez de Cardenas, leader of a band of Coronado's Conquistadores, was directed by the Hopi Indians to the Gorge. They tried for three days to reach the Colorado River, but only managed to get one-third of the way down. Discouraged, they gave up, climbed back to the rim, and left.

Lt. Joseph C. Ives of the Army Topographical Engineer Corps and his party managed to reach Vegas Wash, then crossed over to Diamond Creek. From this point, after an unsuccessful attempt to reach the rim, they concluded that much of the Colorado River would never be visited.

A member of Ives party, John S. Newberry, a geologist, thought otherwise. He convinced John Wesley Powell, a one-armed Major, that it could be done. On May 24, 1869, the Powell Expedition of four boats and nine men left Green River City, Wyoming, and began their hazardous journey. One boat was smashed, scientific instruments and food were lost. Most of the time they were wet, tired, hungry, discouraged, and convinced they would never get out alive. Therefore, three of the men climbed out of the canyon and started cross-country. They were killed by the Indians on Shivwits Plateau. The men that stayed with the expedition survived.

In 1903 Theodore Roosevelt took a trip to Grand Canyon and decided it should be preserved. Unfortunately, people who had other plans for the area blocked its establishment for many years. One

1

group wanted to mine the minerals, another wanted to charge toll on Bright Angel Trail and have a cable car into the canyon. Even today there are groups that would like to dam the Colorado River and flood the park so that tourists could take boat trips to some of the isolated canyons.

GEOLOGIC FEATURES

About 100 miles of the 217 miles of the Colorado River are within the park boundaries. The river carries about a half million *tons* of sediments past any one point each day. It has been often said that the Colorado is too thick to drink but too thin to walk on.

The elevation of the North Rim (8,900') averages some 1200' higher than the South Rim (6,900'). They are only nine miles apart geographically. However, because of the differences in elevation, they are like two separate worlds. The North Rim is characterized by blue spruce, pine, fir, aspen, and the Kaibab squirrel which has a black body, black stomach and white tail. It is typical of the Canadian and sub-Arctic Hudson zone of life. The lower South Rim has mainly juniper and

pinon, typical of the Upper Sonoran Life Zone, and the Abert squirrel with its grizzled grey body, white stomach and grey tail.

About nine million years ago, the Colorado River started to carve the Grand Canyon. It exposed one of the most complete geologic columns on the earth encompassing some two billion years of history. As the Rocky Mountains were folded on either side of the Colorado Plateau, it acted as a buttress and was slowly raised. This permitted the Colorado River to downcut and erode as fast as the area was uplifted. *Mass wasting,* the process of the transportation of material downslope under the influence of gravity, is responsible for the width of the canyon.

There is the possibility that the Colorado drained at one time towards the east. The Kaibab Plateau acted as a barrier preventing the river from entering a drainage basin that drained to the west. Uplift by the Arizona-New Mexico border dammed the river and formed a large lake. *Headward erosion,* the cutting back of the headwaters of a stream by the addition of tributaries, eventually cut back to the lake and began to drain it. This was when the cutting of the canyon began.

GEOLOGIC HISTORY

1. Deposition of sediments which now form the Vishnu schist. (Precambrian X)

During the Precambrian more than 2,000 million years ago, sands, silt and muds were deposited in a shallow marine basin. Debris from nearby volcanoes is found in these beds, along with basic igneous intrusions (some of which may be lava flows). The subsiding basin permitted the accumulation of great thicknesses of sediments and volcanics. There are no indications of any life in this sea.

2. Volcanic activity and formation of Brahma schist.

Some of the basic lava flows were under water as they have a *pillow structure,* with its characteristic ellipsoid shape, fine-grained outer skin and coarse-grained interior. The pillows rest upon each other like a series of flour sacks. Because these flows came from *fissures,* or large cracks in the crust of the earth, instead of from a central volcanic vent, they accumulated to great thicknesses. Today the 15,000 feet of flows are tilted until they are almost vertical. Interbedded between them are *tuffs,* the consolidated fine-grained debris ejected from the volcanoes, mixed with sands and clays.

These sediments are now metamorphosed into quartzites and a quartz mica schist.

3. Formation of the Mazatzal Mountains, folding, faulting, intrusion and metamorphism.

About 1700 million years ago, the beds which are now the Vishnu and Brahma schists were uplifted into *isoclinal folds* with their parallel limbs. Intrusion of granites into the sediments metamorphosed the sands into quartzites and the shales and mudstones into schists. This *contact metamorphism,* because of the heat from the intrusion, formed garnet, staurolite, and sillimanite schists in the Vishnu, and tourmaline, epidote-hornblende schists in the Brahma. This folding and metamorphism formed the Mazatzal Mountains.

4. Faulting, vein filling, additional faulting.

During the folding a series of intersecting faults trending northwest-southeast and northeast-southwest developed. During the metamorphism caused by the intrusion of the granites hydrothermal (hot water) solutions developed. These solutions formed quartz veins as they filled in the fissures produced by faulting.

After the intrusion of the Zoroaster granite, faulting continued, as it cooled into a *batholith* (a large intrusive body with no known floor).

5. Erosion of Mazatzal Mountains and the formation of the Arizonian erosional surface.

The high mountain range exposed to the elements underwent stream erosion and gradually became lower and lower until they were a nearly flat, gently sloping surface called an *erosional surface.* There is evidence that in the last stages of erosion the region was arid.

6. Transgression of the seas and deposition of the Unkar Group unconformably on the Vishnu and Brahma schists 1,000 million years ago. (Precambrian Y)

A sea coming from the west gradually covered or *transgressed* onto the land. As it advanced, the waves smoothed out the old land surface, with the larger fragments forming a gravel in some areas. This formed a *basal conglomerate* and is known as the Hotauta. A basal conglomerate indicates that an *unconformity* or a break in the sequence of deposition has occurred. The Hotauta member of the Bass Limestone consists of igneous and metamorphic fragments derived from the weathering of the Arizonian surface.

Part of the Unkar Group consists of the following formations:

a. Bass Limestone—marine, shallow water, cliff former, 120-340 feet thick, mostly dolomite. (Some sandstone, shale, and algae).

b. Hakati Shale—mostly nonmarine, shallow water, slope former, 580-830 feet thick, mostly shales (thin bedding), mudstones (thick or no bedding) with some sandstones, and possible salt crystal molds.

c. Shinumo Sandstone—marine, shallow water, cliff former, 1,100-1,560 feet thick, cross-bedded sandstone.

d. Dox formation—marine, shallow water, steep slopes, cliff former, 3,000 feet thick, mostly sandstone with some shale.

7. Renewed faulting in the existing system.

Movement once again occurred along the intersecting fault system. Some of these were later filled in by the Rama formation.

8. Volcanic activity with the formation of the Rama formation as lava flows and intrusions.

Part of the Rama formation intruded along any available openings, such as bedding planes and fault systems forming *dikes* (which lie across the bedding planes), and *sills* (which lie parallel to the bedding). It is primarily diabasic (lath shaped crystals in the basalt).

The remaining Rama occurs as basaltic lava flows up to 1,000 feet thick, that erupted from fissures.

9. Deposition of the Nankoweap and Chuar formation.

The area once again was covered by a shallow sea depositing some 300 feet of the Nankoweap on the surface of the Rama formation. The Nankoweap is the last member of the Unkar Group.

The Chuar Group was also deposited in a shallow sea, accumulating to a thickness of 5,000 feet.

10. Fault block mountain building during the Grand Canyon revolution with accompanying thrust faulting and the formation of monoclines.

The block faulting occurred in stages along the old fault lines that were already in existence. First, along the northeast-southwest system blocks dropped down in relation to the two adjacent

Types of Unconformities

Nonconformity

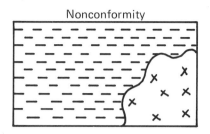

Contact between eroded igneous beds and sedimentary beds.

Disconformity

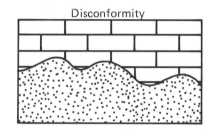

Contact between two parallel beds separated by an erosional surface.

Paraconformity

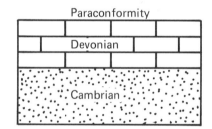

Devonian

Cambrian

Contact between two parallel beds where the unconformity can be distinguished by age of beds.

Angular Unconformity

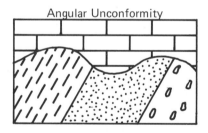

Contact between beds at an angle separated by an erosional surface.

blocks. This is called a *graben fault.* The higher blocks are known as *horsts.*

Secondly, the faulted beds were folded into a single flexure called a *monocline.* Enough stress built up to produce thrust faulting (a low angle reverse fault) along the base of the monocline in a northeast-southwest direction.

Thirdly, *normal faulting* with the hanging wall dropping down in relation to the footwall occurred along the original set of faults. This movement offset some of the folded monoclinal beds. It also permitted further downdropping of several of the graben blocks.

Last of all renewed thrust faulting parallel to the monocline.

11. Erosion of Grand Canyon Mountains to an erosional surface formation of monadnocks.

The mountains underwent erosion, stripping off much of the Chuar and some of the Unkar Group, exposing some of the more resistant beds. Since the beds eroded more slowly, they stood as low hills on the erosional surface and are called *monadnocks.* Features such as the depth of weathering, oxidation, and *decomposition* (chemical weathering) of the feldspar minerals into clay indicate

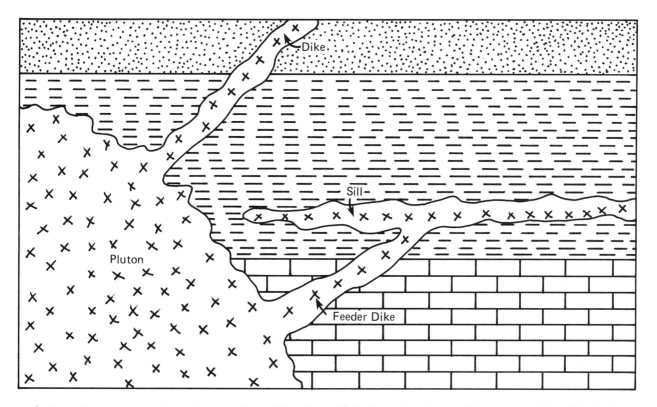

A pluton is any igneous intrusion regardless of its shape. If the intrusion lies parallel to the bedding of the bedrock it is a sill, if it cuts across the bedding it is a dike.

weathering for a long period of time in an arid climate.

12. Deposition of Cambrian beds by a transgressing sea forming an angular unconformity with the underlying Precambrian beds.

The transgressing sea came from the west about 500 million years ago, depositing three formations, each representing a different environment. They also illustrate the principle of *temporal transgression* or crossing of time boundaries.

Imagine yourself standing on a sandy beach. This would become the Tapeats Sandstone. On this beach would be cobbles (256-64 mm in diameter) and boulders (256+ mm in diameter) weathered out from the bedrock and transported to the shore. These would form the *basal conglomerate* at the base of the Tapeats. As you wade out into the water the bottom would become muddy. These muds formed the Bright Angel Shale. Way out in the deep water the *calcium carbonate* or lime ($CaCO_3$) would precipitate out, forming a limy ooze on the bottom. This ooze becomes the Muav Limestone.

Because sea level is gradually rising, the beach and shoreline is gradually shifting eastward. Hence, when the Tapeats and Bright Angel were first de-

posited, they were Lower Cambrian in age, but by the time the last grain was laid down, time had advanced to Middle Cambrian time. Hence, the Tapeats and Bright Angel are one age in one geographic area, and a different age in another area. They have crossed time boundaries.

The sea is known to be a *transgressing sea* because we find the finer grained shale deposited over the coarser grained sandstone. If the sea had been *regressing,* the coarser sediment would lie over the finer.

Since these beds were deposited in a horizontal position on the eroded, folded fault block surface, the beds are at an angle to one another and this type of unconformity is called an *angular unconformity.*

13. Disconformity formed between the Cambrian and Devonian beds because of erosion.

There is no evidence one way or the other if Ordovician and Silurian beds were deposited in the area and removed by erosion, or were never formed. At any rate deep channels were carved by some agency on the top of the Muav such as streams, or perhaps they were formed by marine scour. In these depressions were deposited the Devonian Temple Butte Limestone which is miss-

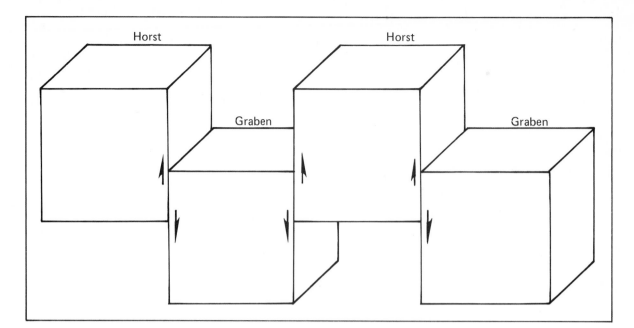

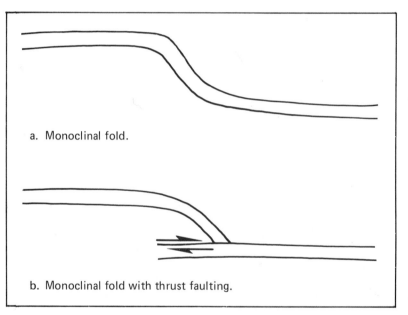

a. Monoclinal fold.

b. Monoclinal fold with thrust faulting.

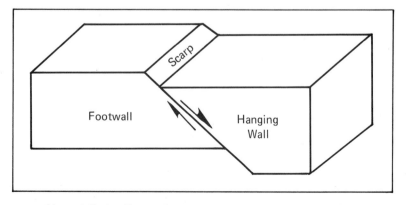

Normal Fault—Footwall moves up in relation to the hanging wall. The cliff formed is called a scarp.

ing in some regions and up to 100 feet thick in others. This uneven surface forms the type of unconformity known as a *disconformity*. Where the beds above and below the erosional surface are parallel but different ages. This occurred around 300 million years ago.

14. Erosion of Temple Butte and Muav formations.

No sooner than the Devonian Temple Butte was deposited, it began to undergo erosion and most of it was removed along with part of the Muav. The only remnants of the Temple Butte that can be found is where it was deposited in some type of hollow and protected by the surrounding Muav. Hence another unconformity is formed between Cambrian-Devonian beds and the overlying Mississippian formations.

15. About 230 million years ago the deposition of the Mississippian Redwall formation, uplift, erosion, karst topography.

This bluish-gray limestone contains chert nodules and was deposited in a shallow sea. It is a cliff former and is usually stained red by the overlying beds. It is about 500 feet thick.

The region was gradually uplifted and the beds underwent erosion. Since limestone is highly susceptible to solution, a *karst topography* developed with many caves and sinkholes.

16. Deposition of the Supai formation 230 million years ago during Pennsylvanian-Permian time.

The Supai is mostly nonmarine and contains footprints of amphibians and perhaps primitive reptiles. These footprints, along with the plant fossils, indicate that the region was a vast flood plain. The lower beds may be marine in origin and are probably Pennsylvanian in age. The upper beds are Permian, nonmarine, and oxidized a bright red color. Total thickness is around 1,000 feet.

17. Deposition of Permian beds around 205 million years ago.

The Hermit Shale is also red in color, about 100 feet thick, nonmarine in origin, and contains ripple marks, mudcracks and footprints. This indicates an origin similar to the underlying Supai. Main fossils are plants, but insect remains are also found.

The flood plain was gradually buried by migrating sand dunes which now form the Coconino Sandstone. The uniformly sized, perfectly rounded, frosted quartz grains that are arranged in a crossbedded pattern all testify to the *aeolian* or wind-

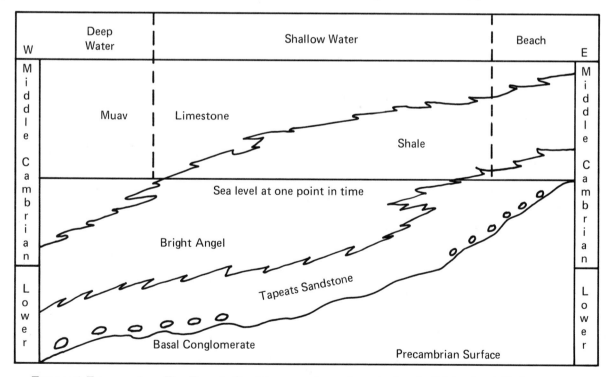

Temporal Transgression. The Tapeats Sandstone was formed on the beach, Bright Angel Shale in shallow water and Muav Limestone in deep water. As sea level rose, the shoreline shifted to the east along with the deposition of the sediments. Thus the Bright Angel Shale is lower Cambrian in age in one area and middle Cambrian in another.

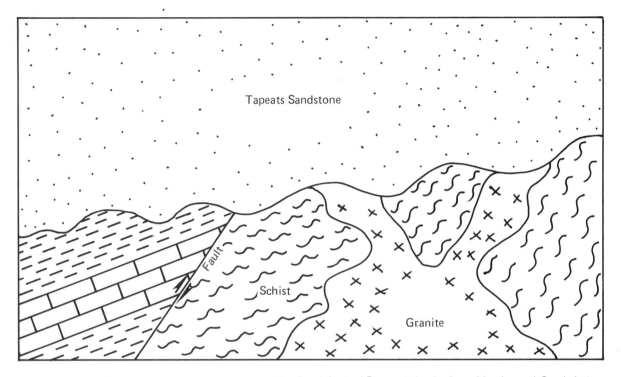

Angular unconformity produced between the tilted and faulted Precambrian beds and horizontal Cambrian Tapeats Sandstone.

blown deposit. The dunes were fairly large, as the Coconino is 400 feet thick. But in spite of their size, animals wandered around the dunes because we can see their footprints even today. The Coconino is the whitish band that can be seen near the top of the Canyon.

The marine Toroweap formation with its sandstones and limestones implies that as the seas advanced, they smoothed out the sand dunes, and gradually buried them as sea level increased. It is not quite 300 feet thick. At one time it used to be considered as part of the overlying Kaibab formation.

Forming the rim of the plateau is the massive, thick bedded Kaibab formation which is over 300 feet thick. The Kaibab is mostly limestone but also contains layers of chert. Some zones have sand and chert nodules. The fossils are middle Permian in age.

18. Withdrawal of the Kaibab Sea, erosion, deposition of the Moenkopi, about 180 million years ago. Possible deposition of other Mesozoic sediments.

When the sea regressed, the streams carved wide low valleys on the newly exposed sloping surface. During the Triassic Period, which is the begin

ning of the Mesozoic Era, the Moenkopi formation was deposited. This is a continental deposit, a mixture of sandstones and shales with gypsum, that are brightly colored. Fish swam in the streams and ponds, reptiles and amphibians lumbered on the land.

Because there are Jurassic, Cretaceous, and Paleogene beds in adjacent areas, there is the possibility that they were deposited in the area but were removed by erosion at a later time. For example, only a small remnant of the Moenkopi has been preserved in the Cedar Mountain area in a graben.

19. Laramide revolution, formation of Rocky Mountains, monoclinal folding, thrust and normal faulting.

Even though beds were being folded, faulted and uplifted into the Rocky Mountains all around the Colorado Plateau, it remained stable and undisturbed as compared to the adjacent areas.

Monoclinal folds with thrust faulting at the base occurred once again, but in a direction perpendicular to the earlier folds formed during the Precambrian.

Normal and *strike-slip faulting* (movement in a horizontal direction instead of vertical direction) offset sections of the monocline. Many side can yons have been eroded along these strike-slip faults.

20. Erosion of Rockies, formation of High Plains in adjacent areas, erosion.

As the Rocky Mountain underwent erosion the sediments were deposited adjacent to the mountains, perhaps as far south as the Colorado Plateau. If they did exist, they have since been removed at a later time.

21. Ancestral Colorado River, development of present drainage system, uplift of Colorado Plateau during the Miocene.

During Miocene-Pliocene time, the ancestral Colorado River formed. It may have originally drained eastward and lost its headwaters to stream piracy, thus changing it to its southwestern flow. Perhaps a lake formed, and the river formed as the result of the overflow. At any rate the Colorado flowed on a fairly low flat surface of soft sediments such as valley fill and soft rocks such as sandstone and shales.

As the river was developing, the Colorado Plateau was uplifted and warped, aiding in the development of the drainage system. There was some local faulting and volcanic activity.

22. During the Pliocene uplift and tilting, superposition of Colorado River.

The uplift increased the gradient of the stream so it began to downcut rapidly and became *entrenched* or sank itself into the bedrock and still maintained its original course.

As it gradually cut down, it became a *superimposed stream* or a stream that is let down onto a preexisting structure and continues to cut through the structure. There was some modification in the eastern portion of the park where the Colorado River encountered some of the monoclinal folds and faults, but not a great deal. By Mid-Pliocene time its final stream pattern was well established. With the entrenchment and superposition of the stream, the formation of the Grand Canyon had begun. During the Late Pliocene volcanic activity covered the plateau with flows. Some of the flows poured down the valleys, especially in the western section.

23. Deepening of the canyon by the river, widening by mass wasting.

A river only removes the material from its bed and the width of its meanders. Weathering and mass wasting removes the rest. The arid climate, being the dominant force, is responsible for mechanical weathering or *disintegration.* Frost action, spalling (peeling off in layers) and landslides broke up the rocks and moved them downslope to the river where it would be carried away. Chemical weathering or *decomposition* is responsible for the solution of the limestone forming caves and pitted surfaces.

Because more rain falls on the south rim than the north, it weathers more quickly and thus the canyon is asymmetrical in profile.

Sandstones and the limestones are more resistant to weathering, so they stand out as cliffs. The shales are weak and weather back as slopes.

The faulted areas were *brecciated* or broken up into angular fragments or rock, hence were more susceptible to weathering, thus forming the tributary canyons.

The Inner Gorge is made up of Precambrian metamorphic and igneous rocks which are of equal hardness so they form a narrow V-shaped canyon. This is a contrast to the overlying sedimentary Paleozoic beds that form a series of steep cliffs, gentle slopes and plateaus.

The Colorado River still has a steep *gradient* or slope and thus is able to continue its process of downcutting and transportation of material downstream to form the canyon that we know today.

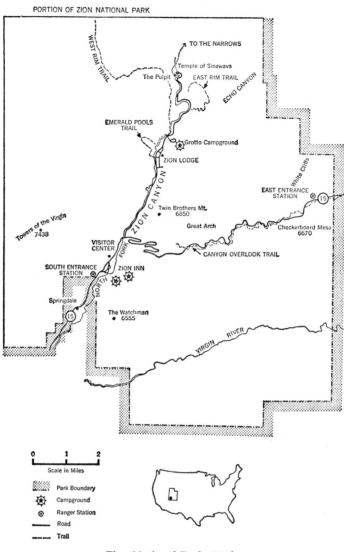

Zion National Park, Utah.

Geologic Column of Zion National Park

Erathem	System	Series	Formation
Cenozoic	Paleogene	Eocene	Wasatch Formation
Mesozoic	Jurassic		Carmel Limestone
			Navajo Sandstone
	Triassic		Kayenta Formation
			Wingate Sandstone
			Chinle Formation
			Shinarump Conglomerate
			Moenkopi Formation
Paleozoic	Permian		Kaibab Limestone

Modified after Gregory, Evans, Newman et. al.

CHAPTER 2

ZION NATIONAL PARK

Location: S.W. Utah
Area: 147,034.97 acres: 1,052 squares miles
Established: November 19, 1919; expansion 1937 and 1956

LOCAL HISTORY

The earliest traces of inhabitants in the park are found in Parunuweap Canyon. The artifacts indicate they belonged to the Basket Maker Culture. They were followed by the Cliff Dwellers or Puebloans who lived in Zion Canyon and Parunuweap Canyon.

The next group to come in were the Piutes, who built no permanent dwelling, but had settlements in the Virgin River Valley. They called the Virgin River "Pahroos," which means "muddy turbulent water" and Zion Canyon "Ioogoon," meaning "arrowquiver," or "come out the way you come."

The first record of white man was the Escalante-Dominquez expedition of 1776. They were two Spanish priests who were trying to find a direct route between Santa Fe, New Mexico, and Monterey, California. Forced to drop down south, they crossed the Virgin River.

In 1825, Jedediah S. Smith, a mountain man with 16 men, left the American Fur Company outpost at Great Salt Lake and headed towards the Wasatch Mountain. In following some of the streams, he encountered a river which he named "Adams River" after the president, John Quincy Adams. During a second expedition in 1827 he renamed it the "Virgin River" after Thomas Virgin, a member of the expedition wounded by the Indians.

The Fremont Expedition of 1843-44 passed through the region also. However, none ever saw Zion Canyon.

Mormons started settling the Virgin River in 1847 with plans to farm and grow cotton. The Mormon search for land for farming was responsible for the Piute Indians leading Nephi Johnson in 1858 to the canyon. Joseph Black explored it thoroughly in 1861 and located several locations suitable for farming. The settlers called the region "Little Zion" and stayed in the area until 1909, when it became a National Monument.

Major John Wesley Powell's party in 1872 crossed over Parunuweap Canyon (water that roars) and Little Zion Canyon, which he named Mukuntuweap (straight canyon). Descriptions and photographs by J.K. Hillers gave the park its fame.

About the same time (1872), Captain George M. Wheeler and G.K. Gilbert (separate expeditions) mapped this region. Additional mapping was done in 1880 by Major C.E. Dutton, with sketches made by W.H. Holmes. Gilbert named "The Narrows" a canyon which is 2,000 feet high and less than 20 feet in places.

Public pressure led to the establishment of a National Monument in 1909, calling it Mukuntuweap (after Powell). In 1918 the name was changed to Zion and in 1919 it was made a National Park.

A railroad spur was built in 1930 to the park, and a wagon road built into the Canyon which was improved to accommodate automobiles. The Pine Creek tunnel is over 5,000 feet long and has six large windows cut into it for views.

GEOLOGIC FEATURES

The Virgin River is the chief agent in the formation of Zion Canyon Section and Timber and Taylor Creeks of the Kolob Canyon Section. As in Grand Canyon, the rivers do the downcutting and mass wasting, the widening. Because the Virgin River has a high gradient or slope that ranges between 50 and 70 feet per mile, it is downcutting more rapidly than it is widening. Some three million tons of rock and sediment are being transported downstream each year.

This results in the tributaries, with their smaller volume of water, being intermittent (do not flow

year round) and having smaller drainage basins, being unable to downcut as rapidly as the main streams. Thus they are in the situation of having their floor way above the floor of the stream they are draining into, and form *hanging valleys* with a waterfall connecting them together.

Zion Canyon is still being formed, for the Virgin River could still downcut another 1,000 feet and have enough energy to transport its sediment to the Colorado River.

The rectangular stream pattern is the result of a closely spaced joint or vertical faults, which permit the broken or exposed rock to weather away more easily, forming valleys. Features such as some of the temples are the result of the unfractured rock between the highly fractured rock being resistant to weathering so it stands out by itself, forming columns and pyramids.

The sapping by springs and undermining of the sand stone cliffs by the removal of the softer underlying shales by gravity has a twofold effect. First, there is the formation of the overhanging cliffs of sandstone, then because of lack of support, large blocks break off and fall to the valley floor. Thus the valley is gradually widened and the

rims of the canyons are gradually developing a scalloped look as they retreat unevenly.

The variety of features are in part due to the wide range of bedding planes and joints and faults. Some are curved, others horizontal or vertical, some close together, still others widely separated. Frost action, rain water and seepage, action of tree roots, differences in the resistance of bedrock to weathering work together to produce Zion Canyon National Park.

The history of Zion Canyon begins where the history of Grand Canyon ends. The rock sequence of Grand Canyon is primarily Precambrian and Paleozoic, whereas Zion is Mesozoic and took some 150 million years to form instead of the two billion it took Grand Canyon.

GEOLOGIC HISTORY

1. Formation of the Kaibab Limestone during the Permian period.

The Kaibab Limestone is not actually exposed within the park boundaries, but is found just beyond the southwest border. This is the same forma-

tion that caps the rim of Grand Canyon. It is a marine limestone deposited in a shallow sea.

2. Deposition of the Moenkopi formation during lower Triassic some 175,000,000 years ago. Forms the Belted Cliff.

The Moenkopi is, therefore, the oldest rocks within the boundaries of the park. It is a mixture of sandstone, shale, limestone and gypsum. The different lithology produces bright bands of colors, thus naming the Belted Cliffs. Not all of these sediments were deposited in the sea. Some were deposited in and along the streams that flowed on the coastal plain adjacent to the sea. The presence of gypsum indicates an arid climate, for it is soluble in water. The marine fossils allow the geologist to date its age.

3. Uplift and erosion of the Moenkopi during Middle Triassic, formation of Shinarump basal conglomerate.

After the deposition of the Moenkopi, the area was uplifted and underwent erosion, forming a surface of *relief* (difference in elevation between the highest and lowest point) of 100 feet or more.

Sands and gravels formed, trees were washed down during floods and became trapped on sand bars and buried. The sands and gravels were later *lithified* (changed into rock by various processes) into a basal conglomerate while the wood underwent petrification. Because the Shinarump is a conglomerate, it is fairly resistant to weathering, and is a cliff former even though it is only 100 feet thick.

4. Deposition of Chinle formation by streams, and in shallow lakes and ponds during Upper Triassic time, forms Vermillion Cliffs. Erosion.

The Chinle is made up of a variety of sediments, it is a mixture of sandstone, shale, limestone and volcanic ash. The latter is of great interest, because it supplies the silica necessary for the petrification of wood in both the Chinle and Shinarump formations. The variation in the lithology and amount of iron oxide cause the brilliant red colors of this 1,000 foot thick formation, hence the name Vermillion Cliffs. Since they are primarily stream deposits, the fossils are mainly river dwellers, fish, amphibians, phytosaurs (resembled crocodile, but not related), and fresh water clams and snails.

After deposition of the Chinle, some uplift and erosion occurred.

5. Deposition of the Kayenta formation as a flood plain and stream deposit. Contains dinosaur tracks and cross-bedding.

Streams meandering back and forth deposited coarse sands and also silt which now make up the sandstones and siltstones of the Kayenta formation. Cross-bedding was produced by the shifting currents and the formation of sand bars. Coming down to the stream for water were dinosaurs, for they have left their footprints in the damp sands along the banks of the ancient streams. The Kayenta has a tendency to be a slope former.

6. Formation of the Navajo Sandstone as migrating sand dunes in a desert. Covered former stream area. Forms White Cliffs.

The climate which was semiarid to begin with grew drier and drier, the permanent streams dried up and became temporary streams that flowed only after a rain storm. Hence the sediments along the banks and on the bottom of the river dried out, and the winds picked them up and deposited them as sand dunes. With no vegetation to hold them down, they began to migrate and the *cross-bedding* that is so prominent in the Navajo developed. The winds came from the north and northeast. At first the dunes interfingered with the Kayenta stream deposits and then finally covered the entire drainage basin. Because of differences in the type and amount of cement, the Navajo varies in color from white to tan to red.

The Navajo Sandstone is a cliff former and forms such features as checkerboard Mesa and the Great White Throne. Dinosaurs also left their footprints in this formation.

Both the Kayenta and Navajo are members of the Glen Canyon Group which includes in other areas the Wingate Sandstone and Moenave formation as lower members. Somewhere in this group is the boundary between the Triassic and Jurassic.

7. Deposition of the Carmel Limestone in a shallow sea which covered the sand dunes during Upper Jurassic time.

About 120,000,000 years ago a shallow warm sea gradually advanced over the area. The wave action flattened the sand dunes and completely buried them. This light colored limestone is between 200 and 300 feet thick and caps many of the mesas on the Kolob Terrace. It also caps the Altar of Sacrifice and some of the Temples.

These are the youngest rocks found in the Park.

8. Deposition of 2,800 feet of undifferentiated beds which form the Gray Cliffs. Erosion.

The beds which overlie the Carmel Limestone now lie outside of Zion's boundaries, for they have been removed by erosion. But the lithology indicates deserts (gypsum) and swamps (sandstone and coal beds). After deposition they underwent erosion until they were buried by Eocene beds.

9. Formation of Wasatch formation and lava flows. Uplift and erosion.

The Mesozoic came to a close with the region under going erosion into Cenozoic time. This ceased in the Eocene epoch with the deposition of the Wasatch formation along streams and in lakes. The beds are about 500 feet in thickness and pinkish in color, hence are referred to as the Pink Cliffs.

On the Kolob Plateau, forming terraces are a series of basaltic lava flows that are mixed with pyroclastic debris from the volcanoes.

With the uplift these beds were removed from the park area, but are found in surrounding areas.

10. Block faulting with uplift of 10,000 feet creating Markagunt Plateau with a tilt eastward of 1-2 degrees.

Starting about 13,000,000 years ago, all of Southern Utah and adjacent areas underwent intermittent uplift over a span of millions of years. Eventually the entire area was elevated some 10,000 feet. This uplifting rejuvenated the streams by forming new base levels, so they once again began to downcut rapidly and form canyons.

The uplifting was uneven so the beds broke into a series of blocks, some of which were miles across. One of these blocks, the Maragunt fault block, has Zion National Park on it. It is bounded by the Hurricane fault on the west, Sevier fault on the east. The faulting tilted the block eastward only slightly, as the slope is only 1-2 degrees. But the uplift was as much as 2,000 feet. So the same beds that cap Elkhart Cliffs are found in the Virgin River below.

11. Dissection of the Markagunt Plateau by the Virgin River and mass wasting forming Zion Canyon.

The faulting and tilting once again increased the gradient of the stream so they began to downcut and gradually the canyon formed. The difference in the lithology of the beds, their reaction to weathering, the formation of the faults and joints have produced the features that we now view.

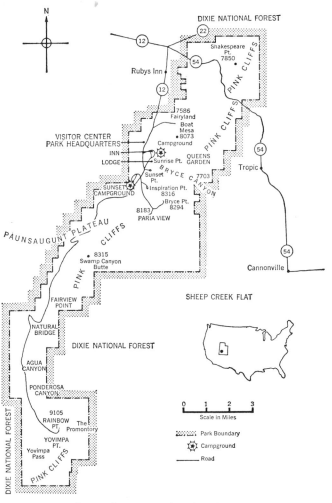

Bryce Canyon National Park, Utah.

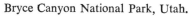

Geologic Column of Bryce Canyon National Park

Erathem	System	Formation
Cenozoic	Paleogene (Eocene Epoch)	Wasatch Formation
Mesozoic	Cretaceous	Kaiparowits Formation
		Wahweap Sandstone
		Straight Cliffs Sandstone
		Tropic Shale
		Dakota Sandstone
	Jurassic	Winsor Formation
		Curtis Formation
		Entrada Sandstone
	Triassic	Navajo Sandstone

Modified after Gregory and Newman et. al.

CHAPTER 3

BRYCE CANYON NATIONAL PARK

Location: S. Utah
Area: 36,010 acres; 50.06 square miles
Established: 1928

LOCAL HISTORY

The first inhabitants were the Basket Maker Indians which were followed by the Pueblo Indians who apparently had no long time settlements. Both cultures lived in the Paria Valley. They were followed by the peaceful Piutes who farmed and hunted in the area, but also did not build permanent settlements here. The Piute name for Bryce was "Unka-timpe-wa-wince-pock-ich" which when translated means "Red rocks standing like men in a bowl-shaped canyon."

The Escalente and Dominquez Expedition viewed the Pink Cliffs in 1776 from a distance, but did not actually enter Bryce. Trappers went through the area in the 1850s and scouts from the Mormon groups from 1851 to 1860.

In 1866, Captain James Andrus led 60 men through the region when they were chasing marauding Navajos. On that expedition, Everett Wash was one of the few white men ever killed by the usually peaceful Piutes.

After his second trip down the Colorado River, Major J.W. Powell initiated a survey to map much of southern Utah in 1871. Under Powell's direction, Almon H. Thompson, a geographer, surveyed and explored the Bryce Amphitheater area. He was responsible for the first scientific description of Bryce. His explorations were followed by Edwin E. Howell, Grove Karl Gilbert, and Lieutenant W.L. Marshall, who were members of a survey party under the direction of Captain George M. Wheeler from 1870 to 1876. Howell mapped the Pink Cliffs for 100 miles and studied them in detail at Table Cliffs. Gilbert went as far as Paunsaugunt Plateau and Paria Valley. John Weyes, an artist from the Wheeler expedition, made the first sketches of Bryce.

Captain C.E. Dutton surveyed the area in 1880 and J.K. Hiller in his party took the first photographs.

The few settlers, mostly Mormons from 1874 to 1891, were unimpressed by the scenery and were more concerned about the lack of soil and water, and losing their cows in the terrain.

Ebenezer Bryce (for whom the canyon is named), kept his cows there in the late 1870s. He is credited with saying it was "a hell of a place to find a cow."

Paunsaugunt Plateau or "home of the beaver" became a National Forest in 1905. When the wagon road was improved (1915-1917), the area became more popular. On June 8, 1923, it became a National Monument, and on September 15, 1928, a National Park.

GEOLOGIC FEATURES

Bryce Canyon is a horseshoe shaped basin carved out of the pink and white Wasatch formation, a mixture of limestone, sandstone, and shale. The basin is 12 miles wide, three miles long and 800 feet deep. It has sculptured terraces and pinnacles that cover the walls and the floor. Streams have carved side canyons along the edges of the cliffs, the rate of cutback is approximately one foot every 50 years.

The region is divided into three features, the Paria Amphitheater, a lowland at the base of the Paunsaugunt Plateau. The Wall which is the rim of the plateau and the Highland at the top of the plateau.

The first beds were deposited in a Permian inland sea some 60 million years ago. Deposition ceased about 25 million years ago. Uplift and faulting formed nine blocks which streams carved into seven plateaus. The uplift produced the cracks and joints. The various systems of cracks or joints create the different features including the arches, windows and natural bridges. The pinnacles are produced by variations in weathering and the erosional resistance of the different types of rocks. The pink color is due to iron oxide.

TABLE 3.1
Weathering

A. *Disintegration or Physical Weathering* is the physical disruption of the rock to form particles of smaller size without change in their chemical composition.

  1. Expansion and contraction causing granular disintegration.

  2. Frost wedging—forcing grains apart.

  3. Frost heaving—raising grains up.

  4. Exfoliation—peeling off in layers.

  5. Spheroidal weathering—weathering in a series of layers. ◎

  6. Organic forms—e.g.: burrowing animals, man, ants, rodents, etc.

B. *Decomposition or Chemical Weathering* is the chemical alteration or decay of minerals of which the rock is composed.

  1. Oxidation—chemical addition of O_2.

  2. Carbonation—chemical addition of CO_2.

  3. Hydration—chemical addition of H_2O.

  4. Solution—dissolves the rock.

The weathering processes are two types in Bryce. First there is disintegration, consisting of frost action and organic activity of plants and animals. Secondly, decomposition in the form of oxidation has produced the various colors. The reds and browns are produced by hematite (Fe_2O_3), the yellows by limonite ($FeO(OH)\cdot nH_2O$ + $Fe_2O_3\cdot nH_2O$), purple by manganese oxide (MnO_2), and white where the beds have been bleached out.

The terraces are formed in the softer rocks such as shale and cliffs in the resistant sandstones and limestones. The many plateaus are produced by various formations: The Pink Cliffs, the Wasatch; Gray Cliffs, The Dakota Sandstone; White Cliffs, the Navajo Sandstone; Vermillion Cliffs, the Wingate Sandstone; and Chocolate Cliffs, the Shinarump conglomerate.

The Paria River and its headwaters plus weathering and erosion is creating the Amphitheater which is in a youthful stage of erosion.

In vivid contrast the Paunsaugunt Plateau is in old age, with slow moving rivers on its surface, meandering widely back and forth in a low shallow

valley. The divide is only five feet wide and separates the *drainage basins* (area a stream drains) which either empty eventually into the Pacific Ocean (east edge) or flow to the Great Basin where it disappears in the desert (west edge).

GEOLOGIC HISTORY

1. Formation of the Navajo Sandstone as migrating sand dunes in a desert.

During the Jurassic Period these migrating dunes covered much of the southwest and are easily distinguished by the well-defined cross-bedding. The Navajo is not exposed within the Park boundaries and can only be found by drilling.

2. Deposition of undifferentiated Jurassic beds.

These beds are a mixture of sandstone and shale, are poorly defined with indefinite contacts and are not exposed within Bryce.

3. Deposition of the Dakota Sandstone in a shallow sea during the Cretaceous.

About 60 million years ago a shallow sea transgressed over the landscape and deposited a *blanket*

sand which is very extensive areally, but usually not more than 50 feet thick. It probably is the Dakota Sandstone, but the lack of a well-defined contact, makes correlation difficult. This is the oldest formation exposed in Bryce.

4. Deposition of the Tropic shale (mostly marine).

There is no sharp separation between the Tropic shale and Dakota Sandstone within the park as they are incomplete and poorly represented. The Tropic was deposited in a shallow sea which occasionally regressed long enough for coal swamps to form. This coal is being mined in adjacent areas.

5. Deposition of the Straight Cliffs and Wahweap Sandstones during the Cretaceous.

These sediments were deposited in a shallow regressing sea and in brackish water lagoons. They have a total thickness of 1,500 to 2,000 feet. A buff colored marker bed of sandstone 30 to 150 feet thick makes them easy to trace, because it forms a vertical cliff. In other areas these two indistinguishable formations are alternate layers of thick and thin sandstones that form a series of steep steps.

6. Deposition of Kaiparowits beds as sandbars, deltas and river flats during the Cretaceous.

The Kaiparowits was deposited on land by meandering streams in several environments. Hence the beds are discontinuous with a variation in composition ranging from pure white sandstones, to arkoses with its sand sized grains of micas, feldspars and quartz. Also argillaceous (clay) shales and calcareous (limy) shales. Lenses of impure limestones, ironstone concretions or sand balls are resistant to weathering so form knobs within the slopes or act as caprock for the pinnacles and buttes. Fossils of crocodiles, turtles, dinosaurs and fresh water invertebrates are found, plus petrified wood and fossil leaves.

7. Erosion of Kaiparowits during Early Paleogene.

The Kaiparowits varies considerably in thickness because much of it was removed by erosion. Therefore the thickness is less than 100 feet to over 1,000 feet. Because of the erosion, an unconformity is formed with the overlying Eocene Wasatch beds.

8. Deposition of Wasatch some 25 million years ago along streams and in quiet water.

The Wasatch is a mixture of thick red limestones, with thin layers of mottled reddish shales and gray sandstones, combined with discontinuous lenses of a calcareous gravel. The lower portion of the Wasatch is a basal conglomerate with rounded fragments of quartzite, quartz, igneous rocks, limestone and sandstone. The gravels and sandstones formed in and along the sides of streams, the limestone and shales in quiet water. It has fossils of primitive mammals, clams, snails, turtles and plants. plants.

Because it is cemented mainly by iron oxide it has various shades of reds and pinks. Thickness varies from 800 feet to about 1,300 feet. It has been covered in some areas such as Red Canyon with basaltic lava flows. North of Bryce, igneous rocks are abundant.

9. Uplift and block faulting elevated the plateau some 2,000 feet.

The uplift of the High Plains of Utah caused it to be broken into nine blocks forming a series of plateaus some 13 million years ago. The Paunsaugunt Plateau has the Sevier fault for its western border known as Sunset Cliffs, separating it from the Maragunt Plateau. To the east is the Aquarius Plateau which is separated by the Paunsaugunt fault. Table Cliffs is a remnant of the former plateau before the uplift and shows a displacement of 2,000 feet along the Paunsaugunt fault. At Bryce Point the same formation on the west side of the fault is at 8,294 feet, while on the east side it is 10,000 feet.

10. During the Holocene erosion of the plateau and formation of Bryce Canyon.

The Sevier River has a gradient of only 15 feet per mile and because it flows on a relatively flat surface it was not affected by the uplift. Therefore it is in a stage of old age. The Paria River, on the other hand, has a high gradient of 1,000 to 1,500 feet per mile and flows over the rim. Thus it was rejuvenated by the uplift which also caused some of its tributaries (e.g., Bryce Creek, Yellow Creek, Willies Creek and Riggs Creek) to steal the headwaters of streams that used to flow north.

Flash floods during late summer are responsible for some of the rapid changes that occur each year, as they undermine and cause collapse of the pinnacles, or columns, and the walls.

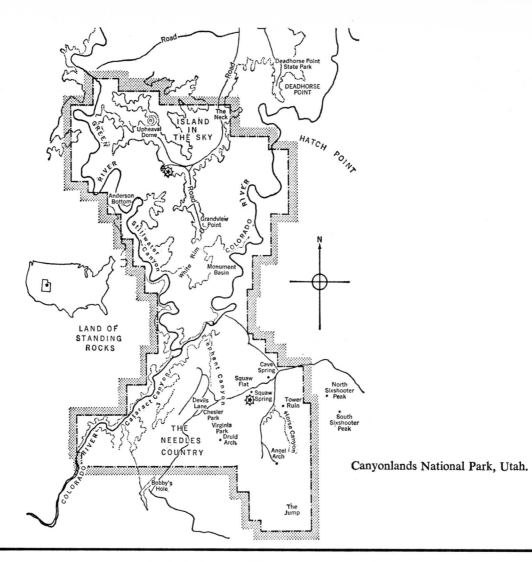

Canyonlands National Park, Utah.

Geologic Column of Canyonlands National Park

Erathem	System	Series	Formation
Mesozoic	Jurassic		Morrison Formation
			Summerville Formation
			Curtis Formation
			Entrada Sandstone
			Carmel Formation
			Navajo Sandstone
	Triassic		Kayenta Formation
			Wingate Sandstone
			Chinle Formation
			Shinarump Conglomerate
			Moenkopi Formation
Paleozoic	Permian	Guadalupe	White Rim Formation
			Bogus Tongue Formation
		Leonard	Cedar Mesa Sandstone
		Wolfcamp	Cutler Formation
			Rico Formation
	Pennsylvanian		Hermosa Formation†

†Includes Paradox member.

Modified after Newman et. al.

CHAPTER 4
CANYONLANDS NATIONAL PARK

Location: S.E. Utah
Area: 337,258 acres
Established: September 12, 1964

LOCAL HISTORY

There is evidence of Indians camping in this area at least 900 years ago. Newspaper Rock (outside park boundaries) is covered with *desert varnish,* a dark coating of iron or manganese oxide on its surface which is formed by the action of lichen on the rocks. Pictographs have been scratched on this surface by the Indians as they passed through the region.

In 1859 Captain John N. Macob took a group of people into the area and thought the region was totally worthless. However, Newberg, a geologist in the group, was awed and delighted by the scenery.

It is still a rough and primitive region which is best traveled by jeep.

In the late 1800s and early 1900s about the only visitors were cowboys with cattle or the sheepherders with their flocks passing through. Since this is desert country, water is precious, as the region only receives five to nine inches per year. Hence people were not attracted to this region for settlement. However, prospectors for gold, silver and more recently uranium, have been lured to this region.

GEOLOGIC FEATURES

Canyonlands is made up of sandstones, with uniformly sized clear quartz grains cemented with an iron oxide cement, giving it its red color. Variation in the resistance to weathering, close parallel vertical joints, and uplift are responsible for most of the features in the park.

Canyonlands has a wide variety of collapse features due to salt domes and igneous intrusions—fins and arches, pinnacles, and graben valleys.

Upheaval Dome was once thought to be a *cryptovolcanic structure,* a round circular area with highly disturbed beds but no evidence of volcanic activity. Geophysical studies have shown that a salt dome is responsible for the feature. During Pennsylvania time the Paradox formation was deposited. It is one of the members of the underlying Hermosa group. Above the Paradox formation lies a mile of sandstone and other materials. Salt is a plastic substance (will flow without rupturing when enough pressure is applied), and therefore the great weight of the overlying sediments causes it to flow. It began to push upwards as a huge plug, arching the beds above it into a circular anticline (beds arched upwards) or *dome.* As the beds are arched they crack, become more susceptible to weathering, and are removed more rapidly. The center of Upheaval Dome has softer beds than the rim. It is for this reason that they have been removed at a faster rate and form a depression in the center. They are removed by a stream that drains to the northwest. It has carved a canyon in the rim called Upheaval Canyon. Some 4,800 feet beneath the Dome is an intrusion of igneous beds that are not of volcanic origin that may also be responsible for the uplifting.

These salt beds are also responsible for the formation of another geologic feature in the park—the graben valleys. As the salt beds shift and flow, the overlying beds develop cracks from the stress and strain. When ground water which now has easy access to the salt beds dissolves the salt, some of the blocks drop down to form the graben valleys. They are located between Cataract Canyon and Elephant Hill.

A rather unique feature of the park is the development of *fins* which are narrow sandstone walls. The close parallel joints are enlarged by both chemical and physical weathering. Water seeping through the joints dissolves iron oxide cement (solution) and carries away the sand grains. Frost action with its alternate freezing and expansion and thawing, breaks up the sandstone which is removed by the streams. Gradually the joints become larger and larger until we have a series of narrow walls called fins separated by narrow valleys. The weathering and additional arching separates the fins even more.

This widening exposes the beds at the base of the fin to weathering, and if they are slightly softer, they will be removed more rapidly than the top. Because these fins are narrow, eventually a hole may be worn completely through, called a *window*. The effects of wind (sand blasting) polishes and rounds off the edges of the windows. The elements of snow, rain and ice break the rock down and gravity provides the energy for rockfalls, thus enlarging the windows until they form a *natural arch*. Certain formations, because of their composition, type of cement, structures, or other features, have a tendency to form arches. Most of the arches in Canyonland have formed in the Cutler formation, which actually has a greater tendency to form standing rocks rather than arches. On the other hand, the Entrada Sandstone, found in nearby Arches National Monument, tends to form arches very readily. This explains why there are only 25 arches in Canyonland and at least 88 arches in Arches National Monument, even though both areas are located in sandstone formations. Eventually the arch will collapse and only the buttresses will remain standing as pillars.

The *needles* and *pillars* (standing rocks) develop when there are two intersecting sets of joints that are spaced closely together. Because of the second set of joints that is at an angle to the first set, fins cannot develop, and in its place are the pointed needles and more rounded pillars. These have a tendency to form in the Cedar Mesa Sandstone, a member of the Cutler formation. Some may be capped by resistant material to form balancing rocks. Many of these features develop when fins develop vertical fractures permitting more rapid weathering along those openings. So eventually the fins break down into a line of columns, pinnacles and spires. This is very evident in Monument Basin.

The Colorado River flows through Canyonlands in a spectacular deep, narrow walled canyon. It is joined by the Green River (its largest tributary) which also has a narrow, deep canyon. Together they flow quietly for about three miles until they enter Cataract Canyon at Spanish Bottom. The 16 mile canyon is appropriately named and exists because of the changes in the bedrock. Both the Green River and Colorado River are *entrenched*. In the geologic past these two rivers were meandering back and forth on a fairly low flat surface close to base level (maximum depth to which a stream may erode) when the area was gradually raised to create a new base level. The uplift was slow enough so that the rivers did not change their path. But it was fast enough that, in spite of their wanderings, they began to downcut until a narrow steep-sided canyon formed, thus becoming entrenched.

This entrenchment isolated the area known as Island in the Sky, a 1,400 foot high mesa that is 6,000 feet above sea level. It is located between the canyons created by the Colorado and Green Rivers.

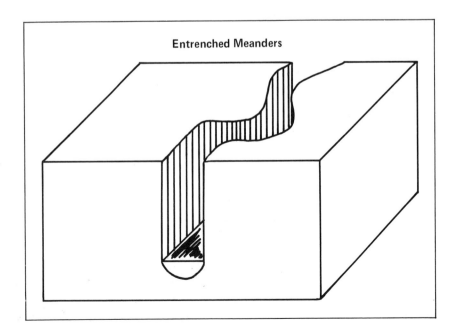

Entrenched Meanders

GEOLOGIC HISTORY

1. **Formation of Cordilleran geosyncline during the Cambrian 600 million years ago.**

   The Cordilleran geosyncline dominated western United States throughout the Paleozoic and Mesozoic until its final destruction during the Laramide Revolution, culminating in the formation of the Rocky Mountains.

2. **During Ordovician-Silurian time, uplift and erosion, Devonian and Early Mississippian—deposition.**

   The seas did not always remain in the geosynclines but for one reason or another, periodically retreated and advanced. Other times a region was temporarily uplifted, subjected to erosion until it was worn down once again.

3. **Middle to late Mississippian time, the formation of the Oklahoma and Colorado Mountains.**

   By the middle of the Mississippian, the uplifting of the areas which were to become the Oklahoma and Colorado mountains began. This caused the seas to retreat westward.

4. **Umcompahgre uplift in Pennsylvanian time, formation of adjacent Paradox Basin and salt beds.**

   About 230 million years ago, a series of uplifts forming mountains with their adjacent basins developed. This produced *closed drainage basins* where the streams would enter a basin, but the only way the water could leave was by evaporation. Hence these basins would be very saline since only the water would evaporate and anything carried in solution would remain behind. When the water became ten times normal salinity, halite (salt) would precipate out. This is how the Rico formation evolved. Layer after layer (total 29) formed until some 3,000 feet had accumulated. Deposition continued from Pennsylvanian to the Permian Wolfcamp time.

5. **During Permian deposition of Rico formation ceased and deposition of Cutler and Cedar Mesa beds began.**

   The Cedar Mesa Sandstone has a tendency to weather with numerous close set parallel joints which were probably formed when the sandstones were broken during crustal uplifts that occurred during fairly recent geologic time. It has the ten-

dency to develop needles, columns and spires. It is found in the Needles area where some of the features are more than 300 feet high. This sandstone formation grades laterally into the Cutler formation and is considered as a member of it.

The Cutler formation has a tendency to form the standing rock. Arches also form in it, but not as readily as other formations, such as the Upper Jurassic Entrada Sandstone in Arches National Monument. Best known arches are Druid Arch, Angel Arch, and Washer-woman Arch.

6. **During Leonard and Guadalupe (Permian) time deposition of Bogus Tongue and White Rim Sandstone, then erosion.**

Outside of Canyonlands boundaries, northwest of the Needles, is the Land of Standing Rocks. The features found here are the narrow fins and balanced rocks on a broad, extensive, bench. It is located west of the Green and Colorado rivers. This area includes the Maze, a canyon network of interlocking gullies and buttes. The White Rim Sandstone forms this intermediate plateau.

7. **Laramide Orogeny and uplift. Formation of Uinta and San Juan Mountains.**

The Laramide Orogeny, which produced the Rocky Mountains, also helped to form other uplifts such as the east-west trending fold called the Uinta Mountains and tilted beds of the San Juan Mountains. All this movement arched the beds and formed the joint systems so crucial to the formation of Canyonlands.

8. **Weathering of joints to produce fins and arches, graben faulting, intrusion of salt domes, formation of Colorado and Green rivers.**

Finally with the exposure of these jointed beds and the uplifting, the rivers formed the canyons, and the weathering the geologic features that make up Canyonlands.

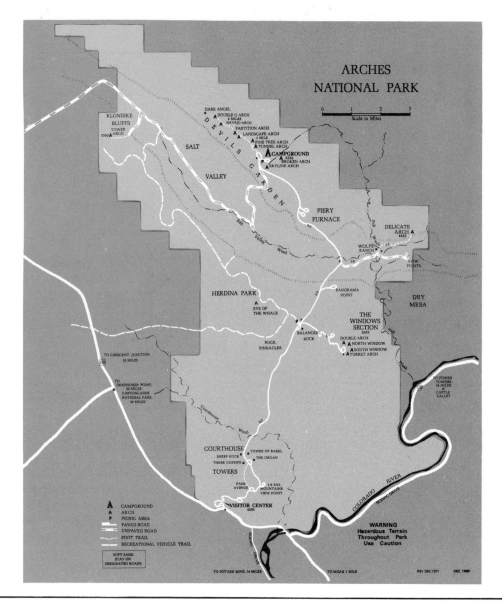

Geologic Column of Arches National Park

Erathem	System	Formation
Mesozoic	Cretaceous	Dakota Sandstone
	Jurassic	Morrison Formation
		Summerville Formation
		Curtis Formation
		Entrada Sandstone
		Carmel Formation
		Navajo Sandstone
	Triassic	Kayenta Formation
		Wingate Sandstone
		Chinle Formation
		Shinarump Conglomerate
Paleozoic	Permian	Moenkopi Formation
		White Rim Formation

Modified after Matthews, Dunbar et. al.

CHAPTER 5

ARCHES NATIONAL PARK

Location: S.E. Utah
Area: 73,233.87 acres
Established: November 12, 1971

LOCAL HISTORY

The region around Landscape Arch shows that Prehistoric Indians used this site for making arrowheads and other tools. It may have been used for a winter camp. They were attracted to the site because of the *chalcedony*, a cryptcrystalline (hidden crystal) variety of quartz used for making points and other tools because of its *conchoidal fracture.* This curved surface is responsible for the scalloped edges found on these tools.

Arches National Park was first discovered by Alex Ringhoffer, a prospector. The barren appearance and lack of water did not attract settlers.

It was proclaimed a National Monument on April 12, 1929. The boundaries were changed three times in 1938, 1960, and 1969. It became a National Park on November 12, 1971. The acreage has been increased from 34,300 acres to over 73,200 acres.

GEOLOGIC FEATURES

The main geologic features are the arches. Thus far some 88 arches have been found. There is a possibility that more may exist in some of the inaccessible isolated portions of the park, such as the Fiery Furnace area and the Dark Angel portion of Devil's Garden.

The arches have formed because the Entrada Sandstone was arched and broken into rows that are 10 to 20 feet wide. Erosion removed the over-

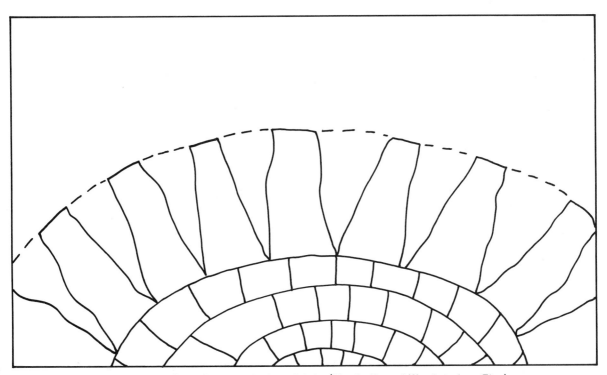

Formation of Fins—Bedrock Arches, Cracks (Vertically and Weathers into Fins)

lying beds, exposing the sandstone to weathering. This is when the formation of the arches began.

Water entering the cracks dissolved some of the cementing material, running water and wind removed the loose sand. The cracks were gradually widened forming very narrow valleys separated by fins. There is more moisture at the base of the fins since it is nearer to the water table and wind doesn't dry them out as quickly. Therefore, they weather first. The rock becomes undercut and small rock falls enlarge the opening even more. Frost action and other forms of physical weathering enlarge the depression until it cuts through, becoming a window. The window is rounded and polished by wind action.

Most of the arches in the park are in Devil's Garden, a grand total of 64. Landscape Arch is the longest natural span in the world, having a length of 291 feet and is 105 feet high.

The arches are in all stages of development, from the faintest suggestion of a depression to arches that have collapsed and have all but weathered away.

Besides the arches, *balanced rocks* have formed because the wind velocity is at its greatest 20 inches above the ground, hence can carry sand grains to that height. *Abrasion,* or the wearing away by friction, removes the softer material protected by the more resistant caprock, forming a pedestal. As it becomes thinner and thinner the heavy balanced rock on top starts to tilt very slowly to one side. This compresses the rock even more, making it more difficult to erode, so the opposite side, not under as much compression, weathers away slightly faster. This causes the rock to tilt to the opposite side compressing it, and the process starts all over again. The rock keeps tilting back and forth on an almost microscopic level until some force finally removes it.

Also found are coves, pinnacles, windows and spires, formed in part by *differential weathering.* This is the weathering of the same rock at different times, mainly because of differences in the intensity of weathering or differences in the composition of the rock.

GEOLOGIC HISTORY

1. **Formation of the Cordilleran geosyncline during the Cambrian, deposition of Paleozoic beds. Erosion.**

The formations that make up the bedrock of Canyonlands National Park were deposited. They underwent erosion at the end of the Permian so now have an unconformable contact with the overlying Triassic beds.

2. Triassic-Jurassic time deposition of the Moenkopi (?) and Wingate Sandstone, then erosion.

These two formations form the upper surface of the plateau and canyon rim. The plateau consists of overlying, twisted and broken rock eroded by wind and rain. The Moenkopi is a continental deposit formed by streams. The Wingate Sandstones were formerly migrating sand dunes.

3. Transgression of the seas, deposition of the Entrada Sandstone during the Jurassic.

The Entrada Sandstones are windblown sands deposited at the edges of shallow seas. The edges of the basins received the coarser sands while the center the smaller sized silt. The sands were derived from the weathering of older sandstone formations. They lie unconformably with the underlying beds.

4. Deposition of Cretaceous beds of sandstone, shale, and limestone.

These beds were deposited in the Rocky Mountain miogeosyncline, each reflecting its specific environment and depth of water. Erosion has removed them from the park area.

5. Laramide Orogeny. Formation of Uinta and San Juan mountains.

The uplifting of the Rockies, San Juans and Uinta mountains put stress and strain on all the rocks forming the joints that are responsible for most of the features in Arches National Park.

6. Formation of features now found in Arches National Park.

The uplift arched the Entrada Sandstone separating the joints even more and making them more vulnerable to weathering. The lithology and structure make this formation an arch former and they form relatively easily as compared to other sandstone formations. The bright red color makes them a very spectacular sight.

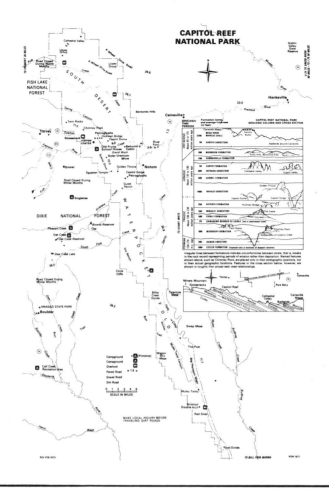

Geologic Column of Capitol Reef National Park

Erathem	System	Formation
Mesozoic	Cretaceous	Mesa Verde Formation
		Mancos Shale
		Dakota Sandstone
	Jurassic	Morrison Formation
		Summerville Formation
		Curtis Formation
		Entrada Formation
		Carmel Formation
		Navajo Sandstone
	Triassic	Kayenta Formation
		Wingate Sandstone
		Chinle Formation
		Shinarump Member of Chinle
		Moenkopi Formation
Paleozoic	Permian	Kaibab Limestone
		Cutler Formation

Modified after National Park Service Pamphlet.

CHAPTER 6

CAPITOL REEF NATIONAL PARK

Location: S.E. Utah
Area: 241,671 acres
Established: December 18, 1971

LOCAL HISTORY

Traces of pre-Columbian Indians (Fremont culture) are found here. They lived in open caves and raised corn. They are also known as the Basket Maker Indians of the southwest. They left petroglyphs, some of which are stained with colors derived from powdering minerals, instead of scratching them out of the desert varnish.

The first white man was Colonel John C. Fremont who was the leader of the Fremont Expedition of 1854. Later in the 1850s, the Mormons passed through this area.

In 1875, one of the members of the Powell Expedition studied this region in detail. The first permanent settler arrived in 1880, a Nels Johnson. He and a few other settlers lived here in a community they called Fruita. His cabin survived until the 1920s. A school and community house was built around 1890 and has been restored by the National Park Service. A cabin built by another settler, Elijah Cutler Behunin, can still be seen today. He also established the route which Scenic Drive follows today. Cohab Canyon, a hanging gorge, was used as a hideout by Mormons, who had more than one wife, and wished to evade the law on polygamy charges.

Capitol Reef received its name because the Jurassic Navajo Sandstone, a former dune formation, with its white color and tendency to weather in rounded domes. This reminded visitors of the dome shaped building found in Washington, D.C. and State Capitols. The "reef" is a term that the miners picked up from Australian sailors during the gold rush. It originally described gold bearing ridges in Australia and Africa, that looked like marine reefs, but its meaning was changed to mean any rocky barrier that made traveling difficult. The word reef, therefore, does not always mean a mound or ridge built on the bottom of the ocean built by some type of living organism.

It was originally a National Monument established in 1937 and did not become a National Park until 1971. Additions in 1958, 1969, and 1971 added the Waterpocket Fold and Cathedral Valley to Capitol Reef National Park.

GEOLOGIC FEATURES

Many of the formations already discussed in nearby National Parks such as Zion, Canyonlands, Arches, occur here, and are exposed in a monoclinal fold. It is not surprising to find, therefore, similar geologic features, such as natural bridges, pinnacles, towers, domes, dikes, sills and narrow canyons abound in this National Park.

A rather unique feature are the basaltic boulders that originally were formed by volcanic activity on Thousand Lake and Boulder Mountains. During the Pleistocene they were removed from these mountains by *valley glaciers* (glaciers formed in former stream valleys). When the glaciers finally melted about 25,000 years ago, their meltwater streams carried them to their present location.

The Waterpocket monoclinal fold is almost entirely located within park boundaries. It begins at Thousand Lake Mountain and extends about 90 miles as a 1,000 foot high wall to the Colorado River.

Mesas, the wide flat topped elevated remnants of horizontal beds, topped with resistant cap rock, protecting the softer beds beneath, developed in some of the formations such as the Mancos shale and Mesa Verde formation. The much smaller *buttes* with their steep sided walls and flat tops also formed.

GEOLOGIC HISTORY

1. Deposition of the Cutler formation and Kaibab Limestone during the Permian. Erosion.

From 270 to 230 million years ago, a shallow sea deposited about 800 feet of the Cutler formation and 300 feet of the Kaibab Limestone. The Cutler formation is best exposed in Canyonlands National Park, while the Kaibab Limestone forms the rim of Grand Canyon National Park. Both of these formations can be seen at the bottoms of some of the deeper canyons. The seas retreated and erosion occurred, forming an unconformity with the overlying beds.

2. Deposition of the Triassic Moenkopi, Shinarump and Chinle formations on an alluvial flood plain.

About 230 million years with the retreat of the early Triassic seas because of the uplift, a very broad alluvial plain formed. This began as a series of *alluvial fans* (land counterpart of a delta) formed at the base of the mountains by their outflowing streams. As the fans grew larger and larger, they gradually coalesced until they formed a continuous sloping plain at the base of these mountains.

First came the red and brown layers of mudstone, siltstone and sandstones, to the depth of 800 feet that is called the Moenkopi. The *primary structural features* (formed when rock formed) such as mud cracks and ripple marks, indicate some were formed in tidal flats, the rest on the flood plains. This too was uplifted and underwent extensive erosion, as is shown by the discontinuous layers of the Shinarump basal conglomerate. The Shinarump ranges in thickness from 0-20 feet and has a yellow color. Uranium salts formed in this layer and was mined in the Oyler Mine. As the marshes and flood plains formed once again, 500 feet of the Chinle was deposited. In this region it consists of volcanic ash and other debris, also siltstone and sandstone. The colors range from browns, to gray to green.

3. Uplift, erosion, formation of the Wingate Sandstone, Kayenta formation and Navajo Sandstone.

Towards the end of the Triassic, another uplift occurred and changed the region into a desert.

Migrating sand dunes advanced over the Chinle surface, burying it with 350 feet of cross-bedded white sandstone. This also produced an unconformity between the Chinle and Wingate. The Castle has formed in these two layers.

The region was tilted and streams deposited the Kayenta formation along its bed and flood plains to a depth of 350 feet. Hickman Natural Bridge forms in the Kayenta.

As the Triassic came to a close about 190 million years ago, the region was once again a desert. Migrating dunes covered the region, accumulating to a thickness of 1,000 feet. The cross bedding is responsible for the weathering of the Navajo into rounded dome shaped hills. The white color above the reddish-brown Kayenta beds is responsible for some of the colorful names for these hills, such as Golden Throne and Capitol Dome. Deposition of the Navajo began in the Triassic, but continued uninterrupted well into the Jurassic.

4. Tilting, erosion, deposition in a sea of the Carmel formation and Entrada Sandstone during the Jurassic.

The land tilted to the south, erosion occurred, the seas advanced and retreated periodically forming the sandstone, siltstones, and mudstones. In the sheltered lagoons, limestone and gypsum precipitated out, forming the Carmel formation to a depth of 500 feet.

The area filled in, producing land locked basins. It was in this environment that the Entrada Sandstone formed. It has a reddish-brown color that is so familiar in the Arches National Park. It is approximately 600 feet thick.

5. Erosion, deposition of Curtis Sandstone, Summerville and Morrison formations during Upper Jurassic time.

An unconformity exists between the Entrada and overlying Curtis Sandstone. It is only 50 feet thick, gray in color and resistant. It caps many of the features in Cathedral Valley. It gradually changes into the Summerville formation which is some 200 feet thick and thin bedded. Both of these were formed in shallow lakes.

With the uplift of the Mesocordilleran *geanticline,* an arch measured in 10s or 100s of miles

with its anticlines (upwarp beds) and synclines (downwarped beds) resembling small wrinkles making up the arch, the Morrison formation developed.

The Morrison is best known in other areas for its dinosaur fossils such as Brontosaurus. The region here became a swampy lake, in other areas an alluvial plain. It is brightly colored with yellows, reds, purples, greens and gray. The type of sediments here caused it to develop *badlands,* small hills that are easily eroded and change their shape after each rainstorm.

6. Erosion, deposition of Dakota Sandstone, Mancos shale and Mesa Verde Sandstone during the Cretaceous.

Between 135 million to 80 million years ago the region underwent erosion. Then the Cretaceous sea crept slowly over the landscape, depositing the thin blanket sand that rarely exceeds 50 feet thickness. Actually the Dakota Sandstone is a mixture of reworked sands and gravel with abundant fossil shells, especially oysters.

During the advance and retreat of the sea, the Mancos shale (and some limestone) formed in the shallow water, while the Mesa Verde formation on the beaches and as deltas. They have a combined thickness of over 3,200 feet and are the youngest beds in the park. Retreat of the seas occurred and then erosion followed.

7. Uplift and formation of the Waterpocket monocline.

On Miner's Mountain the beds are essentially horizontal. They start to fold at the Goosenecks and once again level out by the Caineville Mesas.

8. Volcanic activity and igneous intrusions during Miocene (Paleogene period) time.

About 20 million years ago during the Miocene epoch, *volcanic plugs* (thick intrusions of fairly cool lava intruded into vents to form a dome structure) formed in South Desert and Cathedral Valley. Dikes and sills of basalt intruded the openings of the sedimentary beds (bedding planes and joints).

9. Formation of Fremont River and its tributaries, erosion, weathering, glaciation during the Neogene.

The uplifting which formed the Waterpocket fold also changed some of the streams into *subsequent* or structurally adjusted streams. They flow parallel to the *strike* or trend of the beds and carved steep sided canyons because of rapid uplift and the erosion of soft beds. Mountain glaciation during the Pleistocene transported the basaltic boulders all over the park with its meltwater streams about 25,000 years ago.

The Fremont River is a *consequent* stream and flows to the east over the initial slope of the land parallel to the dip (angle beds form with the horizontal). It and its tributaries have carved the present landscape with the help of weathering.

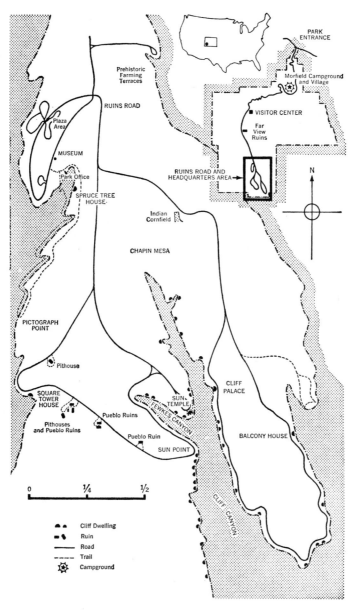

Mesa Verde National Park, Colorado.

Geologic Column of Mesa Verde National Park

Erathem	System	Series	Formation
Mesozoic	Cretaceous	Mesa	Cliff House Sandstone
		Verde	Menefee Coal and Shale
		Formation	Point Lookout Sandstone
		Mancos Shale	
		Dakota Sandstone	
	Jurassic	Morrison Formation	
		Summerville Formation	
		Curtis Formation	
		Entrada Sandstone	

Modified after Newman and Matthews.

CHAPTER 7
MESA VERDE NATIONAL PARK

Location: S.W. Colorado
Area: 52,073.62 acres
Established: June 29, 1906

LOCAL HISTORY

Mesa Verde was probably named by Spanish traders in the middle 1700s. They called it "La Mesa Verde" or the "Green Table."

In 1776 Father Escalante camped along the Mancos River (few miles to the east) during his explorations of the Colorado River area. He apparently did not see the cliff dwellings.

The dwellings were discovered in 1888 by the Wetherill Family. The five brothers plus their brother-in-law Charles Mason explored these dwellings systematically during the winter when the chores on their 160 acre ranch weren't quite as heavy. In the winter of 1889-90 their explorations brought them to Mug House and the other major ruins. In the summer of 1891 John Wetherill helped Gustaf Nordenskiold (23 year old son of a Swedish nobleman) excavate several of the cliff dwellings. In 1893 Nordenskiold published his findings in a book entitled *The Cliff Dwellers of Mesa Verde*.

In 1900, the Colorado Cliff Dwellings Association, a woman's organization, was founded and they began their task of saving and preserving the dwellings. As a result, Mesa Verde National Park was established in 1906, and has been scientifically excavated since then.

The people who were to become the cliff dwellers made their home in this region. They had four main periods of cultural development.

(1) *The Basket Makers* (A.D. 1-400). These were hunters turned farmers. They lived in shallow caves, raised corn and squash, mastered the art of weaving, but not pottery. They made tools out of such materials as shell, bone, stone and wood. The atlatl (primitive dart-throwing stick) was used for hunting. Most of their remains are now covered by the cliff dwellings.

(2) *Modified Basket Makers* (A.D. 400-750). These Indians used the bow and arrow, made baskets and plain pottery, raised beans and turkey. The turkeys were more important for their feathers than for food. They were used for arrows, plus lining blankets and cloaks to keep them warm in the winter. They lived on the mesa top in pit dwellings, which consists of a shallow round pit that had a flat roof consisting of poles and mud. Later the dwellers were more elaborate with various shaped rooms (square, rectangular or D-shaped) which were added to the pit dwellings, complete with ventilation shafts. These homes were clustered on the top of the mesa in villages.

(3) *Developmental Pueblo* (A.D. 750-1100). The pit dwellings were abandoned for living quarters and the Indians lived in rectangular houses with straight sides and flat roofs. These were mostly adobe in the beginning and by adding more and more stone work, the houses ended up being made totally of stone. The individual houses were joined together around a central court. In this court the pit houses gradually evolved into the kiva (club room and secret ceremonial room for men). Pottery improved, the loom and cotton was used. The type of baby cradle utilized resulted in flattened heads in the back. They dug a series of ditches on Chapin Mesa to save and store water.

(4) *Classic Pueblo* (A.D. 100-1300). Maximum development. In the beginning the Indians lived on top of the mesa in pueblos (Spanish—meaning small village) until A.D. 1200. For about the last 100 years they lived in the cliff dwellings, then in A.D. 1300 abandoned them. No one has lived there for the last 700 years. The Pueblo dwellers had well made pottery, used both cotton and yucca fibers for clothing, and had yarn and cord. The jewelry was made of different materials. Their homes were plastered and decorated. They made toys (for children) and games with playing pieces.

No one knows exactly why they left. Among some of the possibilities are a 24 year drought which began in 1276, constant war with neighbor-

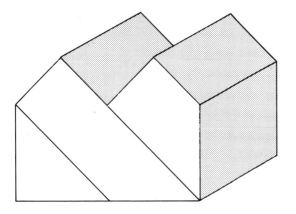

Hogback with Steeply Dipping Beds

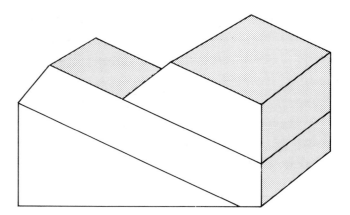

Cuesta with Gently Dipping Beds

ing tribes, or they became tired of living in the dwellings which were hot in the summer and cold in the winter. The depletion of the soil by bad farming was another possibility. Some of the dwellings looked as if the inhabitants simply got up and walked away without taking anything with them.

GEOLOGIC FEATURES

Mesa Verde is a *cuesta,* or asymmetrical ridge that slopes gently on one side (south) and is steep on the other (north). It is an erosional remnant of an ancient *pediment* or sloping rock surface at the base of a mountain covered with a thin veneer of gravel. This pediment once connected Mesa Verde, La Plata and San Juan mountains together. With rejuvenation, the streams became entrenched in the pediment and divided it into its separate components.

McElmo Creek, which flows west, stole or *pirated* the headwaters of the streams that used to flow south and skirted the north and west sides of the cuesta. The piracy occurred because the south flowing streams were at a higher elevation than McElmo Creek. When it extended its tributaries by headward erosion they encountered these streams and stole their waters, thus changing their direction of flow. Along the gentle slope of the cuesta, the Mancos River has cut a 1,000 foot canyon along the east and southern side. The *intermittent streams* which do not flow year round, and drain into the Mancos River, have dissected the cuesta into 15 long parallel steep sided canyons. It is along these canyons and on the intervening plateaus that the dwellings were built.

The sandstones are *resistant* and do not weather away easily while the shales are *weak* and weather back rapidly. This forms alternate layers of vertical cliffs with intervening shallow caves.

GEOLOGIC HISTORY

1. **Deposition of the Mancos shale in the Rocky Mountain miogeosyncline over the Dakota Sandstone during Late Cretaceous time.**

The Dakota Sandstone underlies most of Colorado, it was deposited as an alluvial sediment which is clearly indicated by its crossbredding, and lenses of shale and coal. In this region it is generally 100 feet thick. As the sea floor started to subside the Mancos shale was deposited. It is a marine shale around 2,000 feet thick which contains *concretions* (cement concentrated around a nucleus) and lenses of a yellow-orange limestone. It interfingers with the overlying formation and is the oldest formation exposed in Mesa Verde National Park.

2. **With filling of geosyncline formation of Point Lookout Sandstone (and shale) as a beach and shallow water deposit.**

When the geosyncline stopped sinking it began to fill. This is reflected by the Mesa Verde Group that is made up of three formations—the Point Lookout Sandstone, Menefee Coal and shale and Cliff House Sandstone. Each reflects the changes that occurred.

The Point Lookout Sandstone is divided into two members, the lower portion is alternate layers of sandstone and shale with cumulative thickness of 60 to 140 feet. It interfingers with both the underlying (Mancos shale) and overlying (upper member beds). Deposition occurred in shallow water.

The upper sandstone member is a cliff-former which forms the lower set of cliffs. Thickness varies from 230 to 340 feet and is a yellow-orange, *massive* (no bedding planes) sandstone. It helps to protect the lower member and Mancos shales beneath it. The crossbedding reveals its origin as a beach deposit.

3. Regression of sea, formation of swamp along shoreline forming Menefee formation.

The Menefee is the middle member of the Mesa Verde Group. Thickness varies from 230 to 800 feet. It contains two coal layers, the lower which interfingers with the massive sandstone of the Point Lookout, and the upper that is mixed with a cross-bedded lenticular sandstone that varies from gray to orange in color. Separating the upper and lower coal beds is a carbonaceous shale. This sequence indicates swamps along the beaches. The coal is mined outside the park on a commercial basis. The upper sandstone member interfingers with overlying Cliff House Sandstone.

4. Return of seas and deposition of Cliff House Sandstone with several advances and retreats.

The uppermost member of the Mesa Verde Group is the Cliff House Sandstone. The base is subdivided into three tongues (lower, middle and upper) where it interfingers with the Menefee. It is a cliff-former, the youngest bedrock in the park and averages 400 feet thick. Color ranges from pale to dark yellow orange. The shale zones determine the location of the cliff dwellings since it is less resistant. Water seepage and frost action formed the depressions. This massive sandstone that rims the plateau was one of the last formations to be deposited in a sea in this region. It is unconformable with overlying gravels.

5. Laramide Revolution, domelike uplift of San Juans, La Plata and Mesa Verde.

The uplifting caused this region to dip to the north, south and west beginning at the San Juan Mountains and sloping towards Mesa Verde. Erosion to the southwest by meandering streams transported gravel from the San Juans and La Plata mountains to the top of Mesa Verde. These three regions were connected together as a pediment.

6. Separation of Mesa Verde, La Plata and San Juans by stream erosion.

Further uplift caused the streams on the pediment to start downcutting rapidly, thus creating canyons and valleys that have removed most of the sediments leaving only isolated erosional remnants such as Mesa Verde. Stream piracy had diverted many of the streams and erosion created the 15 canyons.

7. **Seepage and weathering produced shallow caves.**

The Cliff House Sandstone is very porous so the melting snows and rain percolate very easily downwards until they encounter the *impermeable* (does not allow water through) shale beds associated with the Menefee Coal. It moves laterally along the contact until it seeps out at the canyon wall. Constant freezing and thawing breaks up the shale and much of the material enlarging the depressions which the Indians found very convenient for their homes and for protection.

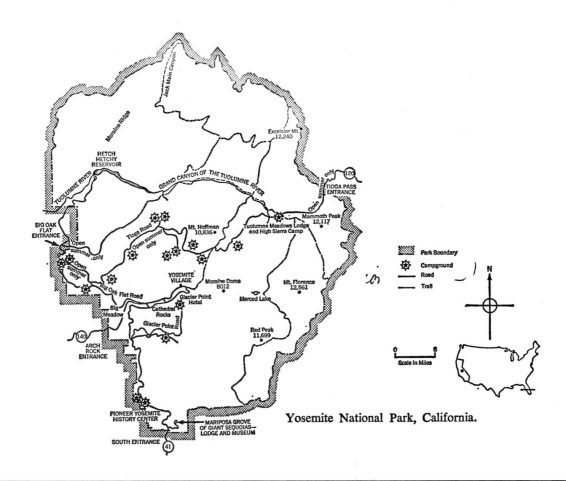

Yosemite National Park, California.

Geologic Column of Yosemite Valley
Yosemite National Park

Erathem	System	Series	Formation	Age M.Y.
Mesozoic	Cretaceous	Tuolumne Intrusive Series	Johnson Granite Porphyry	81.4
			Cathedral Peak Granite	83.7
			Half Dome Quartz Monzonite	84.1
			Sentinel Granodiorite	88.4
		Minor Intrusive Series	Diorite of the "Map of North America"	
			Quartz-Mica Diorite	
			Bridalveil Granite	
			Leaning Tower Quartz Monzonite	
		Western Intrusive Series	El Capitan Granite	92.2
			Granodiorite of the Gateway	92.5
			Granite of Arch Rock	95.3
			Diorite of the Rockslides	
	Jurassic		Sedimentary Rocks Including Schist	
Paleozoic	Permian to Devonian		Calaveras Formation	
	Silurian to Devonian		Metasediments	

Modified after Bateman, Wharhaftig, Clark, Peck et. al.

Geologic Column of Hetch Hetchy Valley
Yosemite National Park

Erathem	System	Series	Formation
Paleozoic	Neogene	Pliocene Miocene (?)	Volcanic Mudflows and Trachyandesite
Mesozoic	Cretaceous	Upper	Aplite
			Half Dome Quartz Monzonite
			Sentinel Granodiorite
			Yosemite Creek Granodiorite of Rose and Quartz Diorite of Mount Gibson
		Lower	Quartz Diorite of Tamarack Creek
	Jurassic	Upper	Alaskite of Ten Lakes
			Granodiorite of Rancheria Mountain
			Granodiorite of Double Rock / Granodiorite of Mount Hoffman / El Capitan Granite } same age
			Quartz Diorite of South Fork of Tuolumne River
	(?)		Diorite
Paleozoic (?)	(?)		Metasediments and Metavolcanics

Modified after Kistler.

PART II
FORMED BY GLACIERS AND WAVE ACTION

CHAPTER 8
YOSEMITE NATIONAL PARK

Location: Central California
Area: 761,320.32 acres
Established: October 1, 1890

LOCAL HISTORY

First inhabited by the Ah-wah-nee-chee Indians. Their name for Yosemite was Ahwahnee which meant "deep grassy valley in the heart of sky mountains."

Joseph R. Walker, one of the "mountain men," was in the area in 1833 and may have seen part of Yosemite Park, but it is uncertain if he ever saw Yosemite Valley.

In the middle of the 1800s the Yosemite tribe occupied this region, led by Chief Tenaya. They were a subgroup called the U-zu-ma-ti, who preferred to keep white man out of their valley which they called home.

Major James D. Savage, who went to California in 1845, had wandered over to this area and had set up Savage's Trading Post. He was a colorful figure who earned his living first as a soldier, then a trapper, and finally a trader. He married into five separate Indian tribes to win the alliance of the various tribes so he could trade beads for gold.

The trading post was located on the Merced River (about 15 miles from Yosemite Valley) and was raided in 1850 along with the rest of the area by the Indians. He organized a group of exsoldiers and miners into a group called the Mariposa Battalion in 1851. It consisted of three companies of men and their plan was to force the Indians out of their stronghold without killing them. Then the Indians were supposed to give themselves up to the Indian Commissioners.

Teneya actually came into the Battalion's camp with some of his men, but most of the tribe stayed behind, so the Mariposa Battalion decided to go in after them. Some of the members of the party were awed by the scenery and that night over the campfire they argued about a name. They finally decided upon Yosemite which may mean "grizzly bear." It is a derivation of the Uzumati subgroup of the tribe.

The enthusiastic letters of the various members of the Battalion to their friends, caused James M. Hutchings (publisher of California Magazine) to bring a horseback party to Yosemite to see its wonders. The group published their experiences which was picked up by papers all over the world, including some of the sketches by Thomas Ayres and others (e.g., Albert Bierstadt and Henry Pratt).

A year after the Hutching's visit a camp (south fork of Merced) was built and tours on horseback were being conducted through the valley.

Horace Greely, editor of the Chicago Tribune, came, but unfortunately picked the month of August when the water level was extremely low. Most of the waterfalls had dried up or had a bare trickle over them so he felt cheated and somewhat disappointed. However the beauty of the region won out and his paper published glowing accounts of it and suggested that the region be preserved for future generations.

Congress in 1864 granted it to the State of California, making it the first public park under the administration of a state. The proclamation was signed by President Lincoln.

John Muir visited it for the first time in 1868 and became captivated by it. He wrote many, many articles about Yosemite that brought world wide attention to it. Muir even convinced Ralph Waldo Emerson to come and see the sights. Unfortunately, in spite of his writings he was afraid of the out-of-doors, so was not impressed, and very glad to get back to the safety of the indoors. Teddy Roosevelt spent three days with Muir in 1903. Yosemite National Park was created around the Yosemite Grant in 1890. To increase the size of the park California turned the Grant back to the Government in 1905.

Travel in the park was by horseback up to 1878. Then wagon roads were put in. The first cars were admitted in 1914 (total of 127 for the year). In 1964 over 54,000 cars entered. To cut down on traffic in the valley itself there is one way traffic to Curry Village and back with free shuttle busses to various areas. Eventually cars may be banned entirely.

In 1914, Stephen T. Mather, an industrialist from Chicago came to Yosemite and was so horrified and angry over the way it was being abused that he wrote a letter of complaint to the Secretary of the Interior. The Secretary in turn offered him the job of running the parks, which he took in 1915. Congress in 1917 created the National Park Service and made Mather its first director. He spent the rest of his life fighting for the preservation of the parks. Sometimes however, the battle is lost, such is the case of Hetch Hetchy Reservoir.

San Francisco in 1903 decided to dam the valley for its hydroelectric power. In spite of the efforts of all the conservation groups, Congress allowed the flooding in 1913. Hetch Hetchy Valley was a minature of Yosemite that could have taken some of the overflow of tourists. Instead, for a good part of the year, it is a mudflat riddled with tree stumps because of the low water level.

GEOLOGIC FEATURES

The two dominant features of the park are the exfoliation domes and glacial features.

Yosemite is located within the Sierra Nevada *Batholith,* a solidified magma chamber. It consists primarily of twelve Mesozoic granitic formations, some Paleozoic metamorphic rocks as roof pendants and overlying Cenozoic volcanoes.

The variation in rock types and composition are responsible for many of the features in Yosemite. They are as follows:

a. *Massive unjointed rock*—El Capitan, Mount Broderick and Liberty Cap.
b. *Vertical joints*—Glacier Point, Upper Yosemite Falls, Ribbon Falls, Washington Column, Sentinal Rock and part of Half Dome.
c. *Joint sets*—Cathedral Rock, Three Brothers.
d. *Sheeting*—Sentinal Dome, Turtle Back Dome, Royal Arches and part of Half Dome.

Bedrock is never absolutely solid. If it is sedimentary it is separated by *bedding planes* or layers. If metamorphic there is usually some type of alignment of the minerals or *foliation* and perhaps *slaty cleavage* where the rock separates along parallel planes. Even igneous rocks that form by crystallization will develop shrinkage cracks or *sheeting* (close parallel cracks).

All three types of rock have a tendency to develop *joints* which are formed when the movement along the fracture is perpendicular to the break. The joints are produced primarily by stress and strain when the beds are uplifted, tilted, folded or faulted. If a series of joints are parallel to one another, they are referred to as *joint sets.* When two or more sets intersect one another at an angle they become a *joint system.*

Sheeting develops in a horizontal direction and the surfaces are frequently curved. It can develop when the weight of overlying rocks is removed permitting the expansion of the newly exposed rock. Sometimes it develops because of the alternate expansion and contraction of the bedrock in response to temperature changes. By either process, the rock gradually peels off layer by layer and *exfoliation* is said to have occurred.

Half Dome, the most famous feature of Yosemite Valley is a classical example of the above processes. Contrary to what many people think, there was never another half to the dome. The vertical face is the result of a set of closely spaced vertical joints, as they weathered, they fell away layer by layer. During the Pleistocene as the Tenaya glacier flowed down Yosemite Valley it plucked away additional layers since it flowed parallel to the joint set. Both of these processes helped to create this impressive wall. The back side of Half Dome weathered more slowly by the process of exfoliation, glaciers never overrode that portion. The backside literally peeled away layer by layer like an onion, forming a dome shaped structure. These features are called *exfoliation domes* because of their origin and commonly form in granite. Sentinal, Turtleback and North Domes are other examples of exfoliation domes.

Not all domes in Yosemite are exfoliation domes, some are the result of glacial action called *roches moutonnees*. These are asymmetrical rock hills with one gentle, smooth side and one steep, rough, jagged side. Because of the hardness of the bedrock, in this case granite, the glacier is forced to override the resistant material scraping and polishing the backside in the process. After it passes over and pulls away from the bedrock it simply plucks away the portions of the bedrock that are frozen to the glacier (controlled in part by the joints) leaving a steep, jagged surface. Liberty Cap and Mount Broderick are both roche moutonnees.

The exfoliation and jointing have created other features, regions that have intersecting joint systems tend to form long pillars such as Cathedral Rocks and Three Brothers. Others that are more closely spaced form pinnacles such as Split Pinnacle, Washington Column (1,700 feet tall) and Watkins Pinnacle.

Exfoliation has also created a series of arches such as Royal Arches, one nested inside of the other. As each curved sheet spalled off, it left part of it as an arc or arch from 10 to 80 feet thick and over 1,500 feet high. The lower portions have been removed by weathering, gravity, running water and glacial action.

The spectacular waterfalls are primarily due to glacial action. Yosemite actually has over half of the highest waterfalls in the United States, as a matter of fact the smallest of the falls is close to being twice as high as Niagara Falls (American 186 feet, Canadian 193 feet). They are as follows:

a. Yosemite Falls–Upper Cascades 1,430 feet, 675 feet of Cascades, Lower California 320 feet–total of 2,565 feet.
b. Ribbon Falls–1,612 feet.
c. Bridal Veil Falls–620 feet.
d. Nevada Falls–594 feet.
e. Illilouette–370 feet.
f. Vernal Falls–317 feet.

Yosemite, Ribbon, Bridal Veil and Illilouette all originate because of *hanging valleys*. As the main large *piedmont glacier* (combination of two

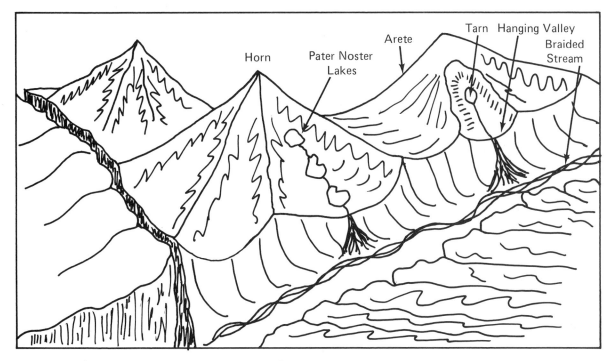

Glacial Features

or more valley glaciers) moved down Yosemite Valley it gouged and scraped deeper than the smaller tributary *valley glaciers* that occupied the side valleys. When the glaciers melted, the floor of the tributary valleys were a considerable elevation above the floor of Yosemite, thus were hanging. The streams that formed were forced to flow over the lip of these valleys creating waterfalls. Some of the hanging valleys are the result of differences in the rates of erosion between the main stream and its tributaries.

Nevada Falls and Vernal Falls originated a different way. They flow over giant *glacial steps* created as a glacier moved down a valley plucking out the closely jointed bedrock and leaving the massive bedrock in place so as to form a series of giant steps 300-500 feet high.

At one time in the head of the valleys where the glaciers originated there existed lakes called *tarns*. These tarn lakes form in the bottom of *cirques* the bowl shape depression formed by glacial action at the head of a glacier. These lakes gradually silted up for form subalpine meadows such as Tuolumne Meadow at an elevation of 8,600 feet.

The flat floor of Yosemite is actually the floor of Ancient Lake Yosemite which was impounded by the El Capitan Bridge recessional moraine blocking the valley and forming a lake 5 1/2 miles long. The *moraine* is the debris deposited by a glacier, along its sides (lateral), in front (terminal or end), between two glaciers (medial) or as it retreats (recessional). Little Yosemite and Tenaya Canyons have similar floors.

GEOLOGIC HISTORY

1. Formation of Paleozoic sedimentary and volcanic rocks.

The oldest rocks are marble, biotite-muscovite-schist and quartzite. They were originally limestone, volcanic rock and sandstone. They are in the form of roof pendants in the Sierra Nevada Batholith both deformed and metamorphosed (hornblende hornfel facies). Not much is known about these Paleozoic rocks.

2. Formation of Jurassic beds.

During the Jurassic a total of six separate formations were intruded. First was the *Quartz diorite of the south fork of Tuolumne River*, this was followed by the simultaneous intrusion of the *Granodiorite of Double Rock*, the *Granodiorite of Mount Hoffman and the El Capitan Granite* which covered a larger areal extent. Next came the *Granodiorite of Rancheria Mountain* and finally the *Alaskite of Ten Lakes*. According to the potassium-argon dating method, the probable time of intrusion was Upper Jurassic.

3. Formation of Mesozoic intrusions.

They are in the form of small intrusions usually less than a mile in diameter, widely scattered and diorite in composition.

4. Formation of Cretaceous beds.

These beds are also intrusive in nature starting with the *Quartz diorite of Tamarack Creek*, which is dated as lower Cretaceous. Upper Cretaceous began with the simultaneous intrusion of the *Yosemite Creek Granodiorite of Rose* and the *Quartz diorite of Mount Gibson*, the former being more extensive than the latter. The *Sentinel Granodiorite* was next, followed by the *Half Dome Quartz Monozonite*. Last of all was the formation of an *Aplite* (fine grained granite).

5. Formation of extrusive rocks.

Sometime during the Miocene (?) and Pliocene epochs of the Neogene period, volcanic activity occurred and a lava flow of *trachyandesite* composition plus *volcanic mud flows* (lahars) occurred north of the Grand Canyon of the Tuolumne River. All of the above are overlain in scattered patches by Pleistocene glacial deposits.

6. During the Eocene or Oligocene time, all land between the Pacific Coast and Rocky Mountains was uplifted, faulted and tilted to form a steep east slope and gentle west slope.

Although the Sierra Nevada batholith formed during the Mesozoic the mountains did not form until the Cenozoic. This was a gradual process which occurred in a series of uplifts. A large block 400 miles long and 80 miles wide was eventually going to develop into the Sierra Nevada Mountains of today. The first uplift was only a moderate height during the first part of the Paleogene, tilting was slightly westward.

7. Formation of Merced River which flowed southwest into the great valley of California Sea.

The second uplift amounted to several thousand feet and included both the land to the east and the block itself, plus increased the slope. Originally the main streams drained in many directions, but the change in slope and uplift caused them to shift their courses and flow westward parallel to each other. Among these was the Merced River that flowed very slowly in a low broad valley, with rolling hills only a couple of hundred feet high. El Capitan (now 8,000 feet high) rose only 900 feet above the floor of the valley. This was to become the now famous Yosemite Valley.

8. Merced River meandered back and forth with a low gradient during the Miocene, then uplift and entrenchment.

The Merced River continued to widen the valley, but as the period wore on, uplift caused it to entrench and form a narrow gorge within the broad valley. It did not stop until it had retrenched 1,000 feet, in spite of the gradually reducing gradient.

9. During Pliocene time the Sierra Block was uplifted several thousand feet, side streams were unable to keep up with the Merced.

With the uplift and mass wasting the valley became a "Two Story Valley." The gentler slopes were at the highest level of the original valley and steeper slopes of the modified entrenched valley. The tributaries that flowed parallel to the faulted edges did not have the gradient increased and hence did not downcut as rapidly as the main stream, so soon developed hanging valleys. As the downcutting of the main stream slowed most of the tributaries managed to catch up to some degree, all except the ones that were flowing on very resistant unfractured granite. The hanging valleys formed were Yosemite, a cascade of 600 feet (2,565 feet today); Bridal Veil 620 feet today and Sentinel Creek Falls (about 400 feet today) were both about 900 feet high; Meadow Brook (1,170 feet today) was about 1,200 feet high and Ribbon Creek (1,612 feet today), was about 1,400 feet.

This uplift caused several major faults along the east side forming Owens Valley which became stable and permitting the block to rise even more.

10. During Pliocene time final uplifting to present 14,000 feet, second entrenchment.

About two million years ago the final uplift and depression occurred permitting Owens Valley and the Sierra Nevadas to reach their present elevation. The increased gradient changed the Merced to a youthful stream again and it became entrenched for a second time forming a "Three Story Canyon."

The stream carved its latest canyon between 1,500 feet to 2,000 feet deep, giving the entire three story canyon a depth of some 3,500 feet.

As before, because of location or type of bedrock, some of the tributaries were unable to keep up and a second set of hanging valleys at a lower elevation were formed such as some of the hanging valleys in lower Merced Canyon. As well as in Illiouett Creek (600 feet), Indian Creek (2,400 feet) and Meadow Brook (2,300 feet).

A few creeks were able to catch up, such as Bridal Veil Creek whose location gave it a steep valley which now forms the lip of the falls, but at that time it was the floor of the valley. The highly fractured rock (easier to erode), direction of flow (southwest), and paralleling the tilt permitted Tenaya Creek to catch up also. The valley then was only 2,400 feet deep (by El Capitan) and narrower than it is today.

11. Pleistocene glaciation, two glaciers more than 2,000 feet thick, one in Tenaya Canyon, the other in Little Yosemite (Upper Merced).

Mountain glaciation occurred at least three times followed by the warmer interglacial periods. The first two advances were much longer and more erosive than the last.

Tuolumne glacier was fed by mountains such as Dana, Gibbs, Lyell, Maclure, etc. It was four to five miles wide and flowed around Mount Hoffman and divided into two glaciers. One flowed in Big Tuolumne Canyon and Hetch Hetchy Valley. The other crossed a divide that was 500 feet high and separated the Tuolumne and Merced basin. It flowed through Tenaya Canyon into Yosemite Valley. The entire valley was under ice. Glacier point which today rises 3,200 feet above the floor was covered with some 500 feet of ice. The peaks of El Capitan, Eagle Peak, Sentinel Dome and Half Dome were *nunataks* (peak of mountain projecting above ice).

At an elevation of 2,000 feet the first two advances melted. The first flowed as far as El Capitan and Cathredral Rocks, the second as far as El Portal. They totally destroyed the inner gorge, cut the projections flush to the sides of the walls. Opposite Bridal Veil Falls the valley was deepened about 600 feet and around 1,500 feet by Glacier Point.

The 2,000 foot thick valley glaciers of Tenaya Canyon and Little Yosemite joined together to form a piedmont glacier and filled the main Yosemite Valley. The smaller valley glaciers shaped their valleys according to their respective joint systems. Tenaya Glacier moved parallel to one set of joints and formed a canyon with a gently sloping floor and was carved some 2,000 feet deeper. However the one flowing down Little Yosemite which is underlain by fairly massive granite, flowed against the joints and glacial plucking formed a glacial stairway over which Vernal and Nevada falls flows.

12. During the last advance the valley was filled to a moderate depth and flowed only to Bridal Veil Meadow.

The last advance was the smallest and the valley was filled about one-third of its depth. It flowed only as far as Bridal Veil Meadow just below El Capitan. It did practically no erosion, and only made the glacial features more prominent at the head of the valley. As it retreated it left a series of recessional moraines, one of these dammed up Ancient Yosemite Lake.

13. About 20,000 years ago glaciers melted leaving present landscape. Appearance of Sequoia Trees.

The glaciers scraped and polished the bedrock forming *glacial polish* on Fairview, Lambert and Pothole Domes. *Glacial pavements* are very extensive especially around Lake Tenaya. The Mono Indians called the lake "Py-we-ack" or "Land of Shining Rocks." Many are found at 8,000 to 9,000 feet. *Glacial erratics* or transported boulders are scattered all over. Glacier Monument is 1,500 feet high and has erratics on it from a mountain that is located 12 miles to the east. The roche moutonnees Fairview and Lembert were formed, and the hanging valleys with their waterfalls came into existence.

Sequoia trees started growing about 2,000 years ago and can be found in three groves:

a. Mariposa Grove—200 trees.
b. Tuolumene Grove—25 trees.
c. Merced Grove—20 trees (also stand of sugar pine).

14. Formation of deltas, stream erosion, exfoliation and landslides.

The streams are still eroding and gradually filling in the shallow lakes with deltas in Little Yosemite and Tenaya Canyons. Landslides in Tenaya Canyon have dammed up the creek and formed Mirror Lake. Other rock slides (perhaps triggered by quakes) have filled in the valley with long slopes.

Exfoliation is continuing forming the domes and arches. About 25 small glaciers can still be found in Yosemite Park. The two largest are Lyell Glacier and Maclure Glacier. All of these are producing glacial features today, only on a smaller scale.

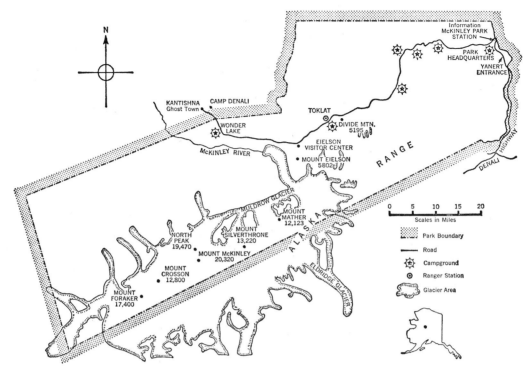

Mount McKinley National Park, Alaska.

Tentative Geologic Column of Mount McKinley National Park

Erathem	Period	Epoch	Formation
Cenozoic	Neogene	Holocene	Alluvium
		Pleistocene	Glacial Deposits
	Paleogene	Eocene	Kenai Formation
		?	
Mesozoic	Cretaceous		Cantwell Formation
			Unnamed Slates and Graywacke
			Unnamed Argillite and Graywacke
	Jurassic		Conglomerates and Other Clastic Rocks
	Triassic		Unnamed Limestone and Basalts
Paleozoic	Permian		Siltstones and Shales Makomen Formation (McCallum Creek Sequence).
	Pennsylvanian to Devonian		Rainbow Mountain Sequence (Probable Correlation with Tetelna and Chisna Formation)
	Silurian	Lower	Totalinika Schist and Shales
	Ordovician	Upper	with Graptolites, Siltstone, Sand-
		Middle	stone, Chert and Limestone
		Lower	Unnamed Limestone and Shale
Precambrian	Precambrian Z		Birch Creek Schist

Modified after Brooks, Churkin, Cobb et. al.

CHAPTER 9
MOUNT MCKINLEY NATIONAL PARK

Location: S.E. Alaska
Area: 1,939,492.8 acres
Established: February 26, 1917

LOCAL HISTORY

Mount McKinley has been known by many names. The Russian traders called it Bulshaia Gora, while the natives of Cook Island called it Traleika. Both meant "Big or High Mountain." The Alaskan Indian referred to it as Denali or "The High (Great) One" or "Home of the Sun." Prospectors that first came to Alaska called it Densmore's Mountain.

W.A. Dicky, a sourdough prospector, in 1896 named it after the presidential candidate for the Republican Party who was strongly for the gold standard—William A. McKinley. This was adapted as the official name.

Its height and position was measured by two geologists of the United States Geological Survey in 1890. They were George H. Eldrich and Robert Muldrow, who determined it was 20,300 feet high, a miscalculation of only 20 feet. Mount McKinley has two peaks, the former North Peak, now officially known as Winston Peak with an elevation of 19,470 feet; plus the high peak located two miles south of Winston Peak with an elevation of 20,320 feet. The former South Peak is now called Churchill.

A United States Geological Survey Party in 1902 led by Alfred H. Brooks, L.M. Prindle, and D.L. Rearburn plus four other men were the first men to reach its base. Brooks climbed to 5,000 feet and left a rock cairn.

In 1903 Judge Wickersham and a party of four led the first attempt to climb Mount McKinley. Unfortunately he chose the north face by way of Peters Glacier which is very sheer and is still today almost impossible to climb.

The present day ghost town Kantishna, sprang up in 1905 when gold was discovered at nearby Kantishna River. In 1910 prospectors William Taylor, Pete Anderson, Thomas Lloyd and Charles McGonogill started out together. Anderson and Taylor climbed Winston Peak and created the "Mudrow Glacier Route." The feat is really amazing because they were really not equipped for climbing. They carried a 14 foot flagpole to show people in Fairbanks that they really climbed it. They thought they had climbed to the top and had actually started out for the other peak but changed their minds.

It was mapped and surveyed again in 1910 by J.W. Bagley and S.R. Capps. Other teams mapped it in 1916, 1919, 1920, 1925, etc. In 1913, Churchill Peak, the true summit was reached by the Hudson Stuck party. (Harry Karstens, Walter Harper [native Alaskan], and Robert Tatum). In 1932 Alfred Lindley, Harry Liek, Erling Strom and Grant Pearson were the first to climb both peaks. McKinley is a very dangerous mountain to climb because of its unpredictable weather ranging from freezing, dense fog to blizzards. Hence between 1913 and 1967 only 39 parties have made it to the top with a total of 116 people (1913-1961). The first woman, Barbara Washburn, climbed to the top in 1947 in an expedition led by her husband. In 1967 a winter ascent was made and one member of the party fell to his death into a crevasse. That same year seven members of a 12 man team died of exposure and fatigue during a storm at 17,000 feet making it one of the worst disasters in mountaineering history. Other climbs have been made successfully since then, but so have there been other deaths such as a three man Japanese team in 1972.

Charles Sheldon, a naturalist and conservationist, while collecting specimens over a three year period for the National museum, conceived the idea that the McKinley area should become a national park, back in 1909. With the help of the Boone and Crockett Club the creation of the park came into being in 1917. It is the second largest National Park.

GEOLOGIC FEATURES

Mount McKinley is the tallest mountain in the world if it is measured from its base. It is some 17,000 feet high. Elevation wise it is about 8,700 feet lower than Mount Everest at an elevation of 20,320 feet.

It is not the only peak in the park, some of its neighbors are Mount Foraker (Denali's wife), 17,400 feet; Mount Silverthorne, 13,220 feet; Mount Crosson. 12,800 feet; Mount Mather, 12,123 feet; Mount Russel, 11,670 feet; Mount Eielson, 5,802 feet; and Divide Mountain 5,195 feet. They are all part of a system that is called a *synclinorium* which is a series of anticlines and synclines that are overall arched downward. They are usually measured in tens of miles. The synclinorium has a major fault on the west side called the Denali fault system that extends some 1,300 miles. It separates the oldest rocks of Alaska from the youngest.

Because of the elevation *mountain or valley glaciers* are found on the slopes. The largest glaciers are located on the south side of the mountains and in the basin of Chulitina and Yentna Rivers because the winds coming off the Pacific Ocean are ladened with moisture. The glaciers on the opposite side (north) do not receive as much moisture and are much smaller as a result of this.

The mountains of the eastern portion of the park are only a moderate height, therefore the glaciers are relatively small, only two or three miles long. The central portion is higher and the glaciers increase in size becoming four to eight miles long. The western portion, the highest, has glaciers that are 30 to 40 miles long. Muldrow Glacier is the largest in the park.

During the Pleistocene, McKinley National Park was at the northern limit of the continental glaciers that left this region some 10,000 to 14,000 years ago, they did not cover the park to any great extent. The glaciers that exist today are mountain glaciers.

The streams in the park have a *braided stream pattern* with intertwining channels that are constantly changing and are rarely any deeper than four feet. They are usually a milky white color because of transporting *rock flour* or finely pulverized rock debris from the glaciers. The braided streams are the result of both a heavy sediment load and a reduced gradient as they leave the mountains.

An *outwash plain* has been formed by these glacial streams as they meander back and forth depositing sand and gravel as an apron at the flank of the mountains and on the valley floor. The lakes are *kettle lakes* formed when segments of retreating glaciers leave a section of ice buried in the outwash. When the ice melts a depression or *kettle* forms which later may fill with water forming a lake.

Deep scratches or *glacial grooves* scour the sides of the glacial valleys, the smaller scratches are called *glacial striations.*

GEOLOGIC HISTORY

1. Formation of Precambrian or early Paleozoic beds. Erosion, deposition and additional metamorphism.

Deposition of sedimentary beds (sandstone, shale, limestone) which have been intruded by igneous beds. Metamorphism changed the beds into mainly a quartz, mica, and chlorite schist and phyllite, with greenstone, marble, serpentine and gneiss occurring in minor amounts. They are named the Birch Creek schist, and are found lying parallel to the axis of the Alaska Range that lies predominately within the park boundaries. Erosion occurred producing an unconformity with the pre-Devonian beds, indicating a possible diastrophism.

Additional sedimentation occurred and these beds were regionally metamorphosed into slates, serpentines, altered acidic (rhyolite) flows. Some contact metamorphism occurred. Faulting has disrupted the metamorphic facies (Totatianika schist and undifferentiated beds). A limestone containing *graptolites* (extinct colonial animal with a primative backbone, resembles a saw blade in appearance) is found nearby and possibly can be correlated with the metamorphosed beds, thus making them Ordovician in age.

2. Deposition of sediments in the Cordilleran geosyncline during the Silurian, Devonian, and Mississippian, folding and erosion. Deposition during Upper Paleozoic.

First shale (Tatina Group?) was deposited, it was possibly Silurian in age. The deposition of chert and limestone was around Middle Devonian time (Tetelna). These are both detrital and volcanic and very siliceous. Dating was possible because of graptolites. Corals are abundant in the limestone. The accumulation was several thousand feet thick and then intrusion by dikes and sills occurred plus volcanic activity. Folding and widespread erosion removed the Late Devonian and Early Carboniferous (Mississippian) beds. Deposition of limestones, volcanics and some sandstones was at the eastern end of the Alaska Range, during Upper Paleozoic.

3. Formation of Skwenta complex during Triassic (?) Jurassic. Uplift, erosion, intrusion of Batholith.

These are found on the south side of the range only, they are both intrusive and extrusive. Lavas, tuffs, graywackes ("dirty sandstone") and some limestones occur. An unconformity is above and below this complex. The beds have been slightly folded, possibly prior to the intrusion of the granitic batholith. Economic minerals such as gold, silver, lead, zinc and antimony were formed.

4. Deposition of Tordrillo series during Middle Jurassic, more erosion.

The series is a mixture of dense feldspathic sandstones, graywackes, limestone and slate (shale). High mountains and volcanoes were the source of these sediments. Adjacent areas were under water. Some of the beds contain plant fossils. Deformation of beds along the axis of the Alaska Range occurred causing additional erosion. Unidentified beds, slates and graywackes were also deposited.

5. Deposition of Cantwell formation and volcanic activity during Lower Cretaceous, unidentified beds above.

The Cantwell formation is a continental gravel deposit made up of well sorted, rounded quartz pebbles. It accumulated to a thickness of up to 4,000 feet, indicating a long period of erosion of the nearby highlands. Lava flows interbedded in the Cantwell indicates volcanic activity. Unidentified beds are found above in other areas, but erosion has removed these beds from the McKinley area.

6. Formation of synclinorium, youngest rocks in the center, oldest on the flanks. Folding several times, intrusion and uplift.

The beds were folded into a series of anticlines and synclines which were overall depressed forming the synclinorium. During and after the folding, granodiorite (intermediate igneous rock) *stocks* or intrusions with an aerial extent of less than 40 square miles, intruded these beds. They form the bulk of Mount McKinley and Mount Mather. Both of which are in the area of highest uplift that is now called the Alaska Range.

7. Erosion and deposition of Paleocene (Eocene) beds.

The uplift was of a long enough duration that a well-defined drainage system formed. Deposition of yellowish clay, and fairly pure sandstone occurred. Swamps permitted the accumulation of vegetation which have now changed to *lignite,* a soft brown coal. The sediments are land deposits about 1,200 feet thick (Kenai formation).

8. Uplift and formation of the Denali fault that elevated McKinley area and depressed the Copper River Basin.

The uplift produced a great deal of deformation. The Denali fault formed as a series of longitudinal faults cutting across the synclinorium which is marked by valleys and passes, especially along the north arm. The major fault at the base of McKinley is the Denali fault.

9. Pleistocene glaciation, retreat of glaciers. Formation of valley glaciers.

Both glacial and interglacial deposits are found. However, most of the streams are of glacial origin. Traces of continental glaciation can be seen but are not well developed since the area was on the edge of the ice sheet. The ice sheet left 10,000-15,000 years ago. Then the formation of the present day valley glaciers which have also left their mark when they were more extensive.

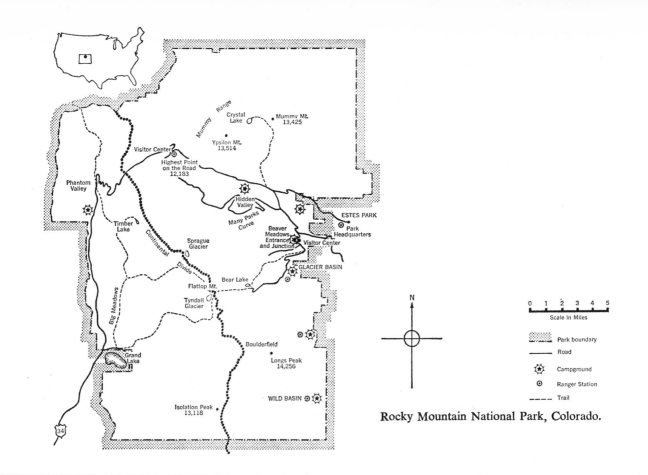

Rocky Mountain National Park, Colorado.

Tentative Geologic Column of Rocky Mountain National Park

Erathem	System	Series	Formation
Cenozoic	Neogene	Holocene	Neoglaciation (Modern)
		Pleistocene	Interglacial Stage
			Pinedale Glaciation
			Interglacial Stage
			Late Bull Lake Glacial
			Interglacial Stage
			Early Bull Lake Glacial
			Interglacial Stage
			Older Intermediate Glacial
		Miocene	Sand, Silt, Volcanic Ash (18 M.Y.)
	Paleogene	Oligocene	Sand, Silt, Volcanic Ash (26 M.Y.)
			Rhyolite, Welded Ash Flows, Obsidian Flows, Pyroclastic Debris
			Volcanic Ash (34-37 M.Y.)
		Paleocene	Andesite Lava Flows and Andesite Plugs (54 M.Y.)
Mesozoic	Cretaceous		Shale (Denver [?]) Formation
Precambrian	Precambrian Y		Gray Granite and Pegmatite Silver Plume Granite (?) (1450 M.Y.)
	Precambrian X		Light Gray Granite Boulder Creek Granite (?) (1720 M.Y.)
	Precambrian X		Gneiss and Schist Basement Complex (1800 Million Years)

Modified after Alt, Hyndman, Richmond et. al.

CHAPTER 10

ROCKY MOUNTAIN NATIONAL PARK

Location: N. Colorado
Area: 262,191.16 acres
Established: January 26, 1915

LOCAL HISTORY

The Indians called this region their home from 8,000 years ago up to about the last 100 years. This was Arapahoe country. An Indian Trail called Taleonbaa, or "Children's Trail," was named because of its steepness. It was so steep that even the children had to dismount their horses and walk. It was used by both the Arapahoe and Utes to cross over the Rocky Mountains and is still easily found, because such deep ruts were formed.

Some 40 Indian campsites are found around the park that were used periodically. They are located around Forest Canyon Pass, Milner Pass and the various meadows found at lower elevations.

Lieutenant Zebulon M. Pike (Pikes Peak was named after him) explored the region in 1906 and referred to the tallest peak as Great Peak but did not go near it. Colonel Stephen H. Long's expedition commissioned by President Madison mapped the region in 1820. His party located the peak that Pike called Great Peak and renamed it Longs Peak after their leader. Other people passed through the region, in 1825 Richard Dodge, 1939 Eliza Farnham, 1839 Frederick Wislizenus, 1840 Kit Carson (trapped game), and also Rufus Sage, who is the first documented person to enter the park. The Fremont expedition went through in 1843-44. Other visitors were Francis Parkman in 1846 and Frederick Roxtan in 1847.

Joel Estes and Milton Estes (his son) in 1859 saw the valley which now bears their name. They built a cabin in 1860 along Willow Creek (located in the foothills). Their claim eventually ended up in the possession of Griff Evans (1867) who later transferred it to the Earl of Dunraven. Dunraven in 1872 tried to acquire the region as a game preserve. One method he used was to pay people to file claims in their name and then transfer the claim to him. However, they were declared invalid. He was useful to the area, as he helped to keep out harmful enterprises that would have ruined the region. Dunraven kept a ranch in Estes Valley and built the first hotel in this region.

In 1864 William N. Beyer tried to climb Longs Peak and failed. Undiscouraged, he kept trying and finally made it on August 23, 1868. Included in the party were Major Powell (Colorado River explorer), N.E. Farrell, S. Gorman, S.W. Keplinger and J.C. Sumner.

The first guide up Longs Peak was an Archdeacon, E.J. Lamb. He climbed it for the first time in 1871 and led other parties up afterwards.

Enos Mills, when he was only 16 years old, built his home (1866) in Longs Peak Valley. He had helped a survey party map Yellowstone in 1891. After this experience, he felt his homesite area should be preserved the same as Yellowstone. He fought very long and hard and finally succeeded, it was established in 1915. He climbed Longs Peak in 1903 by the east face which is a very difficult route.

GEOLOGIC FEATURES

The Front Range is about 200 miles long and from 30 to 50 miles wide. Rocky Mountain National Park is located in the north-central portion. It consists primarily of Precambrian gneiss, schist and granites that were uplifted into an anticline during the Laramide Revolution. The overlying Paleozoic and Mesozoic beds have been removed by erosion exposing the Precambrian beds.

The arching of the igneous and metamorphic beds produced joint systems which have weathered into such features as the Needles and Twin Owls. Exfoliation domes are also present such as McGregor Mountain.

One of the main geologic features in this region is the existence of well-developed erosional surfaces. After the Front Range was elevated by the Laramide Revolution, the erosional forces took over stripping the bedrock and depositing it at the base of the mountains as a series of alluvial fans. The high mountains were gradually reduced to low hills separated by wide shallow valleys. The highest of these stood barely 2,000 feet above the sloping plain. This was the Flattop erosional surface.

At the same time erosion was occurring, volcanic activity was pouring out pyroclastic debris carrying dust and ash everywhere. Eventually this material consolidated into *tuff*. The flows were of intermediate composition and solidified into andesite. Streams flowing on this volcanic surface were carrying their load of weathered rounded boulders of gneiss, schist and granite plus the subangular fragments of the andesite flows and rhyolite tuffs. The streams were overwhelmed by a lava flow that immediately solidified into an *obsidian* or natural glass and consolidated this material into a breccia.

Uplift occurred, elevating the eroded surface. The streams with their new base level started to downcut very rapidly. The only other evidence of this igneous activity is a series of intrusions at Milner Pass around the Poudre Lakes.

Some of the areas were more resistant and today they are mountains with their flat eroded tops left 11,000 feet high as reminders of their past. They are Longs Peak, Taylor Peak, Otis Peak, Hallet Peak, Flattop Mountain, Stones Peak, Trail Ridge, Mount Chapin, Mount Chiquita, Ypsilon Mountain and Mount Fairchild.

Uplift and erosion occurred several times, but never long enough to create another erosional surface until the Pliocene when the Rocky Mountain erosional surface developed. Its erosion remnants are such features as Gianttrack Mountain, Old Man Mountain, Deer Mountain, Castle Mountain, The Needles and Twin Owls.

The last uplift permitted the streams to cut and shape the mountains into the present forms that we are familiar with today with one exception and that is the glaciation during the Pleistocene that added the final touch of forming Rocky Mountain National Park.

There were at least two periods of glaciation, but only the last stage can be accurately mapped. Inside of park boundaries the following glaciers that occurred were Colorado River Glacier, Fall River Glacier, Forest Canyon Glacier, Glacier

Gorge Glacier, North Inlet Glacier, Cache La Poudre Glacier, Monarch Glacier and Middle and North St. Vrain Glaciers. The Falls River Glacier is responsible for Horseshoe Park with its terminal, lateral and ground moraine, while Moraine Park was the melting basin of Thompson Glacier, Bartholf Park for Bartholf Glacier, a basin in North St. Vrain valley between moraines for Wild Basin Glacier, and the Meadows for North Fork Glacier and its tributaries. Several lakes have formed on the moraines such as Burstadt Lake, Copeland Lake, Bear Lake, Dream Lake and Sheep Lake. They are *kettle lakes.*

At the bottom of the cirques left by the Wisconsin glaciers, many lakes have formed called *tarns.* Examples of these are Blue Lake, Black Lake, Shelf Lake, Lake Mills, Lawn Lake, Crystal Lake, Fern Lake, Odessa Lake, Tourmaline Lake and Chasm Lake.

Lakes can form in other ways. For example, *rock basins* form as a glacier moves down sloped and encounters weak zones, plucks the material out, forming a basin. After the glacier melts the basins fill with water. Since it is a series of basins, they are all drained by a single stream like beads on a string. They are called *pater noster lakes* because of their resemblance to prayer beads. Lakes such as the Gorge Lakes, Loch Vale Lakes, Cub Lakes, East Inlet Lakes (including Lake Verna) and Ouzel Creek Lakes.

Other glacial features are abundant, e.g., hanging valleys, examples are valleys in which Lake Nanita, Lake Nokoni and Bench Lake occur. Also Roaring River, (Horseshoe Falls) Chiquita Creek, Sundance Creek and Fern Creek Valley. Roche moutonnees can also be found (Glacier Knobs).

A *stream valley* has a V-shaped profile (Fox Creek, Red Creek and Cow Creek Valleys), the scouring action of glaciers widens and smooths out the floor, changing it to a U-shape. (Fall River Valley, Forest Canyon, Fern Lake Valley, and North St. Vrain Valley).

Glacial erratics, bedrock transported elsewhere by the glaciers, are abundant at Boulder Field and can be found on the sides of the canyon of Fern Creek. *Glacial pavement* occurs around North Inlet. Moraines are scattered all over the park and are easily located since the Lodgepole pine prefers to grow upon them.

A few modern valley glaciers exist in the park today, they are Rowe (Hallett) Glacier, Sprague Glacier, Tyndall Glacier, Andrews Glacier, Taylor Glacier, and Mills Glacier. Scattered snow and ice fields also can be found. In 1906 Sprague Glacier was measured as having movement about 2.5 feet per year.

Solifluction a form of mass wasting (soil flowage is permafrost regions) it is evident on Sundance Mountain and below Trail Ridge Road.

On Longs Peak is a shelf called The Narrows. It was created when movement along a fault dropped a section of bedrock down.

There are 42 mountain peaks that exceed the elevation of 12,000 feet. Some of them are as follows: Mount Evans, 14,260 feet (highest); Longs Peak, 14,256 feet; Mount Meeker, 13,911 feet; Chief's Head, 13,579 feet; McHenry's Peak, 13,327 feet; Mount Lady Washington, 13,281 feet; Taylor Peak, 13,153 feet; Flattop Mountain, 12,342 feet; Ptarmigan Mountain, 12,324 feet; Snowdrift Peak, 12,274 feet; Mount Adams, 12,121 feet; and Battle Mountain, 12,044 feet.

This region is located on the *Continental Divide,* which is the dividing point that will determine if a stream will eventually flow east or west. Streams that are located on the east side of the

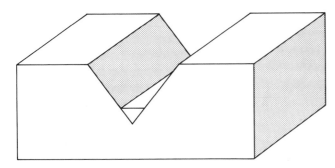

V-Shaped Stream Valley

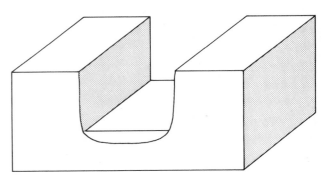

U-Shaped Glacial Valley

divide drain into the south Atlantic Ocean. Those on the west into the Colorado River and eventually into the Pacific Ocean.

GEOLOGIC HISTORY

1. Accumulation of sediments during the Precambrian X.

Over 18 million years ago sandstone, shale, and limestone accumulated in a large linear trough that sank as rapidly as the sediments accumulated, called a *geosyncline*.

2. Destruction of geosyncline by folding, formation of magma chamber and metamorphism.

Eventually the geosyncline stopped sinking because a balance had been reached, so it began to fill up. Once the filling process began, the principle of *isostasy* took over. Isostasy is a condition of balance. For example, if a person dumped into a tub of water pieces of wood all different sizes, the big blocks would sink deeper than the little blocks, because the wood displaces an equal volume of water. If you took one block of wood and placed it on top of another it would sink deeper. If you took it off, the lower block would rise. Basically the same thing happens to the earth. Removing material from one section by the erosion of a mountain range and accumulating the sediments in a geosyncline causes an imbalance that must be adjusted. The crust of the earth is divided into two separate layers according to its composition. The *sial* layer made up of minerals with predominately silicon and aluminum in them and the *sima* layer consisting mostly of the elements silicon and magnesium. The sial layer is much lighter than the sima layer. Thus the continents are made up of sial (main rock granite) and rest or lie upon the sima layer (main rock basalt) beneath, which also makes up the ocean floors. When mountains occur the sial layer is thicker, these are the so-called "roots" of the mountain. As the sediments are removed by erosion, the lighter sial layer is pushed up gradually, (uplift which helped to produce the Flattop and Rocky Mountain Erosional Surfaces) until the entire mountain is eroded away.

Conversely, dumping sediments into the geosyncline causes it to sink until the forces are equalized. The transfer of material needs to be adjusted so forces beneath begin to force up the sediments in the geosyncline. This causes them to be bulged upwards and folded.

As the folding increases, heat builds up because of friction and other causes, some 30,000 to 40,000 feet down metamorphosing these sediments.

The folding produced uplift which took some of the pressure off and the rocks melted forming a *magma chamber*. The magma chamber metamorphosed for the second time surrounding metasediments into gneiss and schist by the heat, pressure and mineralizers about 1,720 million years ago (Precambrian X). These beds can be seen today in the central and western parts of Rocky Mountain Park such as along Trail Ridge Road and Big Thompson Canyon.

3. Uplift produced a mountain range, magma chamber intruded the country rock.

The magma cooled into granite. A second larger intrusion occurred in the same region.

The metamorphosed zone around the magma chamber is called an *aureole*, hot water solutions and other materials are injected into the aureole zone, such as dikes, sills and pegmatites (hydrothermal solutions). This magma chamber solidified into a lighter color granite. The radioactive minerals these rocks contained allowed geologists to date them and they determined this event occurred 1,450 million years ago (Precambrian Y).

4. Erosion of the Precambrian Mountains.

These mountains underwent erosion, from 1,450 to 530 million years they were eroded almost totally away, exposing only the granite in the region of the park. A third possible intrusion may have occurred a basaltic dike called "Iron Dike" that was intruded during Precambrian Y 1,050 million years ago (Mount Chapin and Storm Pass).

5. Transgression of the Cordilleran geosyncline during the Paleozoic, formation of Ancestral Rockies.

Some 530 million years ago the Cordilleran geosyncline transgressed on the landscape depositing sandstone. The higher areas were islands. For 200 million years the Paleozoic sediments formed. About 300 million years ago in the Front Range area, gradual uplifting formed what are called the "Ancestral Rockies." Sometimes they were low, other times a maximum of 2,000 feet high. All the time they were furnishing sediments to the surrounding area. So the Paleozoic beds were stripped away and also some of the Precambrian igneous and metamorphic beds.

6. During the Mesozoic, formation of the Mesocordilleran geanticline and the Rocky Mountain geosyncline.

The Front Range region was a low swampy region 240 million years ago over which dinosaurs roamed on the geanticline. They were only about 500 feet above sea level. Their footprints and remains are found in areas nearby the park. Some coal beds formed in the swamps and near by sand dunes migrated near the inland sea (never came as far as the park).

7. During Late Mesozoic, sea covered area, deposition, then uplift.

The seas finally covered the region completely 100 million years ago depositing sands and clay on the shallow sea floor. Some of these shale beds can be seen east of Lead Mountain, they are about 75 million years old. Eventually the region that was to become the Front Range was gradually raised 70 million years ago to form a low island in the sea. It took the seas some 16 million years to retreat from all of Colorado.

8. Uplift continued, sediments stripped as fast as the island rose.

As the rate of uplift increased (the mountains rose to 2,000 feet) so did the stripping of the beds. The region became low and swampy, coal beds formed once again. The stream stripped the Front Range as fast as they uplifted, gradually exposing the Precambrian beds. These sediments were deposited, forming a new layer of sediments upon the surrounding area. The uplift produced faulting forming north-south trending blocks of the older beds causing the newly deposited sediment covering them to fold into broad folds. Eventually the sediments faulted on the western side and the Precambrian beds either slid down or spread over younger ones.

9. Increased uplifting forms magma chamber, intrusion, extrusion, erosion.

For a second time a magma chamber formed, intruding the metamorphic and igneous beds. Some of the intrusions reached the surface about 68 million years ago and poured out onto the land forming a series of lava flows mixed in with pyroclastic debris. Much of this debris fell into the ocean and was buried with the sediments to become bentonite. Eventually most of the flows were removed by erosion leaving only isolated remnants. The elevation was about 3,000 feet above sea level about 2,500 above the plains. A total of some 5,000 feet of sediments were removed about 67 million years ago, they were deposited as flood plains to the east. These sediments contain dinosaur fossils today.

10. Climax reached, rocks folded sharply, faulting of Precambrian rocks during Laramide revolution. Intrusion and mineralization.

As the mountain range rose another 5,000 feet between 65 and 54 million years ago, the sedimentary beds on its flanks were first tilted gently. This rapid uplift and tilting caused faulting. This in turn produced volcanic activity, one of them erupted where Mount Richthafer (east and south slopes is located). In some localities the faults were almost vertical and beds were simply elevated or lowered. In others *thrust faulting* (low angle reverse fault) caused the older beds to be thrust upon the younger. Along the faults dikes intruded, with them they carried *hydrothermal solutions* (hot water with ions) which filled in the openings in the bedrock pegmatite veins containing valuable mineral deposits such as gold, silver, lead and zinc. Height of mountains between 4,000 and 6,000 feet.

11. Stream erosion shaped the mountains, eroded them and recarved them several times.

At the end of the Cretaceous the Flattop erosional surface formed from 4 to 40 million years ago lowering them to 3,500 feet destroying the *first set* of Rocky Mountains. Weathered residual soil can be found such as along Deer Ridge. Vegetation indicates climate changed from subtropical to savannah to humid forest to hardwood forests. Uplift and stream erosion about 40 million years ago carved the *second set*. Beds were tilted to form hogbacks but were also eroded to form the Rocky Mountain erosional surface. Some remnants can be seen at Big Thompson Canyon. The region was dissected by streams permitting them to become entrenched. About 37-34 million years volcanic activity occurred, burying the adjacent areas with debris, although they themselves since they were 6,000 feet high above sea level. This is when the plant and insect remains were preserved in the Florissant beds (outside of park by Colorado Springs).

Volcanic activity occurred north of the park (Lulu Mountains) some 28 million years ago pouring out the sheets of lava, pyroclastic debris, breccia and obsidian. A volcanic ash flow from that area poured into the park. The remnants of this flow is called Speciman Mountain. These beds can be seen on Iceberg Lake Trail.

Then 26 million years ago the erosional processes stripped many of these sediments from the high areas and deposited them in the lower regions (Grand Lake area). Uplift occurred 18 million years ago with stream erosion creating 200-300 feet deep valleys and in turn burying the surrounding regions in debris. Volcanoes in adjacent areas also contributed material. Mountains were still around 6,000 feet high but seemed lower since surrounding areas were buried. Life at this time were camels, horses the size of ponys and sabertoothed tigers. Vegetation was confined to conifers and some broad-leaved trees.

Uplift once again 18 million years ago to 9,000 feet, stripping the region in the park but burying adjacent areas. This continued up to 5-7 million years ago. This locality became a series of rounded hills and low ridges even though it was 8,000 feet above sea level. Remnants of the old plains from the earlier uplifts still can be seen along the tops of many of the mountains that exist today, e.g.: Flattop Mountain and Castle Mountain. The Front Range appeared to rise as a series of steps, particularly the east side of the continental divide. The sloping uplands of today are not part of this ancient surface for it is possible to find 28 million year old volcanics filling valleys in this surface. It is less than 28 million years old and today forms the roof of the Rockies.

The last and final uplift occurred between five and seven million years ago through faulting elevating the region to 12,000 feet (actual uplift was 4,000 to 5,000 feet). Uplift was along faults in a horst and graben system. The downdropped blocks are called basins or parks (Estes Park), some as much as 1,000 feet. The older and younger erosional surfaces are mixed because some have been lowered, others raised. Erosion was disrupted, streams changed courses, waterfalls developed and the present landscape of today was rapidly forming.

12. Glaciation during the Pleistocene at least twice forming valley and piedmont glaciers.

During the Pleistocene some two million years ago the climate grew cooler and mountain or valley glaciers formed. As they grew they joined together to form piedmont glaciers. They did not cover the entire park extending down to about 8,000 feet. This was about 1.6 million years ago. They were several miles long and several hundred feet deep. They scraped and gouged the land modifying its shape and left their various deposits scattered all over the landscape. This ended about 600,000 years ago. This event occurred at least twice, as seen by the interglacial soils separating the glacial deposits. Intermediate glaciation occurred between 500,000 and 87,000 years ago. Once again there were glacial and interglacial stages. Only the last two stages can be detected. The oldest occurred 160,000 years ago, the younger called Bull Lake was between 127,000 to 105,000 years ago for first advance and about 100,000 to 87,000 (perhaps as late as 70,000) years ago for the second.

The last glaciation called the Pinedale went down the slopes as far as 8,270 feet. It began some 27,000 years ago and finally disappeared around 7,500 years ago. Thus modifying the mountains that we see today in Rocky Mountain National Park.

13. End of Pleistocene, formation of present glaciers, continued erosion.

Thus the Wisconsin (youngest) glaciers melted leaving their features to remind us that they were there. The few isolated glaciers of today formed 3,800 years ago along with a few scattered snow and ice fields (to be a glacier ice must have moved down slope at some time).

Streams are once again busy eroding and depositing materials trying to form another erosional surface. Who knows, perhaps in the geologic future we may have another set of Rockies.

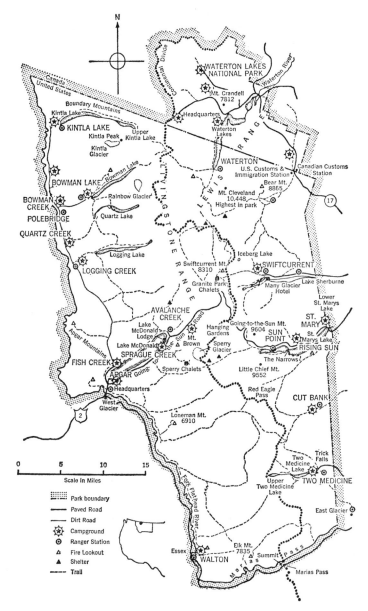

Glacier National Park, Montana.

Geologic Column of Waterton-Glacial International Peace Park

Erathem	System	Supergroup	Group	Formation
Precambrian	Precambrian Y	Belt	Missoula	Mount Shields (Kintla*)
				Shepard Formation
				Purcell Lava
			Unassigned	Helena Dolomite (Siyeh*)
			Ravelli	Spokane Formation (Grinnell*)
				Appekunny Formation
			Pre-Ravelli or Lower Belt	Altyn Limestone
				Waterton Dolomite

*Former Name

Modified after Ross, Alt, Hyndman, Dyson, Alden et. al.

CHAPTER 11

WATERTON-GLACIER NATIONAL PARK

Location: N.W. Montana

Size: 1,013,100.60 acres

Established: May 11, 1910; May 2, 1932 (Peace Park)

LOCAL HISTORY

Indians inhabited this region for about 10,000 years. The most recent tribes are the Kutenai (Flathead) and the Blackfeet who displaced the Kutenai about 200 years ago. They definitely knew about Lake St. Mary and Lake McDonald where they fished and hunted.

The Lewis and Clark expedition of 1806 was in this region and probably saw the Park but made no mention of it, or explored it. The first white man was apparently a Hudson Bay trapper from Canada, Hugh Monroe, who in 1816 visited St. Mary's Lake. His Blackfoot name was "Rising Wolf." Father DeSmet in 1846 named it St. Mary Lake.

In 1853 the editor of *Forest and Stream,* George Grinnell, vacationed there, fell in love with the place, and kept coming back. He was the prime mover in the establishment of Glacier National Park which finally came about in 1910. It was a difficult fight because Congress rejected two bills before it was finally passed (opposition from private interest, Indian problems, congressional committee arguments). In 1932 Glacier National Park in the United States and Waterton Lakes National Park were combined to form the Waterton-Glacier International Peace Park. This came about mainly because of the efforts of Rotary International.

A Major Baldwin in 1889 discovered Marias Pass which up to that time was considered an Indian legend, it was surveyed in December of that year by John Stevens. A road was built in 1895 by settlers to Lake McDonald and the Great Northern Railroad built its line to Glacier in 1892.

Going-to-the-Sun-highway is named after a legendary Blackfoot chief that the Great Spirit sent down to help the Blackfoot tribe to become great

again. When he left, after helping them, he went up the side of the mountain during a terrible storm. When the storm broke, according to the legend, the snow formed a profile of the chief going to the sun, thus the name.

GEOLOGIC FEATURES

The most prominent geologic features of course are the glacial features. The story of glaciers has their beginning in the sky, as can be seen by the outline below.

Water Vapor
↓ Sublimation
Snow Flake (Ice Crystal)
↓ Precipitation
Snow Field
↓ Compaction
Neve
↓ Compaction
Ice Field
↓ Movement
Glacier

The water vapor (a gas) in the air undergoes the process of *sublimation* or changes from a gas directly into a solid, and forms an ice crystal. Ice crystals or snow flakes have six points because they belong to the hexagonal crystal system which has three axes of the same length, lying in a single plane at an angle of 60 degrees to one another. The fourth axis is at right angles to that plane and in the case of snow is much shorter than the other three. Since crystals form along the axes first, a snowflake has six points.

The snow falls and gradually will build up a snow field if more snow falls in the winter than melts in the summer. The snow at the bottom of the field is having the air forced out because of the weight of the overlying beds (*compaction*) gradually changing to a granular snow called *neve*. The neve forms because the pressure on the points of the snowflake causes it to melt and recrystallize in the center, thus forming a sphere. Further compaction forces the air out and changes the material into ice. The ice field builds up and as soon as it becomes large enough that gravity causes it to flow downhill it becomes a *glacier*.

Once the glacier starts to move two zones develop, the bottom half is called the zone of flowage which acts as a *plastic* substance and does not fracture. The upper surface is known as the zone of fracture since it is brittle, this is where crevasses form, which rarely extend any deeper than 150 feet.

The glacial scouring and plucking hollows out the ground beneath the glacier producing a bowl-shaped depression called a *cirque*. The basins that hold the following lakes are all cirques—Avalanche, Cracker, Ellen, Helen, Hidden, Iceberg, Kennedy, Pocket, Ptarmigan, Upper Two Medicine and Wilson lakes.

When two cirques stand side by side there is a thin knife-shaped wall that separates them called an *arete*. Garden Wall and Ptarmigan Wall are examples of these. A *col* forms when the heads of two cirques break down the wall that separates them, forming a saddle-shaped depression. These frequently form passes in a mountain range, Logan Pass, Piegan Pass and Gunsight Pass are no exception. Another col separates Jackson Glacier from Blackfoot Glacier and another one can be found behind Haystack Butte.

A three or four sided pyramid or *horn* forms when the heads of three or more cirques intersect, the famous Matterhorn in Switzerland is the classical example of this. In Glacier National Park many horns also exist such as Clements Mountain, Kinnerly Peak, Little Matterhorn, Reynolds Mountain, Sinopah Mountain, Split Mountain, Mount Wilbur, St. Nicholas and Haystack Butte.

Waterfalls are abundant and have developed for more than one reason. Some occupy hanging valleys such as Avalanche Lake, Bird Women, Dead Horse, and Virginia Falls. Others such as St. Mary, Rockwell, Baring and Florence Falls cascade down glacial steps. The one unique falls is Trick Falls where the top falls flows over resistant beds and the bottom falls usually hidden underneath the top falls, issues out from a cave that formed by the enlargement of the joints in the limestone.

Typical U-shaped glacial valleys are McDonald Valley, St. Mary Valley, Swiftcurrent and Bowman Lake valley.

Ice Caves can usually be found at the base of either Grinnell or Sperry Glaciers. They form where the streams beneath the glacier come out. These streams carry rock flour, (giving them a milky-white color), sand and gravel. They deposit this material as outwash and alluvial fans. These features are responsible for the damming up of St. Mary Lake and Lake McDonald. Both Grinnell and Sperry also show glacial striations and polish.

Other lakes have been dammed up by the various types of moraine such as Bowman Lake, Lake Josephine and Kintla Lake.

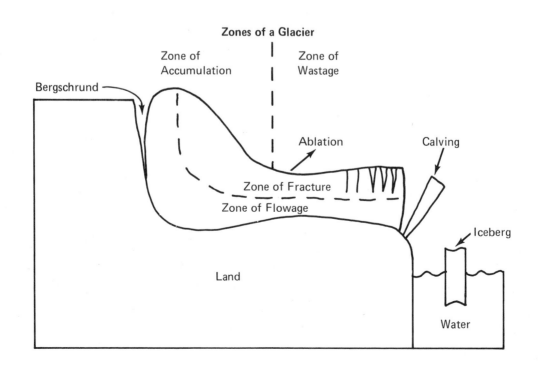

Zones of a Glacier

Zone of Accumulation | Zone of Wastage

Bergschrund

Ablation

Calving

Zone of Fracture

Zone of Flowage

Iceberg

Land

Water

Avalanche chutes are steep sided narrow chutes where snow collects above in pockets during the winter, they become unstable and go rushing down the mountain side bowling over everything in their path.

In these ancient rocks their origin is very evident, for a person can see *ripple marks* in the sandstones formed by wave action in the shallow water. Also *mud cracks*, that formed when the ancient floodplains dried out causing the muds to shrink and crack, are evident.

The most obvious clue to their sedimentary origin is the calcareous algae, they are sometimes nicknamed "cabbage heads" because of their shape. Back during the Precambrian there was no life on land, only in the seas. One form of plant life was single celled plants called *algae*, a blue-green variety. Because they are plants, they use sunlight and carbon dioxide in the process of photosynthesis, and manufacture sugars and starches. Extracting the carbon dioxide from the water causes it to become slightly alkaline immediately around this colonial plant. The alkalinity forces the precipitation of a calcium carbonate ($CaCO_3$ or limestone) crust over the spherical colony. So it will

not suffocate or smother it grows through the crust and over it forming another layer. So layer after layer it builds up, one calcite shell after another. This is what is preserved and not the actual plant. There is some argument as to the origin of the different shapes, some believe that it is due to different species, others state that variations in living conditions such as temperature causes the various forms. Irregardless these primitive plants are collectively known as *stromatolites* and had to develop in shallow water where they could receive abundant sunlight.

Triple Divide Peak is rather unique, inasmuch, as water drains in three directions instead of two as most divides do. Water may drain to the Gulf of Mexico starting its journey down Atlantic Creek. It may go the right and eventually end up in Hudson Bay via Hudson Bay Creek. If it goes to the left it flows down Pacific Creek, and eventually ends up in the Pacific Ocean.

The Lewis Overthrust is a huge thrust fault that underlies all of Glacier National Park and has placed older Precambrian beds on top of younger Cretaceous beds. Chief Mountain just outside the park is the classical example of the thrusting. The

upper third is horizontal Precambrian Pre-Ravelli (Altyn) limestone while the lower two thirds disturbed and undifferentiated Cretaceous beds separated by the fault surface. Erosion has isolated Chief Mountain so it consists of older beds completely surrounded by younger ones and is called an *inlier*.

The Lewis Overthrust is easily seen on the face of Wynn Mountain, it underlies Trick Falls, Swiftcurrent Valley by Many Glacier and the entire East Livingston Range. The vertical face of Appekunny marks the scarp of the Lewis Overthrust.

GEOLOGIC HISTORY

1. Formation of the Belt geosyncline during the Precambrian.

Most of the time the region was under water but occasionally above sea level. The water was shallow and warm, as indicated by the calcareous algae, ripple marks and mud cracks. The algae formed reefs as high as 60 feet high.

2. Deposition of the Pre-Ravelli or Lower Belt (Waterton beds) during the Precambrian (Precambrian Y).

It is a dolomite that weathers gray, tan or reddish brown. Its fresh surface is red-brown or gray and is not exposed in Glacier National Park, but is in Waterton Park. It is usually classified as the lower member of the Pre-Ravelli (Alytn) formation. Thickness is unknown since its base is not exposed.

3. Deposition of the Pre-Ravelli or Lower Belt (Altyn formation), some 2,300 feet of sandy dolomite and limestone, weathers light buff, a ridgeformer (Precambrian Y).

The layers of sandstone and conglomerate at the base, interbedded with Pre-Ravelli (Altyn) limestone indicate that these beds were originally derived from a granite source that bordered the sea. Lower layers contain stromatolites (Appekunny Falls, Rising Sun Camp Ground, Divide Mountain Firetower). Thin layers of quartz that have filled in minute fractures and weather out in sharp relief, thus making it easy to identify. It also contains mudcracks and ripple marks.

The Pre-Ravelli (Altyn) forms *The Narrows* of St. Mary Lake, the ledge over which Swiftcurrent Falls tumbles, and the rim upon which Many Glacier Hotel rests.

4. Deposition of the Appekunny formation, 3,000 feet of greenish shales and argillites (similar to shale but harder). (Precambrian Y).

The Appekunny contains no known fossils, but mud cracks, crossbedding and ripple marks are common. They were derived from streams that traveled long distances over granitic bedrock before depositing their load into the Belt Sea. Most of the beds are a green mudstone deposited underwater where it never had a chance to oxidize. The reddish layers were exposed to the air and became oxidized. Occasionally beds of well sorted clean sand were deposited which now form white stripes in the Appekunny. It usually supports a heavy forest cover.

McDonald Falls flows over the Appekunny, Old Squaw Mountain and Yellow Mountain contain these beds and Aster Creek has carved its gorge in it. It is used as flagstone for walks in the park.

5. Deposition of the Spokane formation (Grinnell formation), 3,000 feet of red argillites interbedded with white quartzite (Precambrian Y).

The upper portion of the Appekunny gradually changes into the Spokane (Grinnell). There is no sharp contact between them. The Spokane (Grinnell) is predominately red in color, but has layers of white sandstone and green mudstone.

It contains a few fossils, but mud cracks, raindrop impressions, and ripple marks are more commonly formed in the same environment as Appekunny but exposed to air more often. Between its thickness and location it makes up the bulk of Glacier National Park.

Avalanche Gorge and Red Rock Canyon is cut into the Spokane (Grinnell). McDonald Creek flows through it forming many *potholes* or hollows formed by abrasion of the bed load of the stream. Sperry Glacier, Grinnell Lake and Falls, Mount Henkel, Ptarmigan Creek, Rising Wolf Mountain, Ptarmigan Tunnel, and many other features lie within its domain.

The Appekunny and the Spokane are members of the Ravalli Group or Middle Belt Supergroup.

6. Deposition of the Helena dolomite (Siyeh) 4,000 feet of limestone, contains algal colonies and diorite sill. (Precambrian Y).

The Helena (Siyeh—pronounced See-ah) was deposited in clear, shallow water. This is the reason that it has so many algal zones. There are three main zones, the lowest is *Collenia symmetrica* which grew on a mud cracked surface. It can be

seen on Garden Wall, by Lake McDonald and Go-
ing-to-the-Sun Highway east of Logan's Pass.

It also contains a series of dikes and sills that
were intruded near the top. They have a metagab-
bro or diabase composition and range in thickness
from 10 to 200 feet. Some copper was emplaced in
the limestone but not enough to mine. Contact
metamorphism has occurred, bleaching the beds
above and below them, and changing them to
marble. They are easily traced all over the park
near the tops of mountains such as Clements
Mountain and Mount Gould, or in the Natural
Amphitheater. They weather more easily so form
deep grooves and will often act as an avalanche
chute.

7. Formation of Purcell lava, a 50 to 275 foot basaltic pillow lava flow formed underwater.

The Purcell flow was probably formed at the
same time as the dikes and sills since the composition is the same, but a direct connection has not
been found between them. It was extruded on top
of the Helena dolomite (Siyeh) while the overlying
Shepard was forming. It was extruded underwater

as is shown by the *pillow structure* with its cooled
glassy skin and coarser interior. Chemically the
flow is the same composition as the intrusions, but
since it cooled more quickly it became basalt
which is finer grained than diabase. The bubble
holes or *vesicles* have become filled with calcite
forming *amygdules.* Some mineralization has occurred, and radioactive dating places the extrusion
at 1,080 million years.

Granite Park is the best locality to see the lava,
it was misnamed by a person who did not know
the difference between granite and basalt. There is
no granite in Glacier National Park.

8. Deposition of Shepard formation, 600 feet of buff colored limestone.

The Shepard was forming at the same time the
Purcell was being intruded so there is some inter-
bedding between them. It is a mixture of siltstone
and limestone with some dolomite. It contains
algae as fossils, also ripple marks, mud cracks and
some channel fillings. These indicate the sea was
becoming quite shallow.

9. **Deposition of the Mount Shields (Kintla formation), some 860 feet of bright red argillites. Contains casts of salt crystals.**

The Mount Shields (Kintla) was formed in shallow lagoons where the water became very salty. It is the youngest bed found in the park. Sometimes the lagoons were exposed as mudflats, other times covered with water. The salt or *halite* would crystallize out when salinity was 10X normal. The bright red color is because oxidization occurred when the beds were exposed to the air. It also contains algal zones. Outcroppings can be seen on the peaks on many mountains such as Running Rabbit Mountain, Mount Custer, Boulder Pass and Hole-in-the-Wall Basin.

Both Shepard and Mount Shields (Kintla) are considered as members of the Missoula Group that are found outside of the park boundaries.

10. **Uplift, withdrawal of seas, erosion of some of the Belt beds, deposition of Paleozoic beds.**

After the deposition of the Belt beds in Glacier National Park, the sea retreated for a while, and erosion occurred, removing some of the soft sediments. They returned during the Paleozoic because the mountains located both north and south of the park are of this age. They are mostly marine and perhaps could be found in Glacier by drilling.

11. **Deposition during the Cretaceous about 100 million years ago.**

During the Cretaceous period, as it was drawing to an end, an inland sea extended into the Glacier Park region. It was shallow and contained marine fossils, other area beds of the same age contained dinosaur fossils. Because the Cretaceous beds were never deeply buried they were never as strongly lithified as the Precambrian rocks, hence they have a tendency to weather rather easily. They form a gently rolling landscape in contrast to the mountains and steep slopes of the Belt series. These beds outcrop on the eastern side of the park and contain oil, there are natural oil seeps in North Fork Valley. Commercial fields are found in Alberta and Montana in these beds.

12. **Formation of Rocky Mountains during the Laramide Revolution.**

The beds deposited during the Paleozoic and Mesozoic were gradually being uplifted. Minor folds were produced as the mountain building forces increased in their intensity, these can be seen in the park along the trails (along the roads the beds appear horizontal).

The region was hilly, the uplifting increased the velocity of the streams so they cut large stream valleys.

13. **During Late Cretaceous-Early Eocene time the Lewis Overthrust occurred, forcing Precambrian rocks over Cretaceous shales.**

Gradually a broad arch (crest 100 miles west) was uplifted vertically several thousand feet. This tilted the sediments gradually eastward. Some of the rocks along the weaker zones began to slide and slip creating the Lewis Overthrust. Movement was as much as 30 miles. The tougher Precambrian beds easily slide over the weaker Cretaceous beds.

Pull apart valleys formed during the slippage such as North Fork Valley, opening a gap of about ten miles wide.

14. **Vertical faulting in Livingstone Range, erosion, uplift, more erosion.**

After the North Fork Valley formed the southern end was dropped down. Late Eocene sediments were deposited in the newly created valleys. Erosion isolated sections of the thrust fault creating rootless mountains such as Chief Mountain, so the older Belt beds are completely surrounded by the younger Cretaceous beds. This changed the drainage system completely.

A single river used to flow to the northwest from Bear Creek, through Middle Fork Valley, over McGees Meadows and then on to Canada. Today two branches of the Flathead River flow to the southeast, suddenly swing west and finally join southwest of the Apgar Mountains. Because of the change in drainage some of the valleys that were poorly drained formed swamps, which today have some coal in them but not of a very high grade.

Erosion stripped the beds from the park region and produced the Blackfoot erosional surface.

15. **Pleistocene glaciation occurred covering park with 3,000 feet of ice, only tallest peaks showed above.**

Three million years ago the mountains had been worn down to low hills. Eastward they sloped as part of the High Plains Surface at an elevation 1,000 feet above present day streams, the streams flowed to the northwest.

To the west were desertlike conditions, the North Fork Valley filled with valley fill because there were no streams to remove the material. The mountains were sharp and angular because physical weathering is dominant in a dry climate.

With the beginning of the ice age more water was available and the streams began to downcut rapidly or become permanent where they used to be temporary. Material was moved rapidly, the streams cutting their valleys about 1,000 feet deeper, thus forming high mountains. Remnants of the former Blackfoot surface was only left on the high divides.

Continental glaciers flowed southward from Canada and almost, but not quite, entered Glacier National Park, this was the Keewatin Ice Sheet.

Instead it encountered the valley glaciers that buried the park under some 3,000 feet of ice. The two main glaciers were the Two Medicine Glacier that flowed east and St. Mary Glacier that flowed north, both encountered the Keewatin Sheet. North Fork Valley and Flathead Valley on the other side of the continental divide were filled with ice also. The earlier ice ages were of greater extent than the later ones. They flowed further eastward onto the plains and filled the valleys 500 to 1,000 feet higher with ice.

Glaciers never erode valleys of their own, they modify preexisting stream valleys. Therefore the *main* features such as location of mountains, valleys and divides was already established. Glaciation modified the landscape to a great extent, widening valleys, creating lakes, changing drainage.

16. Complete retreat of Pleistocene glaciers about 10,000 years ago, formation of present landscape.

When the glaciers left the park, they left essentially the same landscape that we see today. Only minor changes have occurred since, such as the small gorges cut by streams. Sunrise Gorge and Gorge at Hidden Falls are the result of the enlargement of the joint system.

Avalanche Gorge with all its potholes was also created. These took between 9-10,000 years to cut.

Alluvial fans, some as old as 12,000 years have dammed up several lakes such as Waterton Lake, St. Mary Lake and Lower Two Medicine Lake.

Talus slopes are beginning to bury the foot of the mountain ranges of today. Landslides, mudflows, rock falls, avalanches, soil creep and rock falls are also contributing to the landscape.

17. Formation of glaciers that exist today.

About 4,000 years ago, conditions were such that the present glaciers developed. In 1910 about 90 glaciers were active, today only 60 or so are active. Grinnell Glacier was the largest until recently, but now Sperry Glacier is. They move between 12 and 50 inches per year.

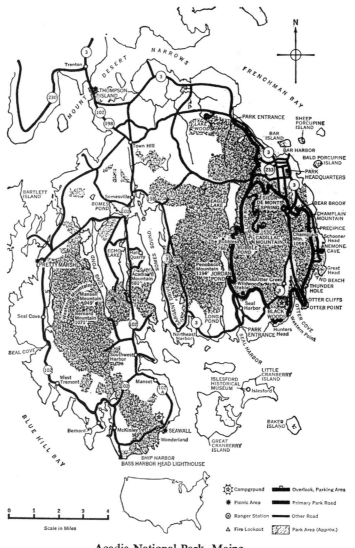

Acadia National Park, Maine.

Geologic Column of Acadia National Park

Erathem	System	Series	Formation
Cenozoic	Holocene	Pleistocene	Glacial Deposits
Paleozoic	Devonian		Nondifferentiated Granite
			Shatter Zone and Pegmatites
			Recrystallized Granite
			Medium-Grained Granite and Dikes
			Coarse-Grained Granite and Dikes
			Fine-Grained Granite and Dikes
			Diorite
			Bar Harbor Series
	Silurian		Cranberry Island Series
	Ordovician Cambrian		Ellsworth Schist Series

Modified after Chapman and Shaler et. al.

CHAPTER 12

ACADIA NATIONAL PARK

Location: Maine (mostly Mount Desert Island)
Area: 41,642.41 acres
Established: February 26, 1919

LOCAL HISTORY

Indians inhabited the island some 6,000 years ago. They were fishermen in canoes, catching mostly tuna. A second group inhabited the area about 3,000 years ago. They are called the Red Paint People because of the habit of sprinkling the bodies of their dead with red and yellow ochre. The Red Paint People used slate tools and made pottery.

Some 16 years before the pilgrims settled on Cape Cod, Samuel de Champlain in 1604 discovered the island which he named "Isles de Monts Deserts" (Island of Barren Mountain, or Solitary Mountain).

Mount Desert Island was the first French missionary colony in the United States settled by the Jesuits in 1613. Their mission was to teach the Abnaki Indians, who were forest Indians and used birch bark canoes. The Indians lived off of shell fish and cranberries during the summer, which they found on Mount Desert Island. The site for the colony was selected accidently when a heavy fog prevented them from landing where they had planned to (mythical city of Norumbego). A few months later an English Captain stationed in Virginia, Samuel Argall, wiped out the settlement Saint Sauveur in a surprise raid. The Indians finally left Pemetic (their name for Mt. Desert Island) permanently in 1840. From then on warships of both the French and English remained in the area in a long battle for supremacy and no one was able to settle there because it was too dangerous.

Louis XIV gave the island to Sieur de la Mothe Cadillac in 1688. He and his wife dwelled there for a short time on its eastern shore. He went on to become the founder of Detroit, Michigan, and later the Governor of Louisiana.

In 1713, after Louis XIV was defeated, it became English property. However, battles continued until it was given as a Grant by George III to Francis Bernard in 1759. Bernard was the last English Governor before the American Revolution, his settlement was located on Somes Sound. During the Revolution his mansion was confiscated and reverted to the Providence of Massachusetts, because Bernard sided with the King. Upon his death his son John requested that ownership be restored to him since he was loyal to the colonies. A one half interest was granted to him, he received the western half of the island.

The granddaughter of Cadillac—Marie de Cadillac and her husband (Messieur de Gregorie), armed with a letter from Lafayette, requested their rights to the island for services rendered. The General Court of Massachusetts gave them the eastern half.

Bernard, as soon as it was surveyed and the boundaries established, sold his share and went back home to England. Cadillac's granddaughter sold it off a piece at a time to settlers.

When the French and Indian War ended, permanent settlers came in 1761, they were James Richardson and Abraham Somes.

In 1820 Maine became a state, but the island stayed a fishing hamlet and was fairly remote. This isolation was reduced somewhat when roads were built and the island was finally connected to the mainland by a bridge and causeway. In 1844 Thomas Cole, a painter, came, fell in love with it, told his friends about it, and it soon became an artist colony. The rich followed shortly afterwards.

The Boston and Bangor Steamship Line established a stop there in 1868 and thus helped the area to develop into a resort area for the rich, especially around Bar Harbor by 1890.

In 1901 the "rusticaters" or summer residents who owned most of the island were afraid that commercial enterprise, especially lumbermen, would come in and ruin the island, so they formed a corporation to protect the area. It received its

charter in 1903, and then a gift of one square rod of land (16 square feet), which was the beginning of the acquisition. Some 6,000 acres were acquired by 1913, located on Mount Desert Island, Schoodic Peninsula and Isle au Haut "High Island" (which can only be reached by boat). Two of the prime workers were George Bucknam and Dr. Charles Elliot. One of the prime donators was John D. Rockefeller. An offer was made to donate the land to the Federal Government in 1914, but it took until 1916 to fulfill the requirements. It became the *"Sieur de Monts National Monument"* on July 8, 1916 (President Wilson), then on February 26, 1919, the first National Park in the east, *Lafayette National Park,* which means "place of rest and delight in nature."

GEOLOGIC FEATURES

Acadia National Park has the only true fiord in North America. A *fiord* is a U-shaped glacial valley that has been drowned by sea water. It has almost vertical sides and the water is quite deep. This is possible because all of the land north of Chesa-

peake Bay is submergent, that is, depressed by the weight of glaciers below sea level. This explains the irregular coastline.

The island was depressed by as much as 600 feet, but has only managed to *rebound* or spring back some 250 feet, hence all the headlands and coves are former stream valleys with their divides.

Glacial features are all over. The Bubbles are roche moutonnees, resting on top of them are large glacial erratics. Many of the lakes occupy the bottom of glacial valleys, some such as Long Pond which has been dammed up by glacial debris.

The direction of ice flow in the beginning was from north to south, due to the south trending passes in the ridges. Therefore, as the ice moved south, it slid over the north slopes and polished them smooth, but as it passed over the southern slopes glacial plucking occurred, forming rough jagged slopes.

Glacial striations, glacial grooves, glacial pavement and *chatter marks* (crescent-shaped fractures in a series) are found everywhere.

There are two other geologic processes that are dominant here, they are igneous intrusions and, since this is an island, wave action.

The island is made up of a series of igneous intrusions producing various types of granites usually distinguished by the size of the crystals. The entire island is also criss-crossed by a series of basaltic dikes that are the classic example of the *Law of Cross-Cutting Relationships.* This law states a bed is younger than the bed it cuts across and it is through the use of this law that much of the geologic history has been worked out. Pegmatites with the giant crystals of quartz, feldspar, epidote and other minerals are abundant.

Volcanic activity occurred and the so-called Ovens are domed caves eroded in the volcanic tuff by weathering and wave action.

Xenoliths or sections of local bedrock, are found completely surrounded by the granitic intrusions, especially around Otter Point.

The dikes frequently weather away more quickly than the surrounding granite forming narrow depressions.

Sheeting is very evident around Cadillac Mountain where the granite has been broken into large sheets, one resting upon the other as the granite expanded in response to the overlying rock being removed by erosion.

Vertical fractures also developed as a result of stress and strain which formed the joint systems. Weathering of these joints produces such features as The Beehive.

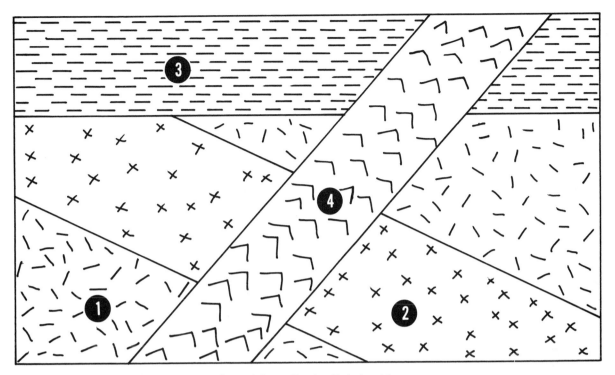

Law of Cross-Cutting Relationships

Wave action enlarging the joint system has produced Thunder Hole. The noise is the result of the trapped air becoming compressed by the wave action and emitting a thundering sound. The debris collecting at the bottom of Thunder Hole is helping to enlarge the joint even more by abrasion.

Natural *sea walls* form as the result of large blocks of bedrock becoming detached during storms and tossed up onto the beaches, forming long ridges. During less violent storms rounded stones called cobbles (spherical) or shingles (disc-shaped) are tossed over the seawalls by waves, become trapped and form *cobble and shingle beaches.*

In protected regions at the mouths of bays, *sand bars* and *spits* form because of the action of *longshore currents* flowing parallel to the shoreline and depositing sand in the quiet water. If the bar is attached at one end to the land it is called a spit. The quiet *lagoons* form between the land and the bar, permitting sandy beaches to form. Eventually the lagoons will become totally filled in with sediments.

The effects of weather can be seen on the bedrock itself. For example, much of the granite has *weather pits* where the higher temperature, less resistant minerals weather out, leaving a depression. Or *differential weathering* occurs because of differences in composition of the bedrock and intensity of weathering due to location.

On Isle Au Haut a slightly different feature can

be seen. *Cross-bedded gravels* are present, left by glacial meltwater streams as they deposited their loads into bodies of quiet water. They can be seen in the scattered gravel pits all over the island.

Wave cut cliffs such as Otter Cliffs were formed when wave action undercut the base of the cliff forming a *nip.* The cliff then collapses by landslides because its support has been removed and recedes back. The floor of the nip then becomes part of the *wave cut bench.* If a joint is enlarged by wave action, it becomes a *sea cave* such as Anemone Cave. The resistant rock that extends out into the ocean is called a *headland* (e.g., Alter Point, Schooner Head), and the less resistant rock recedes back to form a protected area called a cove, such as Seal Cove or Ship Harbor.

GEOLOGIC HISTORY

1. Formation of the Ellsworth schist (Barlett Island series) folding, uplifting and erosion.

About 450 million years ago (around Cambrian time), New England was covered with water in a sea called the Acadian geosyncline. Sediments of sand, silt, and mud were being carried by streams and deposited into the ocean. Nearby volcanoes contributed pyroclastic debris to the sediments which were mixed in by wave action or accumulated in thick enough layers to form stratified beds of their own. Probably several thousand feet of material accumulated.

Diastrophism, or crustal disturbance on a large scale occurred and metamorphosed these sediments into a schist that was rich in mica, layered gneisses and some quartzites. The beds were folded and uplifted above sea level probably into a mountain range.

2. Formation of Cranberry Island series about 420 million years ago. Folding uplift and erosion.

The newly metamorphosed beds underwent erosion until they were once again submerged. On top of this eroded disturbed surface the Cranberry Island series formed uncomfortably.

This time volcanic activity had increased considerably and the majority of the beds are volcanic mixed in with some sediments. The volcanic beds are mostly *tuff* made up of dust, ash and cinder sized debris, but some fairly larged sized volcanic blocks are also incorporated. They are intermixed with felsite lava flows that are either porphyritic or amygdaloidal. The *porphyritic* rock has two distinct sizes of crystals, while the *amygdaloidal* rock has filled in vesicles.

For the second time the region underwent folding and uplift, and tilted some of the volcanic beds to a nearly vertical position. Both units were metamorphosed slightly.

Erosion took over and started stripping away these disturbed beds to near sea level once again. Wave action formed sea cliffs which collapsed and permitted much of the material to be dumped into the ocean. Wave action quickly rounded these in cobbles (256 mm to 64 mm) and pebbles (64 mm to 4 mm in diameter) thus covering the sea floor with a gravel.

3. Submergence and deposition of the Bar Harbor series about 375 million years ago.

As the gravel from the remnants of the Ellsworth schist and Cranberry Island beds formed on the sea floor, the region was once again submerged and completely covered by water. The sand and silts became mixed with the gravel and was lithified into a basal conglomerate. Sandstones and siltstones were deposited on top and were mixed in with some volcanic ash beds. The sandstone and siltstone splits apart fairly evenly and can be used as flagstone. These events occurred during the Devonian period. The Bar Harbor series produces an angular unconformity between the two older deformed formations.

4. Intrusion of diorite as a series of dikes, sills and lava flows, contact metamorphism.

The magma chamber that probably supplied the earlier volcanics moved up towards the surface and intruded the metamorphic and sedimentary beds. It spread out in the Bar Harbor series (because they were softer sedimentary beds) as sills up to 100 feet thick. One intrusion is 3,000 feet thick and has formed between Bar Harbor and Ellsworth, possibly moving along the old eroded tilted surface of the Cranberry Island series. It is so thick that it is referred to as an *intrusive sheet*. The heat of the intrusion produced contact metamorphism. The surrounding rock especially the shales and siltstones were baked into *hornfels*. Contact of the magma with the bed rock caused rapid cooling and a finer grained igneous texture resulted. Towards the centers of the intrusions cooling was slower and the rock is coarser grained. Two groups of minerals crystallized out, the basic dark colored minerals of olivine, pyroxenes and amphiboles (mainly hornblende) plus the lighter colored, more acidic minerals such as plagioclase feldspars. Combined together, the rock formed is called a diorite, which has an intermediate composition and is made up of approximately half light and half dark-colored minerals.

5. Intrusion of fine-grained granite, country rock shattered, formation of xenoliths. Uplift, erosion, and depression.

A second intrusion occurred soon after the diorite intrusion cooled. A granitic magma intruded the older beds and made contact with both the diorite and Bar Harbor series. The force of the intrusion shattered the surrounding country rock, so the molten material filled in the cracks and partly assimilated some of the local rock, thus forming *xenoliths*. The fine-grained texture indicates fairly rapid cooling.

During Late Devonian this region was apparently uplifted and underwent erosion then was depressed and covered with sediments that have since been removed.

6. Intrusion of coarse-grained granite, collapse of roof, formation of ring dikes, metamorphism.

Gradual intrusion of a granite which cooled at a slower rate and therefore became coarse-grained. The bedrock over the liquid intrusion began to sag and fracture since it lacked support then gradually sank into the magma chamber to become assimilated. Naturally the liquid magma filled in the vacated space and eventually crystallized.

The falling in of the country rock was not clean, so it caused some of the surrounding rock to shatter. This permitted the magma to force its way into these cracks surrounding the chamber forming *ring dikes*.

All this intrusion produced contact metamorphism and baked the rock, especially the Bar Harbor series into hornfels.

7. **Permeation of hydrothermal solutions into fractured rock, collapse of portion of roof, intrusion of medium-grained granite.**

As fractures developed *hydrothermal* (hot water) solutions permeated through the bedrock and recrystallized some of the granite. Additional fracturing occurred with the formation of ring dikes (such as on Bartlett Island).

A second section of roof rock plus part of the solidified chamber started to break off and began to subside. Magma move in and cooled to form a medium-grained granite. Some faulting and shattering of bedrock occurred. Hydrothermal solutions formed veins and intrusion formed dikes.

8. **Erosion of sediments by Cretaceous time. Mount Desert Island became a monadnock, intrusives were exposed by erosion.**

After the intrusion of the medium-grained granite, and for the next 350 million years, the record is not complete. There may have been additional sediments deposited, if so no trace or clue of them remains today. About 135 million years ago most of the eastern United States was stripped to an erosional surface (Schooley). Only very resistant areas, especially those made up of igneous rock stood as low hills or *monadnocks* above the surrounding plains. The erosion stripped the sediments, exposing the igneous core of Acadia National Park.

9. **Paleogene uplift and tilting seaward, rejuvenation and dissection of land, subsidence and drowning of coastline.**

About 60 million years ago (Paleogene) Mount Desert Island and the surrounding areas were part of the mainland which had an east-west trending ridge and streams running parallel north and south of the ridge.

Uplift tilted the region seaward, the new base level permitted the streams to carve new valleys between low hills.

When the area subsided the hills became headlands and the valleys formed coves, isolated hills became islands, forming Mount Desert Island, Isle Au Haut, Baker Island and others.

10. Glaciation during the Pleistocene, island covered by 5,000 feet of ice.

Between one-two million years ago the Pleistocene began and the ice crept slowly southward. Some 20,000 years ago they reached Mount Desert Island flowing south and encountered the east-west mountains. The glaciers then piled up on the north flank and were deflected around both ends.

Eventually it overrode the mountain and changed the divides into deep saddles. Through these they moved, scraping, gouging and plucking. They enlarged the passes to valleys cutting across the mountains.

Eventually this east-west trending ridge was dissected into a series of isolated peaks with north-south U shaped valleys separating them. This is why the north side of the ridges is smooth with the ice gradually sliding over, and the south sides steep as the overriding glacial ice plucked away weak beds. Some of the valleys were cut below present day sea level such as Somes Sound which is a fiord. Others became scraped depressions that now form Long Pond, Eagle Lake, Jordan Pond and others.

11. Melting of ice, area remains submerged, wave action and weathering forming present landscape.

With the retreat of the glaciers the processes of weathering and wave action now dominate. Exfoliation of the granite is building up talus slopes, and differential weathering is enlarging joints and removing dikes and sills.

Wave action is cutting back the headlands, forming wave cut benches and filling in the coves with sandbars by forming sea walls and beaches of sand and cobbles. Enlargement of joints or weathering of rocks are forming caves. All these processes combined have made Acadia National Park a lovely place to visit.

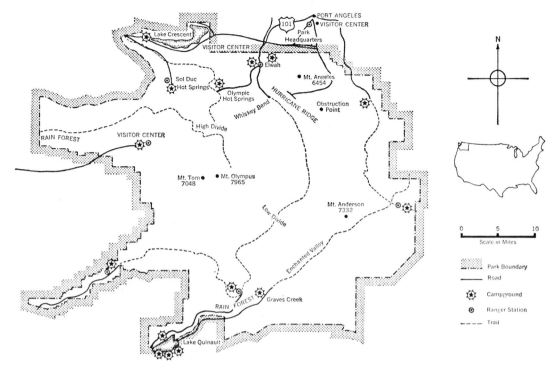

Olympic National Park, Washington.

Geologic Column of Olympic National Park
Along the Coastline of Pacific Ocean

Era	System	Series	Formation
Cenozoic	Neogene	Holocene	Alluvium
		Pleistocene	Glacial Outwash Glacial Deposits
	Paleogene	Pliocene	Quinault Formation
		Miocene	Hoh Rock Assemblage
		Eocene	Broken Sedimentary Rocks (Point Grenville Area)
			Fractured Volcanic Rocks and Marine Silts—Pt. Grenville Area

Modified after Rau.

Geologic Column of Olympic National Park
Axial Facies or Core Rock

Era	Period	Epoch	Formation
Cenozoic	Paleogene	Miocene to Middle Eocene	Crescent Formation
		Eocene	"Soleduck" Argillite and Graywacke or Metchosin Volcanics (?)

Modified after McKee, Tabort and Stewart.

Geologic Column of Olympic National Park
Marginal Facies (Peripheral Rocks) Along
Strait of Juan De Fuca and East Side of Park

Era	System	Series	Formation	
Cenozoic	Neogene	Holocene	Alluvium and Marine	
		Pleistocene	Glacial Drift	
	Paleogene	Early Miocene to Late Eocene	Twin River Formation	Upper Member
				Middle Member
				Lower Member
		Late Eocene	Lyre Formation	
		Middle Eocene	Aldwell Formation	

Modified after McKee, Tabort and Stewart.

CHAPTER 13

OLYMPIC NATIONAL PARK

Location: N.W. Washington
Area: 896,599.10 acres
Established: June 29, 1938

LOCAL HISTORY

According to a legend, the Greek God Jupiter, disgusted with the attitude of his people, left Mount Olympus and moved to the New World to a mountain on the west coast. Perhaps it was Mount Olympus in Olympic National Park.

At any rate the Olympic Peninsula has had inhabitants for a long time. The aboriginal tribes were part of the northwest coast culture. Fish was the staple food, especially salmon and smelt. They dug clams and went out to sea in large dugouts (30 to 40 feet long) to catch seal, porpoise and whale. They hunted deer and elk and supplemented their diet with roots and berries.

Cedar was used in making their canoes, houses, clothing, etc. The houses were permanent homes made out of planks, some single, others large enough for several families. The homes were clustered in villages by the sea or mouths of rivers.

One custom that still exists today, although in a modified form, was that of potlatch. They gave away possessions during a feast, with the wealth being measured by the amount given away. The recipient was required to give an even larger gift in return, so the giver was not left destitute.

The first white men were Spanish explorers in the sixteenth century. Supposedly Juan de Fuca (a Greek pilot in Spain's service) was supposed to have entered the strait named after him in 1592.

In 1774 the Spanish renewed their efforts in exploring, after hearing rumors that the Russians were moving south from Alaska. A Spanish sea captain, Juan Perez, saw what is now Mount Olympus and named it "La Sierra de la Santa Rosalia" or Saint Rosalie Mountain. In 1788 Captain John Mears, a trader, saw Perez's mountain and named it Mount Olympus, since he thought he was the first to see it. This was the name recorded by Captain

George Vancouver when he mapped the area in 1792.

In spite of exploration by the English, Americans, and Spaniards, the Spanish were the first to actually enter the park area. Two Spainards, Juan de la Bodegay and Bruno Heceta made a voyage in 1775. Heceta made a landing by the mouth of the Quinault River.

Captain James Cook in 1778 named Cape Flattery. His crew obtained furs from this area and sold them later in China at a good profit. This resulted in bringing both English and American fur companies to the northwest.

Juan Francisco de Eliza in 1791 entered the straits and called the harbor "Puerto de Nuestra Senora de Los Angeles," translated means Our Lady of the Angels. In 1792 the Spanish established a colony at Neah Bay but it only lasted five months. Also in 1792 Captain Walter Grey, an American, and Captain George Vancouver explored this area.

By 1810 fur traders were following the trails of such men as Alexander Mackenzie ((1792-93), David Thompson (1806-09) and Lewis and Clark (1805-06). From 1821 on, this became Hudson Bay Fur Company territory.

The American population kept increasing in the area and in 1846 Great Britain finally gave in and the 49th parallel boundary was established.

Nearby, Port Townsend was established in 1851, but prior to that (1849) John Everett and John Sutherland, two trappers operated trap lines in Lake Crescent (formerly Lake Everett) and Lake Sutherland, that are now located in the park.

Mount Olympus, elevation 7,965 feet, was first climbed in 1854. However, the first group to really explore Olympus was in 1885 led by Lieutenant Joseph P. O'Neil (14th Infantry).

In 1889 the city editor of the "Seattle Press," Edmond Meany (27 years old), printed an article about the Olympic Mountains suggesting its exploration. He convinced the paper to sponsor an expedition which left in December 1889. The group thought the Olympics they saw was a rim which surrounded a protected valley. The plans were to cross over the rim, winter in the valley, and cross over the other side the following spring. Six months later they came out with a rough map, photographs and sadder but much wiser men. They endured many hardships and managed to name some 50 landmarks. Press Valley was named after the paper, Bailey Range after the owner, Mount Meany after the city editor and Mount Christie after the leader of the expedition and Mount Barnes after the narrator of the group.

In 1890 a second group went in sponsored by the Oregon Alpine Club who supplied the scientific staff and, most important of all, the money. The army furnished soldiers and Lieutenant O'Neil to lead the party. They named O'Neil Creek and O'Neil Pass after their leader, Mount Bretherton after the naturalist-cartographer and Mount Henderson after the botanist. O'Neil suggested the interior of the Olympic mountains is useless and would make a good National Park.

In 1897 the Olympic Forest Reserve was established and mapped by Dodwell and Rixon (1897-1900). In 1904 Representative Cushman tried to establish Elk National Park, but it failed. Both in 1906 and 1908 Representative Humphrey tried to make the area a game refuge and it failed. Humphrey asked Teddy Roosevelt to create Mount Olympus National Monument two days before his administration ended, which he did and it became official on March 2, 1909.

At first it was administered under the Forest Service because it was within a National Forest, but was transferred in 1933 to the National Park Service. Through the efforts of Representative Wallgreen after quite a few failures it was approved as a National Park in 1938, but only after Franklin Roosevelt had visited it and agreed to its establishment. It is the fourth largest park (Yellowstone, McKinley, Glacier).

GEOLOGIC FEATURES

One of the main features of Olympic National Park is the amount of rainfall and its distribution. Moist winds coming off of the Pacific Ocean are cooled as they ascend the Olympic Mountains. The majority of the peaks are around 5,000 and 6,000 feet in elevation and none exceed 8,000 feet. The moisture condenses into clouds and it rains up to 140 inches per year on the western side of the slope, on the eastern side it is only 17 inches per year. Therefore one side is a rain forest where chemical weathering dominates and the other a

semiarid climate where physical weathering dominates.

In this section the glacial features are common, such features as U-shaped glacial valleys such as Soleduck Valley that contains the glacial Lakes Crescent and Sutherland, and the Strait of Juan de Fuca that is also a former glacial valley.

The glaciers that exist today are the result of geographic location and high elevation in the mountains, they are also found at one of the lowest altitudes in the United States (exception is Carbon Glacier at Mount Rainier). They are small valley glaciers. The six glaciers on Mount Olympus, Blue, Hoh, Hubert, Hume, Jeffers, and White have a total area of ten square miles. The largest are located on the protected north slope. The remaining 54 glaciers combined, equal about 15 square miles and are found on Mounts Anderson, Carrie, Christie, the Bailey Range. The remaining areas are snowfields. Around these areas can be seen cirques, medial, lateral and end moraines and glacial lakes. Also the *bergschrund* or the crevasse between the valley wall and the head of the glacier may be observed. The crevasses on the surface of the glaciers are frequently covered with snow bridges.

Olympic National Park also has a 50 mile stretch along the ocean where the various wave cut features may be seen such as the coves and headlands, wave-cut cliffs, wave-cut terraces and sea caves that are also found at Acadia National Park.

Other features have been produced by abrasion and *hydraulic action* caused by the waves battering the sedimentary rocks with violent blows, shattering the jointed rock and washing the pieces out to sea. They have formed *sea arches* and *sea stacks*. A narrow headland is battered from both sides by the waves, eventually a tunnel is drilled through the headland forming the sea arch. When the unsupported roof of the arch collapses a pillar of rock is left standing in the ocean that once formed the

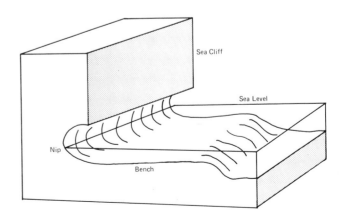

seaward end of the arch. These are called sea stacks.

Barrier bars, or ridges of sand, have been built across the mouth of coves deposited by the long-shore currents, if the bar is attached at one end to the land it is called a *spit.* Sometimes the longshore currents and waves deposit sand on the land forming a *beach.* There are even two *hot springs* where the temperature of the water is above 98° F, they are called Olympic Hot Springs and Sol Duc Hot Springs, remember this is not a volcanic area, but are heated by the *geothermal gradient* of the earth which increases in temperature 1° F for every 60 to 100 feet in depth. This means if the ground water travels down deep enough it is heated when it reaches the surface.

Although not specifically a geologic feature the rain forests formed on the western slopes must be mentioned. They receive about 133 inches of rain per year and contain some of the largest trees in the world. There is Western Red Cedar 230 feet high with a diameter of over 21 feet; Silka Spruce 248 feet high, a diameter over 18 feet; Douglas Fir 221 feet high, a diameter over 14 feet and Western Hemlock and Black Cottonwood, both over 175 feet tall. There is not a square inch of ground that doesn't have something growing on it, mosses and ferns cover the ground. Everything is covered by moss, even the trees.

GEOLOGIC HISTORY

1. Formation of core of the Olympic Mountains.

A portion of the oceanic crust plus its overlying sediments that were originally called the "Soleduck" formation are a mixture of Eocene and Oligocene sediments such as argillite and graywacke mixed in with basalt flows, volcanic breccias and water deposited pyroclastic debris. When the Juan de Fuca Plate began moving away from the Juan de Fuca Ridge during Cenozoic time this portion of the ocean floor encountered the resistant beds of the Crescent formation.

The Crescent formation is primarily a *pillow basalt,* that is, a lava flow that solidified under water. Intermixed with the pillow lavas are massive flows (diabasic and fine-grained gabbro or coarse-grained basalt), breccias that are either a flow breccia or tuff breccia, sandstone, siltstones, argillites or conglomerates with basaltic pebbles.

The Crescent was originally part of the continent, such as a basaltic flow extruded onto the continental shelf, perhaps *seamounts* (underwater mountains) in a chain or even a series of offshore islands.

The encounter between the two formations caused the "Soleduck" to be underthrust beneath the Crescent uplifting the region into the Olympic Mountains. The "Soleduck" broke up into a series of thrust plates one stacked upon another en-echelon style. Shear zones developed.

Sections of both formations have undergone a low rank metamorphism in the green schist facies about 29 million years ago during Oligocene time.

The metamorphism increases eastward changing from the zeolite facies to the prehnite-pumpellylite facies. A *metamorphic facies* indicates the conditions under which the metamorphic rock forms and not the composition of the rock in question. The two main factors are temperature and pressure. Specific minerals will occur within a facies creating *metamorphic zones.* For example, laumontite is a mineral that belongs to the zeolite group. The zeolite facies boundary can then be marked where laumontite no longer occurs.

The faulting, shearing and metamorphism has created regions of so badly fractured rock that it is almost impossible to assign a formation name to, and in some instances an age.

Maximum deformation occurred during middle to late Miocene as late Miocene (?) and Pliocene beds overlie unconformably the earlier Miocene beds.

2. Formation of sediments on the northern flank of the Olympics which are upper Eocene to middle Miocene in age.

First was the deposition of the Aldwell formation resting apparently conformably upon the Crescent formation. The Aldwell is middle Eocene in age, up to 2,900 feet thick. It is a marine siltstone, mudstone, sandstone with a greenish gray cast. The lower half of the beds are basaltic flows and flow breccias while the layer directly above has lenses of basalt cobbles and other material.

Above it lying conformably is the marine Lyre formation, late Eocene in age and up to 3,280 feet thick. It is a conglomerate and sandstone. The subrounded to subangular pebbles in the conglomerate have the following compositions: argillite (hard shale), chert (amorphous form of quartz), metavolcanics (volcanic rocks metamorphosed) and quartzite (metamorphosed sandstone). The sandstone grades into the conglomerate indicating the source area was getting progressively higher allowing the streams with their increased velocity to carry larger fragments. The short transportation distance is indicated by the fact that the pebbles were only beginning to become rounded.

An unconformity exists between the Lyre formation and the overlying Twin River formation. Age wise, the Twin River is from late Eocene to early Miocene. up to 9,000 feet thick and is divided into three members.

The Lower Member has channel deposits at its base where streams flowed over the Lyre surface. The sediments are conglomerate (gravel) and a coarse sandstone (containing pebbles and granules). Above this are marine sandstones and siltstones with a greenish cast. It is very compact, well bedded and calcareous in some regions.

The Middle Member has a conglomerate lens in the Eden Valley area, (containing basaltic boulders) otherwise it is mostly siltstone, mudstone, sandstone and occasional concretions. It has calcareous lenses up to five feet thick. Frequently conglomerate is found at the base of the sandstone layers.

The Upper Member consists of siltstone, mudstone, claystone, it is poorly cemented and sorted, and has a gray-green color. It contains beds one-twenty feet thick of calcareous sandstone (massive). Mudstone layers have concretions of various shapes (odd-shaped, round or cylindrical).

In some regions the Clallam formation of middle Miocene age exists, in others, the Pleistocene deposits form an angular unconformity. It is less than 2,500 feet thick, and consists of marine and nonmarine sandstone and siltstone. Some conglomerate layers exist and coal beds.

3. Formation of the Eocene to Pliocene sediments along the west coast of the Olympic Mountains.

About 45 to 50 million years ago during the middle Eocene according to the marine microfossils, submarine lava flows occurred. Some of this formed obsidian or pitchstone, others cooled a little slower forming basalt. Frequently the hot lava shattered and exploded upon contact with the cold water and formed a volcanic breccia. In the intervals between eruptions, sediments such as siltstone was deposited. Much of it is calcareous and in various localities has concretions. These unnamed volcanic beds can be seen at Point Grenville south of the Quinalt River.

Unconformably resting on the Grenville Point, volcanics are alternate layers of sandstone and siltstone that were deposited in deep (1,000 feet plus) water. They have been tilted, folded and faulted. Perhaps even faulted on top of the volcanics. They were formed 35 to 40 million years ago during the late Eocene.

No sedimentation occurred during Oligocene time, the next record is during Miocene with the formation of the Hoh Assemblage. According to some authors, this is the equivalent to the "Soleduck" of the northern belt of sediments.

There are two major groups, a "tectonic melange" of a chaotic assemblage of greywacke, basalt in a matrix of claystone and siltstone (broken) and a sequence of sandstone and siltstone (layered) that are faulted, tilted and overturned. They were formed in a marine basin. Fossils indicate the sequence occurred 15 to 22 million years ago.

Some of these are *turbidites* or sediments that settled out of muddy water and are *graded* or sorted according to size becoming progressively finer. These turbidites possibly formed as unstable sediments on the shelf "landslided" to deeper water and resettled. These form the upper layer.

The lower "tectonic melange" is the result of extreme deformation and crushing. Some petroleum is mixed with these giving them the local name of "smell rocks."

The Hoh Assemblage is probably a section of the Juan de Fuca plate that is being shoved against the North American plate and instead of sliding under was "skimmed off" and was crushed against the rigid contact between the two plates.

There is some *piercement structure* where the plastic rock was injected into the surrounding bedrock which is called the Quinault formation.

The Quinault is Pliocene in age, marine in origin and moderately tilted. It is well bedded and a mixture of conglomerate, sandstone and siltstone. It has been tilted, was deposited between 7 to 1.5 million years ago, and is about 6,000 feet thick. It may contain oil.

4. Formation of Pleistocene deposits.

After the Quinault was deposited, it was uplifted and tilted with streams forming valleys. The Pleistocene glaciation lowered the sea level so many of these valleys are below sea level today. This was complicated by the fact that the land was depressed by the weight of the ice and then rebounded, possibly more than once. Just as the sea level fluctuated with the advance and retreat of the glaciers.

Hence are formed a younger and older piedmont surface and a wave cut bench about 50 feet higher than today. The piedmont surfaces are remnants of coalescing fans formed by the meltwater streams of the glaciers.

Along the west coast the record is as follows: Quinault formation uplift and erosion, glacial deposits (tilted), marine erosion, glacial deposits, erosion and modern alluvium.

Along the north belt of sediments is an angular unconformity with the bedrock, "Vashon" age glacial deposits from both alpine and continental glaciers, erosional surface and modern alluvium. Much of the debris in the Vashon glacial material (till and stratified drift) is of local origin.

Between one and two million years ago the climate grew colder and great ice sheets advanced and retreated at least four times. The continental glacier from British Columbia barely touched the park area. One lobe flowed south into Puget Sound stopping ten miles past Olympia, Washington, and against the eastern edge of Olympic Mountains, reaching a height of only 750 meters. The other lobe went into the Straits of Juan de Fuca, went through Crescent Lake and into the northern portion of Soleduck Valley.

It picked up granite from British Columbia and deposited it as erratics upstream and as high as 3,000 feet on Klahhane ridge (near camp Wilder). They left about 11,000 years ago.

5. Modern glaciation carved most of the features.

The Olympics are actually a mountain dome some 60 miles in diameter cut into a series of peaks, ridges, and intervening valleys that have been modified by glaciers. These valley glaciers are more responsible for the present day features than the continental glaciers, for they never actually eroded the mountains proper.

The present day glaciers, because of the slight change in climate that has caused western Washington to become cooler and moister, and has actually caused some of the glaciers to advance slightly. This is more obvious in Blue Glacier because it has been the subject of close scientific study for the last few years. It undergoes wastage a slightly different way than the rest of the glaciers in the park. Most of them melt and evaporate at the tip, a process called *ablation*, and streams carry away the water. Blue Glacier ends at a steep cliff, so as it moves forward pieces break off or *calves* and goes thundering down onto the rocks below. The sound can be heard for miles.

Surprisingly enough the meltwaters do not contain rock flour, so the streams are quite clear, making it a very unusual park indeed, ranging from a rain forest to almost desert, to a sea shore.

The Olympic Mountains are the result of plate tectonics and have a very complicated history indeed.

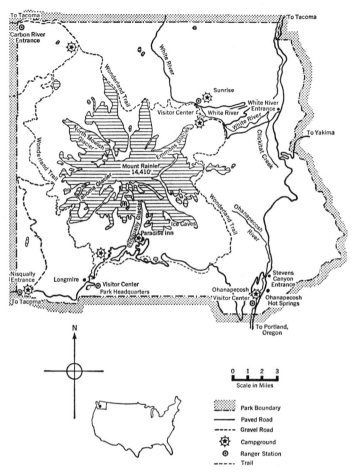

N

0 1 2 3
Scale in Miles

▒ Park Boundary
— Paved Road
--- Gravel Road
✹ Campground
◎ Ranger Station
----- Trail

Mount Rainier National Park, Washington.

Geologic Column of Mount Rainier National Park

Erathem	System	Series	Formation
Cenozoic	Neogene	Holocene to Pleistocene	Andesite Lava Flow Plus Breccia and Mudflows
		Miocene	Katoosh Granodiorite and Quartz Monzonite
			Fifes Peak Volcanics
			Stevens Ridge Formation
	Paleogene	Oligocene	Ohanapecosh Volcanics
		Eocene	Puget Sound Group

Modified after Barnett, Crandall, Matthes et. al.

PART III
FORMED BY IGNEOUS ACTIVITY

CHAPTER 14
MOUNT RAINIER NATIONAL PARK

Location: S. Central Washington
Size: 241,992.0 acres
Established: March 2, 1899

LOCAL HISTORY

According to a Nisqually Indian legend, Mount Rainier moved to the other side of Puget Sound when it became too crowded because of the rapid growth of the other mountains. She changed into a monster who tried to suck in and trap anything that came near by. Eventually (in the form of a fox) the Changer challenged the mountain, but first had taken the precaution of tying himself to a nearby mountain. So the monster Mount Rainier tried and tried to suck the Changer in, but had no success and finally broke a blood vessel in the effort which poured down her side and then she died. This probably describes a volcanic mudflow that occurred about 5,000 years ago.

More than a dozen Indian tribes lived in the vicinity of Mount Rainier and they all had legends about it and considered it a holy place. They used only the lower slopes for hunting but feared the god who lived upon it. Their name for Mount Rainier was "Tahoma" or "Snow Mountain." Some of the early tribes there were the Cowlitz, Klickitat, Nisqually and Yakima.

Rainier might have been seen by the early

Spanish and English explorers, if they did, they did not record it. The official sighting was on May 8, 1792 by Captain George Vancouver (English Royal Navy) who named it after Rear Admiral Peter Rainier (close friend) who never saw the mountain named after him.

In the early 1800s both Americans and English were exploring this country and establishing outposts that were barely able to survive. By 1818 the United States and England had a joint occupancy agreement.

Fort Nisqually was established in 1833. One of the men assigned to the post was a young Scottish physician William Frazer Tolmie who worked for the Hudson Bay Company. He was an accomplished botanist and in August of that year with five Indian companions became the first white man to enter the park. They climbed Hessong Rock (NW section by Spray Park), and then a peak that is now called Tolmie Peak (5,939 feet).

Settlement in the area continued and the boundary between the United States and Canada was established in 1846. Because of trouble between the Indians and settlers, Fort Steilacoom was established in 1849 (south of Fort Nisqually). By 1858 the Indian trouble was over. According to a local newspaper four men (Bailey, Ford, Edgar and Shaw) climbed Mount Rainier in 1852. In 1855 two white men got Alex Saluskin, an Indian to guide them up the mountain. They left him high on the slopes and continued upwards on their own. When they returned they spoke of a lake and steam vents at the top.

In 1857 Lieutenant August V. Kautz with Dr. R.O. Craig, four soldiers and Wapowety, a Nisqually Indian, almost made it to the summit but was turned back by weather and high winds.

In August of 1870 Hazard Stevens and Philemon B. Van Trump plus an Indian guide named Sluiskin started up the peak. When the guide realized they really meant to climb it he refused to go because an evil spirit lived in a lake of fire, so instead he started chanting about the mountain. He promised to wait three days and asked for a letter to vindicate him if they did not make it back. They made it to the peak but it took longer than planned and could not make it down that night because of weather and darkness. They thought they were going to freeze to death on the mountain when they discovered the steam vents and ice caves and spent the night in the caves. It was an uncomfortable night roasting by the vents and freezing if they went back into the caves, but they survived and came down the next day. The route they took is called the Gibraltar route.

James Longmire was an early settler, he helped to search for a route over the Cascades to the east. He helped to guide Stevens and Van Trump, later on in 1870 Emmons and Wilson (they gathered geologic information). Finally he climbed it himself in 1833. On that trip he discovered a mineral springs on the southwest side. He developed a resort and hotel by the springs creating the first settlement in the park.

A fight was raging in 1890 about the name, should it be Tohoma, Tacoma or Mount Rainier, the latter won out.

Because of the vandalism with the increase of tourists it was decided it should be made into a National Park to protect it. Thus, in 1899 it became the fifth National Park to be created.

It was under the administration of the U.S. Forest Service until 1916 when the National Park Service was created under the Department of the Interior.

GEOLOGIC FEATURES

Mount Rainier is a *dormant or quiescent composite* volcanic cone. It is dormant because there is a possibility that it may erupt again. Composite, since the cone or *edifice* is made up of alternate layers of lava flows and pyroclastic debris.

In 1955 Jim and Lou Whittaker after climbing Mount Rainier decided to explore one of the ice caves located at the summit. It had a 30 degree slope, and they went down some 300 to 400 feet and stopped for fear of being overcome by the hydrogen sulfide gas. When they rolled rocks down ahead of them they heard splashes, indicating a lake of a fairly good size under the snow cap.

In 1967 a mudflow from South Tahoma glacier, without warning, buried the White River Campground under two to three feet of debris. That winter instead of snow accumulating to the depth of 20 to 30 feet it stayed bare of new snow. Instead of advancing as normal it remained in place or became *stagnant.* This was because it was becoming so thin due to the heat beneath it.

In 1969 Emmons Glacier had a section of 105 acres of ice suddenly collapse, once again because of heat from below (in the middle of the winter). A *debris flow* (mixture of mud, rock and water) poured out of the collapsed area and moved downslope for some two miles before it ceased. The greatest debris ever recorded for Emmons Glacier. New steam vents have also formed.

In 1969 an infrared imager (heat sensing device) has shown along the east rim a new field of magma has moved in.

Micro-earthquakes (cannot be felt, but can be recorded by a seismograph) in 1969 have increased from four to six times normal. This type of quake indicates changes are occurring in the *vent* or lava pipe, that the magma travels up on its way to the surface. All these indicate that Mount Rainier may erupt again, but geologists cannot say exactly when.

Pumice is a volcanic rock that is so light that it is the only rock that can float on water. It is filled with *vesicles* or gas bubbles and resembles a sponge. It was formed when lava filled with gas was thrown into the air, the gases expanded and the rock solidified before it struck the ground. Pumice is found in layers at Mount Rainier which is not surprising. However not all of the pumice came from Mount Rainier. Because of its light weight it can be carried by the wind for long distances. There are eight pumice layers which have been assigned letters, from youngest to oldest. They are X, W, C, Y, D, L, O and R. Ones erupted by Mount Rainier are X, C, D, L and R, the respective ages (dating from 1968) and 100-150, 2,150-2,500, 5,800-6,600, 8,750-11,000 (?) years ago.

The two eruptions from Mount Saint Helen which is some 50 miles southwest of Rainier are eruption W about 450 years ago and eruption Y about 3,250-4,000 years ago. Last of all is layer 0 from Mount Mazama (Crater Lake) about 6,600

years ago. Mount Mazama is located about 250 miles south.

None of the layers are thick, most never exceed eight inches, they vary in color from white, to shades of green, brown, yellow or orange. The layers thicken as they approach the volcano that spawned them.

Landslides are a type of fast mass wasting. They are divided into slump, rock slides and debris slides. *Slump* is the downward and outward movement of a section of material, *rock slides* are rapid downhill movement of consolidated material downslope and *debris slides* are made up of unconsolidated material. All three varieties can be found. Slides can be found at Slide Mountain, another dams Ghost Lake (Klickilat Creek), and still another is in Grand Park called Scarface. All are located in the northeast section of the park. Several are on or near Backbone Ridge, and one is near Paradise Park.

A spectacular series of *rockfalls* (large segments of rock detach from a cliff and drop down) occurred in 1963 and fell from Tahoma Peak onto Emmons Glacier.

Avalanches are landslides primarily of snow but can be mixed with other material. Both modern and ancient ones can be seen. A modern one made up of reddish-gray rock debris is located in White River Valley near Little Tahoma Peak.

An ancient one occurred between 6,600 and 5,800 years ago and covered Paradise Park and Paradise Valley with as much as 600 feet of material. Erosion has removed most of it for it only averages about 15 feet thick of yellowish orange debris. Some of it crossed Mazama Ridge and filled the basin that the Reflection Lakes rest in. Part of it was liquid enough to be called a *mudflow* which is another type of fast mass wasting where a mixture of mud and water flows downhill.

The mudflow moved down Paradise River Valley and Nisqually River Valley some 18 miles down stream. Its thickness is measured in hundreds of feet.

About the same time another mudflow filled the floor of White River Valley some 30 miles beyond Mount Rainier to a depth of several hundred feet. Buried in this flow are huge rocks 5-35 feet high and up to several feet in diameter. They appear as mounds or hills covered with debris.

A slightly younger mudflow (5,800 years ago)

called the Osceola Mudflow, flowed down West Fork Valley and White River Valley joined together and flowed an additional 15 miles down slope into the Puget Sound lowland. It covered 100 square miles with mud some 70 feet deep. Part of it even flowed into Puget Sound. Evidence indicates it originated above Steamboat Prow, probably from the missing portion of the summit.

Sunset Amphitheater was formed by a mudflow 3,000 years ago. Another filled South Puyallup River and Tahoma Creek about 2,800 years ago, and also deposited some 400 feet of debris at Round Pass. A relatively recent one (occurred 600 years ago) buried the area with 15 feet of debris where the town of Orting is. It began at Sunset Amphitheater and moved down Puyallup River Valley.

The Kautz Creek Mudflow occurred in 1947 because of heavy rains saturating the region around Kautz Glacier (six inches in a few hours). Ice shifted to block and dam up material. The saturated mud broke through the dam and poured down the stream valley, went through a forest destroying it, bowled over trees five feet in diameter, buried a road with 50 feet of debris and finally stopped where Kautz flows into Nisqually River. In 15 hours some 48 cubic yards was moved. There is evidence of six other flows in the area.

Hot Springs and mineral baths are found by Ohanapecosh and Longmire. At Longmire there are about 20 springs with temperatures 50°-90° F. Other springs are located on Summit Creek, Cowlitz River, Tuton River, and Khikitat River.

Glacial features such as *ice caves* form because Paradise has stopped flowing and become a "dead" glacier (lack of enough ice). Meltwater streams that flow beneath the glacier have hollowed out tunnels or caves. Air flowing down crevasses have enlarged them into caves. Some of the caves are partially enlarged by steam vents nearby.

Glacial cirques are abundant such as Willis Wall (Carbon Glacier), Sunset Amphitheater (Puyallup Glacier) and head of Paradise Valley. *Aretes* are Cathedral Rocks, Liberty Ridge; *horns* such as Little Tahoma; cirque lakes or *tarns* such as Crescent, Tipsoo, Mowich Lakes; and *hanging valleys* over which Christine Falls and Comet Falls flow. *Cleavers* are volcanic ridges with sharp faceted angular faces, they form as ridges between glaciers, examples are Cowitz, Wapowety, Success and Puyallup cleavers. Also terminal, medial, end and recessional moraines are easily found by Flett, Frying pan, Nisqually, and Emmons Glaciers.

There are 26 glaciers in the park, occupying 37 square miles of land. They form in three areas:

1. On the summit (10,000 feet plus).
   a. Kautz
   b. Tahoma
   c. Winthrop
   d. Emmons
   e. Ingraham
   f. Nisqually

2. On the slopes (7,000-10,000 feet).
   a. Russell
   b. North Mowich
   c. Edmunds
   d. South Mowich
   e. Puyallup
   f. South Tahoma
   g. Wilson
   h. Paradise—"dead"
   i. Cowlitz
   j. Whitman
   k. Frying Pan
   l. Ohanapecosh
   m. Inter
   n. Carbon
   o. Pyramid
   p. Van Trump
   q. Stevens

3. North side of other mountains (around 6,000 feet).
   a. Sarvent
   b. Pinnacle
   c. Unicorn
   d. "Unamed" on Burroughs Mountain
   e. "Unamed" on Mother Mountain
   f. "Unamed" by Spray Park

Emmons Glacier is the largest and moves on the average of 50 feet per year, however, between 1952 and 1957 it moved 102 feet per year. Even though Nisqually Glacier moves foreward at 25 feet per year, it retreats back 70 feet per year so is shrinking.

Since 1950 many of the glaciers have started to advance e.g.: Carbon Glacier (8-26 feet per year), this change to cooler temperatures and increased snowfall actually began around mid-1940. Prior to that they were all shrinking and covered only two thirds the area that they occupied in the mid-1800s.

Glaciers during the Pleistocene were more extensive and existed before Rainier did as is evidenced by lava flows over glacial debris. Glaciation occurred over one million years ago, Mount Rainier built up to maximum size only 75,000 years ago.

The first two glacial episodes covered entire flanks of Rainier and adjacent mountains. The last episode 25,000 to 10,000 years ago was much less extensive.

Todays glaciers exist because of the amount of precipitation averaging 50 to 100 inches per year (depending upon location). Record depth of snow was in 1972 when Paradise Valley received 93 feet, it averages 50 feet per year.

Besides regular glaciers forming *rock glaciers* also form. They are a mixture of mostly rock with ice in the spaces or interstices between. The expansive and contraction during freezing and thawing moves these glacierlike bodies down slope. Examples can be found between the Palisades and Hidden Lake or in a cirque on Mount Fremont on the southeast side.

Mount Rainier used to be approximately 2,000 feet higher than it is today. Its original summit which formed some 4,000 years ago. An explosion formed a summit caldera around two miles in diameter, remnants of these are Point Success (14,150 feet) and Gibralter Rock (12,679 feet) and the slope by Liberty Cap (14,133 feet). It was then filled in by two smaller volcanoes that lie within it, first the western cone that is not completely preserved, then the eastern cone that has a circular crater some 1,200 feet in diameter (the one that the tourists visit) called Columbia Crest. The third cone is Liberty Cap.

GEOLOGIC STORY

1. Flat lowland during Eocene period built on ancient lava flows.

Before the Eocene epoch there used to be mountain ranges in the location of Mount Rainier today. They were formed and destroyed several times, including lava sheets several thousand feet thick.

About 60 million years ago an arm of the Pacific Ocean extended to just north of Seattle, Washington. To the east low hills existed which gave way to higher mountains that supplied the sediments. The area was low and swampy, with coal beds forming on the coastal plain swamps. The shoreline fluctuated while sands and clays were deposited in the shallow water of the gradually sinking basin. These are known as the *Puget Sound*

Group and are a mixture of shale, sandstone and coal. They took 20 million years to form.

2. During Late Eocene-Early Oligocene, formation of Ohanapecosh formation.

The coastal plain region slowly sank beneath sea level, the volcanoes that had formed became islands. This happened about 40 million years ago. Eruptions occurred both from the islands and underwater, mixing with the accumulating sediments. Thus the Ohanapocosh formation is a mixture of lava, mudflows, grayish-green sandstones and volcanic breccia. They accumulated to a thickness of 10,000 feet. These are the oldest rocks seen in the park.

3. Uplift, folding, volcanic activity, erosion Oligocene-Miocene time.

Western Washington was uplifted and the Puget Sound Group and Ohanapecosh formation were folded into gentle northeast-southwest anticlines and synclines which underwent erosion and weathering forming reddish soils in the stream valleys that were several hundred feet deep.

About 30 to 25 million years ago volcanoes (outside park boundaries) had nuee ardentes type eruptions which solidified into *welded tuffs.* Some were 350 feet thick. These tuffs are called the Stevens Ridge formation.

Following this central vent eruption, fissure eruptions poured out layer after layer of andesite lava flows from 50 to 500 feet thick until they accumulated to a depth of 2,500 feet. These flows are called the Fifes Peak Volcanics and formed 20 to 30 million years ago, they are cliff and mountain formers.

4. Uplift, folding and faulting, intrusion of grano-diorite and quartz monozonites, erosion.

After the Fifes Peak volcanics formed the entire region was uplifted and folded once again, parallel to the first set of folds, (Unicorn Peak syncline). Faulting accompanied uplift forming a mountain range. About ten million years ago a granodiorite and quartz monzonite intruded. (Katoosh granite). Some was extrusive and formed volcanoes nearby. Erosion wore down the mountains to low hills.

5. Uplift of Cascades and formation of the Cascade Mountains.

While the Rainier area was undergoing erosion to the west the Olympics Mountains were being formed and to the east the Cascades. The streams dissected them into the mountains about 6,000 feet high with the intervening valleys. This occurred during the Cascadian Revolution.

6. Formation of Mount Rainier on granodiorite platform during early Pleistocene.

Somewhere between 1 to 1/2 million years ago molten rock reached the surface of the Unicorn Peak syncline one of the folds that formed the Cascade Mountains. The flows first moved down the valleys as far as 15 miles away filling them in. Eventually a volcanic cone built up from a central vent with alternate layers of pyroclastic debris of bombs, blocks and cinders plus andesite lava flows. This is why Mount Rainier has a symmetrical shape. There were long spans of times between the eruptions because glacial sediments were deposited in between some of the volcanic layers. At about the same time in adjacent areas the volcanoes Mount Adams, Mount Baker, Glacier Peak and Mount Hood were forming.

7. Glaciation of Mount Rainier during the Pleistocene some two million years ago.

Much of the region was covered by the Cordilleran Ice Sheet, with the exception of the highest peaks. Mount Rainier was covered by large alpine glaciers that filled the valleys with 2,000 to 3,000 feet of ice. They were up to 40 miles long. This was during the formation of Mount Rainier. With the last major glaciation some 25,000 years ago the glaciers were not as extensive. They retreated for the last time about 10,000 years ago.

8. Holocene time destruction of original summit.

About 5,000 years ago Mount Rainier was much higher according to the angle of the slope by as much as 2,000 feet. Part of it was probably removed by avalanches, landslides and mudflows. Possibly *phreatic explosions* occurred when ground water turned to steam very quickly and violently, triggering the slides and mudflows. The Osceola Mudflow of 5,800 years ago may have marked the final destruction of the original summit.

9. Formation of present glaciers about 3,000 years ago.

From about 3,000 years to the present is often referred to as the "Little Ice Age." The glacier that are now on Rainier formed at that time. The old crater or caldera that was 1 1/2 miles across has been filled in with two minor cones that make up the present summit cone, these formed 2,000 years ago. Columbia Crest Cone, Point Success, Liberty Cap and Gilbralter rocks form the rim of the old crater.

First formed was the western crater which is about 1,300 feet across (part of the rim is missing on the eastern side). Then the younger eastern cone developed, it is about 1,200 feet in diameter and is perfectly formed within the central crater. Combined they form the *Columbia Crest Cone,* and have the floor filled with snow (depth unknown). Steam vents issue from them just inside the rim and form caves. These are unglaciated since they formed after the Pleistocene.

10. Last major eruption 2,000 years ago, possible minor eruptions between 1820 and 1854.

The last major eruptions created the "C" layer of pumice that covered about two-thirds of the park mostly in the northeast section with a layer of brown pyroclastic debris from one to eight inches deep.

A localized layer "X" probably occurred between 1820 and 1854, it is light olive gray and only one inch thick. Other reports in the late 1800s were probably dust clouds formed when landslides and avalanches occurred. However, since Mount Rainier is quiescent, that situation may change any day.

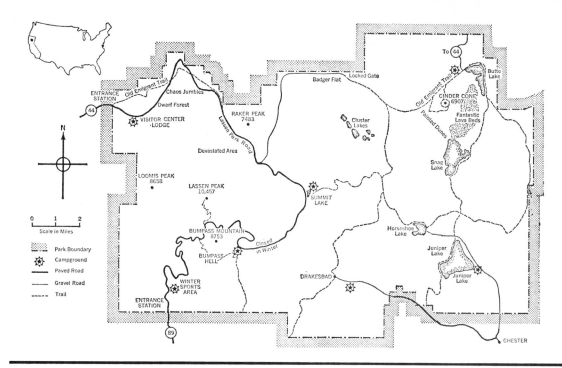

Lassen Volcanic National Park, California.

Tentative Geologic Column of Lassen National Park

Erathem	System	Series	Formation	
Cenozoic	Neogene	Holocene	Rockfall Avalanche Deposits of the Chaos Jumbles	
			Dacite of the Chaos Crags Dome	
		Pleistocene	Moraines of Late Tioga Age	
			Dacite of Lassen Peak Dome	
			Tioga Till	
			Tahoe Till	
			Pre-Tahoe Till	
			Lassen Peak Dacite	
			Pre-Lassen Dacite	
			Rhyolite and Dacite Flow	Cones at Edge of Central Plateau
			Basalt and Andesite Flow	
			Eastern Basalt	
			Flat Iron Andesite	
			Juniper Lake Basalt ⟩ Twin Lakes Andesite	
			Willow Lake Andesite	
		Pliocene	Tuscan Formation	Unaltered Sedimentary Rock
	Paleogene	Eocene	Auriferous Gravels	
			Montgomery Creek (Ione Fm)	
Mesozoic	Cretaceous		Chico Formation	
	Jurassic		Bend Formation	Altered Sedimentary Rock
			Cedar Formation	
Paleozoic	Pennsylvanian Mississippian		Robinson Formation	
			Calaveras Formation	
	Devonian		Arlington Formation	
	Silurian		Grizzley Formation	

Modified after Diller, Warren, Loomis et. al.

CHAPTER 15
LASSEN VOLCANIC NATIONAL PARK

Location: California
Area: 106,933.78 acres
Established: August 9, 1916

LOCAL HISTORY

This was the ancestral home of four Indian tribes: the Atsugewi, Mountain Maidu, Yahi and Yana. This was a summer camping ground where they arrived in late spring, hunted and fished and left in late fall for the regular homes in the foothills and valleys. They were peaceful tribes that got along together well enough for intermarriage to occur.

The Indians called Lassen Peak "The Sweat of the Gods" for they believed that one day it would blow itself up into pieces, and that all the fire and water would go away, thus the gods would come back and all would be right with the world again.

Around 1770 the total population of all four tribes was around 4,025. By 1950 there were only 385 persons left. Around 70 of them live near Lassen Peak. The last member of the Yahi Tribe died in 1913, his name was Ishi.

It was orginally called St. Joseph Mountain, or San Jose, by Captain Luis Arguello, a Spanish explorer who discovered it in 1821. Lassen Peak was named after Peter Lassen, who was born in Denmark around 1900. He immigrated to the United States and worked for a while, as a blacksmith in the east, before coming to California in the 1830s. He received a land grant from the Governor of California (Mexican) and started a 26,000 acre ranchero he called Rancho Bosquejo. He laid out the town of Benton on it, then went to Missouri to bring settlers to it in 1848.

He guided the immigrants into Sacramento Valley from the Overland Trail across the Cascade Mountains and used San Jose as a landmark. It was supposedly a short cut, but the group suffered so many hardships they referred to it as "Lassen's Folly." With the discovery of gold, Benton was abandoned.

He led parties into Sacramento from about 1848 to 1850. Unfortunately, he frequently got lost himself. He couldn't even tell the difference between St. Joseph and Mount Shasta, and used them both as the same landmark. According to one legend, one of the parties he was supposed to be guiding forced him to climb St. Joseph's Mountain (Lassen) at gunpoint to locate himself. By 1850 the word got around not to take Lassen's Trail. Instead they used the Nobel Trail, parts of which can be seen in the park today as wagon tracks.

Lassen went to Susanville in June of 1853 with some prospectors and struck gold. He settled in a cabin along Lassen Creek in 1855. In 1856 he was elected surveyor of Nataqua—a short lived attempt of establishing a capitol of California territory. He was finally killed at the age of 66 on April 26, 1859 by Indians. There are two monuments erected to him near Susanville.

The United Geographic Board officially named it after Lassen in 1915. However, in 1907 it was declared Lassen Peak and Cinder Cone National Monument. After the eruptions of 1914, it was changed to a National Park in 1916.

Bumpass Hell is named after Kendall V. Bumpass who worked on a ranch in 1864. He was hunting for stray cattle and broke through the crust and scalded his feet. When his boss asked him what took him so long to get back and where had he been, his reply was "In Hell." The following year he was guiding a Red Buff editor through it and had just finished warning him to be careful in this dangerous ground when he broke through the crust. The boiling mud burned his leg and he was so irked that he forgot to swear, which was very unusual for him.

Lake Helen has been named after the very first woman to ever climb Lassen Peak in 1864, her name was Helen Tanner Brodt. It fills a glacial basin.

GEOLOGIC FEATURES

Lassen Peak is a *plug-dome volcano* that formed when a thick, stiff mass of dacite lava was squeezed out of a vent to form a irregular dome-shaped bulbous mass that plugged up the vent it was forced out of. As the dome was elevated, the broken crust and debris flanked its slopes. The vent is located on the northern slope of an ancient extinct volcano called Brokoff (Tehama). An unusual feature of Lassen is the crater at the top (not usually found in plug domes) formed by violently escaping gasses. It was originally 1,000 feet in diameter with a depth of 360 feet, now is presently filled with lava. Other plug domes are Bumpas Mountain, Chaos Crags, Eagle Peak and Vulcan's Castle, they formed around 200 years ago. They formed in three stages—first formation of cinder cones, second intrusion of rhyolite dome, and third partial destruction of dome by explosion. The first stage can still be seen on the slope. The crater was 600 feet in diameter and 60 feet deep. The Chaos Jumbles at the base of Chaos Crags are the result of an avalanche and volcanic explosion from the third stage, when the three plugs that form the Crags literally blew their top. The violence of the blast undercut the cliffs resulting in three landslides sliding over the wet ash. Some of the slide blocked Manzanita Creek, thus forming Manzanita Lake.

Other types of volcanoes can be found such as the *pyroclastic* or *cinder cones* made up entirely of pyroclastic debris. Cinder Cone, northeast of Lassen Peak is an example of a pyroclastic cone, it is 700 feet high and last erupted in 1851, the composition of material is quartz-basalt.

The Fantastic lava beds and Fantastic dunes formed during the 1851 eruption, the dunes are made up of pyroclastic debris (ash and cinders) and lava beds that are an *aa* flow (rough jagged flow). These became colored as gasses passed through the debris and oxidized forming the reds, oranges, and yellow brown colors.

Another type of volcanic cone is the *lava cone* or *shield volcano* made up entirely of lava flows. They are Mount Harkness, Prospect Peak, Raker Peak, and Red Mountain. Mount Harkness and Red Mountain since then, have had the addition of cinder cones upon them.

Sulfur Works is a mixture of hot springs, steam vents or *fumaroles* and *mud pots*. The mud pots develop when the hot springs are unable to remove the mud so it accumulates, ending up with mostly boiling mud and very little water. The Sulfur Works are located over the old vent system of ancient Brokoff Mountain (Mount Tehama). Similar features can be seen at Boiling Springs Lake, Bumpass Hell, Devil's Kitchen, Little Hot Springs Valley, and Cold Boiling Lake. A straight line could be drawn through all these features. They all probably lie above a fault trending northwest-southeast that

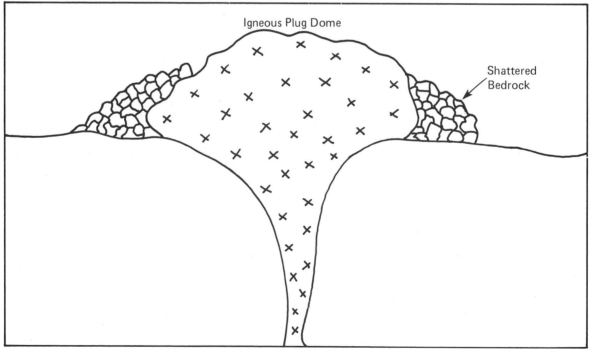

Plug Dome Volcano

is located in Brokoff's Mountain (Mt. Tehama's) caldera. Bumpass Basin is a crater shaped basin that developed its thermal features when the hot sulfuric acids dissolved holes in the lava flows forming its various thermal features.

The Devastated Area resulted from a geologic feature called a *nuee ardentees.* This a super-heated (above boiling point but does not boil) cloud of steam and droplets of lava. Since this cloud has weight, it rolls downslope. They have been clocked at speeds as high as 60 miles per hour. This fiery cloud roared down so fast that it bowled over trees as far as three miles away uphill and sandblasted the broken tops smooth. For the trees that somehow remained standing, the side facing the crater had the bark sandblasted off and sides polished smooth. The effects died out after a distance of five miles. Sand grains were driven in an inch into the trees. Debris was tossed five miles high in a huge cloud, which winds picked up and blew as far away as Nevada. Falling stones punched holes in the roof of the Lookout (it was destroyed by a later eruption).

The edges of the ancient caldera of (Mount Tehama) Brokoff Mountain, can still be seen, they are Brokoff Mountain, Mount Piller, Pilot Pinnacle, Mount Conrad (moving clockwise). Diamond Peak is some of the rock that congealed in the vent and Sulfur Works represents the vent area. Brokoff Mountain (Mount Tehama) probably had a 15 mile diameter base, was about 11,000 feet high and had a 3,000 foot diameter crater. It formed about 11,000 years ago. The sides collapsed forming a caldera and about 1,000 years ago Lassen Peak formed on its flank, with its three craters that are funnel shaped (one has a lake) and 35 steam vents at the summit.

Emerald Lake is a cirque lake or tarn, Lake Helen is of glacial origin along with most of the lakes in the central portion of Lassen such as Feather, Silver, Big Bear, Little Bear, Twin (upper and lower) Swan and Echo Lakes. Summit Lake is dammed by moraine.

GEOLOGIC HISTORY

1. **Formation of the gold-bearing beds of Paleozoic and Mesozoic time, total thickness 4.5+ miles.**

Cretaceous	Chico formation—sandstone, conglomerate, shale
Jurassic	Bend formation—slate, sandstone, conglomerate
	Cedar formation—slates, sandstone, some conglomerate
Carboniferous	Robinson formation—conglomerate, volcanic sandstone, tuff
	Calvareas formation—quartzites, slates, limestone
Devonian	Arlington formation—sandstones, slates some conglomerates, Lithostrotian, Fusulina
Silurian	Grizzly formation—slate, quartzites and limestone.

This is a mixed group of both sedimentary and extrusive igneous rocks that have undergone changes in structure and position since their formation.

Interbedded are lens-shaped masses of limestone that dates it Mississippi-Pennsylvanian according to the fossils. Some of these slates are Paleozoic, the rest are Mesozoic. Most of the rocks were originally shales. Folding first occurred in the Jurassic. The land then subsided and seas covered the area. The Chico formation was deposited at the base of the Sierra Nevadas.

2. **Extrusion of peridotite (olivine lavas) interbedded with shales, weathering alters to serpentine.**

Ultrabasic flows of almost pure olivine formed, thus the rock is called a *dunite* or *peridotite.* Chemical weathering or decomposition occurred, converting olivine into tremolite if it was unprotected. The remainder stayed olivine.

3. **While shales were forming they became interbedded with limestone, metamorphism.**

As the Mesozoic began, limestones were deposited instead, and more volcanic eruptions occurred. Metamorphism changed the shales to slates, limestones and peridotite to greenstones. The degree of metamorphism varied. Diorites, gabbros, peridotites and layers of lapilli and volcanic sand also were formed.

4. **Formation of Cenozoic formation.**

	Auriferous Gravels—gravel
Paleogene (Eocene)	Montgomery Creek (Ione formation)—conglomerate, sandstone, shale, coal.

The region was uplifted and tilted at the end of the Cretaceous with the deposition of the Montgomery Creek (Ione) formation in an estuary or lake. It separated the Klamath Mountains from the

Sierra Nevadas and covered the Lassen area about 70 million years ago. This was called the Lassen Straits (Cascade Mountains today). The surrounding land areas were being worn down to an erosional surface. Streams meandered back and forth, unable to carry very much of the insoluble material. When the area was elevated the streams began to downcut and also concentrate the insoluble material, such as the placer gold, where it became mixed in with the gravels. The elevation was produced by the Laramide Revolution that was creating the Rocky Mountains.

The uplifting relieved pressure on the rocks deep below and they were able to melt and work their way to the surface and erupt as volcanoes. The activity began 11 to 12 million years ago and formed what is now the Western Cascades (low rolling hills). These are the Tuscan beds. To the north, basalt was pouring out in Washington and Oregon, forming the Columbia Plateau. To the east the Modoc Plateau formed, with its thinner flows (basalt). This also continued until 10 or 11 million years ago and formed the High Cascades. These beds were uplifted, faulted, and for the next 10 million years built up shield type volcanoes with the usual basaltic composition.

5. Formation of Tuscan formation between two and three million years ago.

For about one million years a series of flows, *mudflows* or *lahars* covered some 2,000 square miles. These were derived from three sources: First Mount Yana (south) was the oldest major source. Mount Yana was some 10,000 feet high and 15 miles in diameter. The second source was Mount Maidu which was at the time it contributed to the Tuscan formation was about half its size (it eventually became bigger than Mount Yana). Mount Maidu is where the town of Mineral is today. The third major source (north of Latour Butte) contributed some material. Other possible sources (minor) were the dikes by Inskip Hill, Campbell Mound (north of Chico) and Hatchet Mountain Pass (located to the northwest of Burney Mountain).

6. At the same time basaltic flows within the park created a flat lava plain.

The first eruptions within the park occurred at Willow Lake (just south of boundary). They were basaltic but the source is unknown.

Shortly after the Juniper Lake Flows (southeast corner of the park) developed some four miles westward. The composition was a pyroxene andesite. The vents were to the east of the park and today are probably covered up by later material.

At the same time as the Juniper Lake flows were occurring the Twin Lakes andesites, a black (porphyritic) flow, extruded from a group of vents on the Central Plateaus. They covered an area of 30,000 square miles. This particular andesite contains phenocrysts of quartz and today lies next to Cinder Cone which is a very recent addition that also has these quartz crystals in it but is of basaltic composition. At a much later time in the same region Crater Butte, Hat Mountain and Fairfield Peak, three cinder cones formed.

The Flat iron andesites, a pyroxene andesite, poured out of the Reading Peak area (central portion of the park), and traveled some five miles to the head of Warner Valley.

All these flows combined, transformed the originally mountainous region into a flat lava plain. There was practically no pyroclastic debris ejected.

Just east of the park the Eastern basalts which are pyroxene basalts covered up the Juniper Lake andesites. These are the hills found on the east side of the park today that have been created by the erosion of these hills. Towards the end of the eruptions they changed to pyroclastic debris. Around this time Badger Mountain and Table Mountain areas (just north of boundary) region had active cones with the andesite and basalt lavas.

7. Formation of Brokoff (Mount Tehama) Mountain.

All this time, after Mount Yana started to erupt and before Mount Maidu began a volcano was forming immediately after the eruption of the Flatiron andesites. It eventually reached a height of 11,000 feet with a diameter of around 15 miles and was known as Mount Tehama but recently its name has been changed to Brokoff Cone. The sequence of eruptions compositionwise was augite andesite, hypersthene andesite with tuffs and breccias interbedded near the peak. The main vent was in the locality of Sulfur Works today, a second vent (which did not produce lava) formed on the eastern side of Little Hot Springs Valley. (Note: Bumpass Hell is outside of the caldera, it was not a main vent.)

8. Formation of shield volcanoes.

Some time later four shield volcanoes formed, one located at each of the Central Plateaus corners. Going clockwise at 1:00 was Prospect Peak, 5:00 Mount Harkness, at 7:00 Red Mountain and 10:00

Raker Peak. By the time these had erupted the Juniper and Flatiron flows were dissected by erosion, so the lavas from the shield volcanoes began to fill the valleys first. Prospect, Red Mountain and Harkness erupted pyroxene basalt, Raker Peak pyroxene andesite. Eventually all developed cinder cones within their craters. After they were all dormant some were intruded by plugs. Rhyolite forced its way into the north side of Red Mountain, while hornblende-mica dacite intruded the west side of Raker.

9. New vent opens on Brokoff Mountain formation of Lassen Peak and other plug domes, glaciation.

Just about the location of Lassen Peak on what was the northeast slope of Brokoff Cone another vent opened and quietly extruded some 1,500 feet of the pre-Lassen dacites. Today they form the beds with columnar jointing that circle the base of Lassen Peak. These pre-Lassen dacite flows were the gas-rich lavas that managed to escape, leaving the gas-poor magma to slowly well up with its burden of fragments of the crust of the magma cham-

ber (basic with hornblende). The lava was half solid and the liquid portion extremely thick, so as the mass rose the sides of the dome became polished, and then broken into a pile of talus all around the sides.

Other similar domes formed but on a smaller scale. They are Bumpass Mountain, Eagle Peak, Helen Ridge, Reading Peak and Vulcan's Castle, all south of Lassen. Crescent Crater occurred before Lassen. The vents were the same ones that supplied the Flat Iron Flows, and the domes near Lost Creek (upper limit of 10,000 years ago). All formed rapidly. The entire region underwent glaciation forming many of the lakes in the park and leaving other glacial features such as striations and moraines.

Three periods of glaciation occurred; Pre-Tahoe, Tahoe and Tioga plus moraines of late Tioga that occurred after the Dacite of Lassen Peak Dome. These tills have been separated by leached and weathered soil layers. Lassen Peak erupted and formed between Tioga and late Tioga time about 11,000 years ago.

Pre-Tahoe oxidized zone may be as old as

100,000 years and forms to the depth of 15 feet. Original rock pebbles have weathered to clay, larger fragments to the depth of four inches.

The Tahoe Drift is oxidized to depths of three to eight feet and is between 30-40,000 years old. Fragments of rock are weathered only a slightly (1-2 mm) forming a thin rind. These beds include more than one advance. The till itself is at least 70,000 years old.

The Tioga glaciation is divided into two stages—the Tioga and late Tioga separated by the formation of Lassen Peak. The Tioga stage was deposited by an icecap variety of glaciers. The late-Tioga by small valley or cirque glaciers that formed at 8,000 feet.

Tioga Drift is oxidized to a maximum depth of 2.5 feet and incorporated bedrock does not have a weathered coating or rind.

The late Tioga drift is only oxidized to a depth of two feet and ranges in age from 10-15,000 years. Incorporated fragments are unweathered and the moraines are covered by a 1,200 year old pumice (Choas Crags). They are also covered by mudflows and alluvium by Lost Creek, and were probably deposited by glacial meltwater.

10. Destruction of Brokoff Cone, formation of pyroclastic cones along fault, formation of Chaos Crags.

All the activity and possibly the removal of so much dacite caused Brokoff Cone to collapse and form a caldera of which Pilot Pinnacle, Mount Conrad, Diamond Peak and Mount Diller forming the rim. Most of the hot springs of today are located within this caldera.

Before Lassen Peak had developed there was a string of dacite pumice cones trending northwest from Crescent Crater. About 1,100 years ago Chaos Crags evolved as plug domes in the same locality destroying completely all but a half of one of the cones (the most southerly). Around 200 to 300 years ago steam explosions (in all probability) triggered off an avalanche that was cushioned by air. It traveled about 100 miles per hour, dammed up Monzanita Creek creating Manzanita Lake and finally collapsed to form Chaos Jumbles an area of 2.5 square miles. Depressions in these debris filled with water and we call them Lily Pond and Reflection Lake.

11. Formation of cinder cone, Painted Dunes and Fantastic lava beds.

About 500 years ago on the floor of what was called Lake Bidwell (now called Butte Lake) a cinder cone built up and in the process blanketed the surrounding 30 square miles with debris. The ashes that fell on the lava pouring down the east slope we now call the Painted Dunes. Another flow of blocky quartz basalt (several hundred years later) poured into Butte Lake, dammed the drainage system and Snag Lake formed. Last eruption was in 1851. Gases seeping through the flow and oxidizing are responsible for the color, giving them the name Fantastic lava beds.

12. Eruption of Lassen Peak in 1915.

On May 30, 1914 an eruption began that formed a 25 by 44 foot crater and fissures that radiated 360 degrees. Both pyroclastic debris and lava was ejected and a cone built up. This continued about a year. Winds deposited some of this material as far east as Nevada.

On May 19, 1915 lava poured out from two notches (southwest and northwest) with a tremendous explosion. The southwest flow traveled about 1,000 feet before it hardened. The northeast flow melted snow and triggered a mudflow. This debris started downslope, encountered Raker Peak, split into two. Half went down Lost Creek, the other half flowed over a 100 foot divide and down Hat Creek. It destroyed everything in its path including trees, carried 20 ton boulders some five to six miles before dropping them. One boulder sizzled in the water for several days and is called Hot Rock today. It's about 18 X 20 feet for length and width and 14 feet high. Smaller flows occurred on the west and north sides.

13. Nuee ardentees forming devastated area.

Three days later on May 22, 1915 a nuee ardentees first rose 40,000 feet, some of it sank and went roaring down the mountain side. This superheated cloud of steam and droplets of lava bowled over trees on Raker Peak (five million board feet), sandblasted the bark off on the crater side and polished and rounded the tops of the stumps. Some sand was actually driven into the trees to the depth of one inch. Trees three miles away were toppled, but have since decayed or are so covered with vegetation that they can't be seen. Almost no fires occurred so the temperature at the crater must have been fairly low.

Several small explosions occurred and finally in 1921 all activity ceased and so far has remained that way.

Lassen Peak is considered as quiescent but it may erupt sometime in the future.

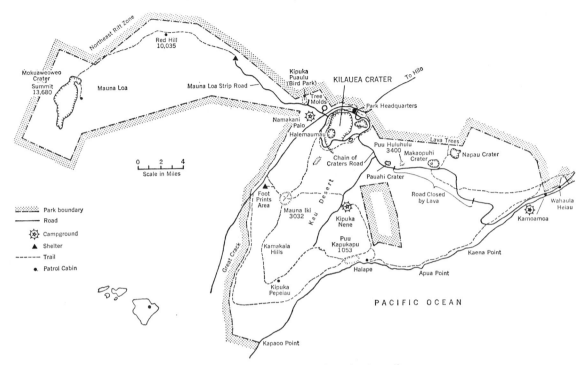

Hawaii Volcanoes National Park, Hawaii.

Geologic Column of Hawaii National Park

Erathem	System	Series	Formation	
			Mauna Loa	**Kilauea**
Cenozoic	Paleogene	Holocene	Kau Volcanic Series of 1832-1950 including Mudflow of 1868	Puna Volcanic Series (1790-1974)
			Dunes	Dunes
			Prehistoric Member of Kau Volcanic Series	Prehistoric Member of Puna Volcanic Series
		Pleistocene	Pahala Ash	Pahala Ash
			Kahuku Volcanic Series	Hilina Volcanic Series
			Unconformity	
		Pliocene	Ninole Volcanic Series	

Modified after Sterns and Wright.

CHAPTER 16
HAWAII VOLCANIC NATIONAL PARK

Location: Island of Hawaii, Hawaii
Size: 229,615.87
Established: August 1, 1916

LOCAL HISTORY

According to legend Pele came somewhere from the south, after having a quarrel with her sister Namakaokahai, looking for a new place to live. She carried with her a magic digging tool called a Paoa. When she used it on the earth a volcano sprang up. First she tried Kauai and formed Puu-ka-Pele ("Pele's Hill"), but her sister found her there, they fought and Pele was left for dead. When she recovered she went to Oahu and tried digging a volcano but it filled with salt water (Keealipaakai Salt Lake), the debris from the hole formed a hill (White Bird). Unable to find fire she moved to the coast and tried again (Diamond Head) but she struck water once again and it put out the fire she discovered. Therefore, she left and went to Molokai and dug Kauhako, but again encountered water, so she went on to Maui and dug Haleakala. When her sister saw the smoke at Haleakala she knew Pele was alive. They fought again and Pele was killed. Her sister scattered Pele's bones along seashore forming Kaiwi o Pele (Pele's Bones). While Namakaokahai was celebrating her victory she looked over to Hawaii and saw Pele's spirit over Mauna Loa (as smoke and fire) and realized she could never defeat her so she left never to bother Pele again. Pele dug Halemaumau, "House of Ever-lasting Fire," (in crater of Kilauea) to be her final home and brought her relatives to live with her to this day. This is why Kilauea is still so active. According to legend, she is temporarily residing in whatever volcano is active.

The Polynesians came to Hawaii some 2,000 years ago. They worshiped Pele the volcano god-dess. Captain Cook came in 1778 and officially discovered the island. The natives worshipped him at first but after one of his men was killed, they realized that they were not gods and Captain Cook was killed at Kealakeakua Bay.

Captain George Vancouver (Mount Rainier discoverer) visited Hawaii in 1792, 1793 and 1794. Some of his men climbed Mauna Loa to study it. He knew the islands as the Sandwich Islands instead of Hawaii. His visits and the visits of others including the missionaries in the 1820s changed the lives of Hawaiians forever.

Reverend William Ellis in 1823 was the first white man to enter what is now the Hawaii Volcanic National Park. He also wrote the first account of an eruption of Kilauea.

Wahuala Heiau (near east edge of park) is most likely the oldest temple in Hawaii and was the last one used in the ancient Hawaiian religion which included human sacrifice. Paao, a priest from Tahiti, built it and formed a very strict religion which included human sacrifice. Hewahewa, a descent of Paao, did not agree (he was a high priest of King Kamehameha) and in 1819 managed to overthrow the religion and destroy their temples.

In 1790 when Kamehameha was trying to become the first king of Hawaii, his cousin Keoua was the last of the warriors to oppose him. Keoua and his army (plus their women and children) had camped by Kilauea which erupted that night. They thought that they had offended Pele in some way and tried to appease her the following day but were without success. The following day they split up into three companies and started to leave. After the second group passed Kilauea by (some seven miles away) it exploded with a steam blast (phreatic explosion with poisonous gases) that killed all the people in the second company and some of those in the first. Some of their foot prints can still be seen today in the pyroclastic debris that had fallen earlier and became welded by the heat.

GEOLOGIC FEATURES

The Hawaiian Island chain are all volcanoes that had their origin on the ocean floor as submarine volcanoes. The island of Hawaii is the youngest and has two active volcanoes on it, Mauna Loa ("The Great Mountain") and Kilauea ("Rising Smoke Cloud"). Mauna Loa is 13,680 feet above sea level, its lava flows occupy 2,035 square miles. The cone measures 75 miles long by 64 miles wide. It occupies about 50 percent of the island of Hawaii. Its caldera is 1 1/2 miles wide, 3 miles long and is around 600 feet deep and is called Moluaweoweo Crater (even though it is a caldera). It formed when the withdrawing magma no longer supported the sides of the edifice, so it collapsed inwards.

Kilauea is 4,088 feet high, it is smaller and younger and is the most active. Kilauea is only 51 miles long and 14 miles wide and occupies about 14 percent of the island. Its caldera is 2 1/2 miles long, 2 miles wide and about 450 feet deep. Some of its steep walls are *fault scarps* (cliff formed by faulting) formed when the caldera collapsed. On the floor is the vent Halemaumau. A wall separates the main crater from a smaller pit crater called Kilauea Iki or "Little Kilauea," it has been the more active of the two more recently.

Both volcanoes are *shield volcanoes* or *lava cones*, since they have been built up by successive layers of lava. This type of volcanic cone is characterized by a very low broad profile with a slope of four to ten degrees. Size wise they cover 10s or 100s of square miles in area. Their shape in profile is the reason for the name shield volcano.

Fire fountains form during eruptions, because of the expansion of the trapped gasses in the lava. As the lava moves upwards towards the surface it is under less pressure permitting the gasses to expand. These gasses exert tremendous pressure since the lava is confined within the vent. With relatively little pressure from above, the greater volume of lava is forced out as a fire fountain. Some fountains have been recorded as erupting as high as 1,900 feet.

Both types of flows are found here, the rough jagged *aa* flow and the smooth ropy *pahoehoe* flows. Occasionally in these flows *lava tunnels* or lava tubes will form such as Thurston Lava Tube. They form when either a lava flow or the side of the edifice breaks open and the liquid lava comes pouring out through the opening. Sometimes the liquid lava will splash onto the ceiling and solidify while dripping onto the floor, thus forms *lava statactites.*

Not all volcanoes will have the vent in the center. During the 1959 eruption of Kilauea Iki a 1,000 foot high cinder cone formed behind one of the fire fountains because the prevailing trade winds blew the material behind the vent.

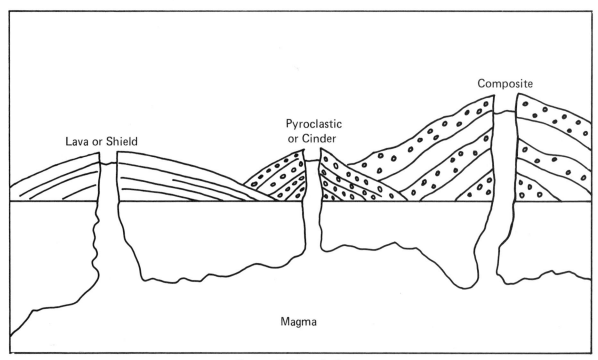

Types of Volcanic Cones

Tree molds will form at the fringes of the flows, since the lava is chilled upon contact with the tree. A crust will form around it, from one to six inches thick. If the lava remains, a hole is left in the flow with only the carbonized impressions of the tree remaining. If most of it flows away a mound is left, sometimes with the charred tree inside.

Spatter cones may form when clots of lava with their chilled glassy skins accumulate in a cone at the base of a fire fountain. A *spatter rampart* is a wall along a fissure eruption with fountains. It is formed the same way.

Halemaumau is a *pit crater* or *volcanic sink.* It is a steep walled, circular depression formed by collapse, that does not have a mound or wall of lava surrounding it. Additional pit craters can be found along the east rift zone called the Chain of Craters. Up to 1924 there was a lake of molten lava that rose and fell periodically in them. Occasionally the lava dropped completely out of sight. When this happens the sides of the pit crater will landslide in because of the lack of support. If ground water seeps in through the cracks it is immediately and violently converted into steam producing an explosion of steam called a *phreatic volcanic explosion.* There is no pyroclastic debris involved and the temperature is usually quite low, thus differing from the nuee ardentees.

Pele's tears and *Pele's hair* form in the basaltic flow when exploding gas bubbles throw droplets of lava into the air that are picked up by an updraft and solidified into glassy globules (Pele's tear). Sometimes a thread hangs from the globule and breaks off (Pele's hair), or jets of steam from ground water table is blown into the lava during fountaining, forming the delicate threads.

When larger fragments of lava solidify they form *volcanic bombs.* There are several varieties such as the *spindle bombs* that are tapered on each end, *bread crust bombs* that have a cracked crusty exterior formed by shrinkage while cooling, some are simply irregularly shaped or round bombs with a relatively smooth exterior.

The *rift zones* are narrow zones of fractures on the side of the volcano that extend from the crater to the foot of the cone or beyond. Kilauea has an east rift zone and southwest zone, Mauna Loa a south rift zone and a northeast one. The lava is called a *tholeiitic basalt* which indicates it has very little olivine, but other crystals are present as the result of separating and settling out.

The lavas from Mauna Loa do not show any correlation as to when the eruption occurred or where. However, the ones from Kilauea are separated into three groups, those before 1750, the ones between the eighteenth and nineteenth century, and those of the twentieth century. Apparently in Kilauea's lava, the crystals which have formed, are being removed from the reactive part of the chamber by *fractionation* or settling out.

Hawaii Volcanic National Park has several extremes, from snow on Mauna Loa for a good part of the year to the Kau Desert in the rain shadow of Kilauea and to tropical jungles on the windward side of the island.

GEOLOGIC HISTORY

1. **Fissure forming on floor of ocean some 1,600 feet long, formation of five shield volcanoes.**

Millions of years ago during an earthquake some 18,000 feet below sea level on the floor of the ocean a fissure opened up. Hot gases and primitive basalts (undifferentiated magma from sima layer-parent magma) poured out and began to the formation of a chain of volcanic islands.

Layer after layer built up with some of the flows traveling 20 miles along the floor of the

	Types of Pyroclastic Debris	
Name	Description	Size
Bomb	Rounded, May Be Elongated at Both Ends	32mm+
Block	Angular, Part of Vent or Edifice Torn Loose	32mm+
Cinder or Lapilli	Usually Vesicular Scoria or Pumice	32–4mm
Ash	Fine Grained May Be Flaky	$4-\frac{1}{4}$mm
Dust	Dust-sized Particles Gritty	$-\frac{1}{4}$mm

ocean before congealing. These were pouring out of four separate vents which joined together to form the island of Hawaii. They were Kohala, Hualalai, Mauna Kea, and Mauna Loa. Mauna Loa and Hualalai formed over the Western Fundamental Fissure while Kohala and Mauna Kea formed over the Eastern Fundamental Fissure during early Pleistocene. When they finally rose above sea level they formed a mass 70 miles long and 31,000 feet from the floor of the ocean to the summit.

To form these shield volcanoes, a magma chamber formed some 35 miles within the earth's mantle. The magma began to rise because it is lighter than the overlying material. About two miles beneath the surface in the crust the reservoir built up. The lava found its way to the surface through a set of fissures and began its eruptive process.

Today most of the eruptions occur within the summit or along the rift zones on its flanks.

2. Formation and eruptions of Mauna Loa.

By the end of early Pleistocene, the Ninole Volcanic series had erupted. Then a series of en eche-

lon faults formed on along the southeastern flank of Mauna Loa. This was followed by weathering carving deep canyons. Seacliffs were formed. Stream erosion removed much of the material, especially on the southeastern slope. This has produced a great unconformity between the Ninole and overlying Kahuko Volcanic series. The Kahuko series are mainly mid-Pleistone in age and were forming at the time that Kilauea evolved. It filled in the earlier formed canyons.

The island underwent submergence (maximum depth 2,400 feet), then emerged very rapidly about 1,500 feet.

Wave action cut a sea cliff that presently is 300 feet above sea level. Stream erosion started stripping sediments. The sea level dropped 400 feet again. Kilauea blanketed the island with the Pahala ash. The early Kau volcanic series formed during late Pleistocene. The shoreline continued to fluctuate, dropping to 60 feet below sea level, then rising 85 feet above (sea caves and benches formed at 25 feet above sea level). Sea level dropped once more and sea caves and benches formed five feet above sea level. This time Mauna Kea developed glaciers on its crest, and Mauna Loa, sand dunes. Glaciers

melted and sea level rose to present elevation. The modern or Historic Kau flows began in 1832 and ceased eruption in 1950.

The areas of active vents are the southwest rift, northeast rift, northwest rift and Mokuaweoweo Crater. Most of the eruptions have occurred in the Crater (18 from 1849 to 1950), only one in 1859 at the northwest rift, seven at the northeast rift and eight at the southwest rift. Some start in the crater for a day or two and move to the rift zone. Occasionally eruptions last close to a year such as in the years of 1855, 1859, 1873 and 1880. The eruptions average every 3-3 1/2 years apart. It is the world's largest active volcano. Some of the more famous eruptions were a flank eruptions that approached the city of Hilo in 1881. Eruptions also endangered Hilo in 1935 and 1942. Both of these times, bombs from planes, were used to try to seal off the openings but had little or no effect. In 1926 a flow destroyed the village of Hoopuloa.

The biggest record flow, volume wise, was on June 1, 1950 when in less than a month's time some 600 million cubic yards was extruded. This is enough to pave a four lane highway around the world four and one half times. It took the flow

that came from a 13 mile fissure some three hours (at a speed of 5.8 miles per hour) to reach the sea. A village and some buildings were destroyed by this flow. This was its last eruption.

3. Formation and eruptions of Kilauea.

According to the latest evidence Kilauea is the youngest and most active of the five volcanoes. Its birth occurred after the other four cones that make up the Island of Hawaii had developed. Kilauea formed on the flanks of Mauna Loa at the intersection of the Eastern Fundamental Fissure and fault zone on the southeastern side of Mauna Loa. Its creation formed the Hilina Volcanic series. Eruption of the Pahala ash from Kilauea blanketed the entire island with a 0-50 foot layer and overlies the early Pleistocene beds of Mauna Loa, Mauna Kea and Kilauea. It probably developed as the caldera was forming and was created by phreatic explosions.

Towards the end of the Pleistocene, sea level fluctuated and the Prehistoric Puna Flows formed. The close of the Pleistocene buried these beds also with dunes. The Historic Puna series began in 1790 and is still forming today. The other three (excluding Mauna Loa) are considered extinct. Its summit has collapsed to form a caldera and within the caldera is the pit crater Halemaumau, which has been active from 1911 to 1924. Between 1911 and 1924 there was a permanent lake of lava. There has been 12 additional eruptions from Halemaumau since then.

Kilauea caldera has had 14 eruptions since 1790 the east rift zone 17 eruptions, the southwest rift zone three eruptions and Kilauea Iki has had two and Keana kakoi only one. Both aa and pahoehoe flows have occurred.

The continuous lake in the pit crater which existed for 13 years was destroyed by a phreatic eruption. The crater at that time was enlarged from a 1,300 foot diameter to almost one half a mile and became 1,300 feet deep.

Kilauea's most spectacular eruption occurred in the Kilauea Iki Crater where a rift opened up along the side of the Crater about half way up its side. Lava poured into the lake to a depth of 414 feet. It was a combination of extrusion, and then filling and backflow into the vent. One fountain reached the height of 1,900 feet and a huge cinder cone built up behind it. Today the lake has dropped down to 360 feet and has a hardened crust that has cooled enough to walk on, but it will be many centuries before it will have completely solidified.

The lava moved along the east rift zone, and produced a graben fault in Kapoho (28 miles from summit). A string of fire fountains developed in the fault zone and completely destroyed the village. It reached the sea and added some 500 acres of new land, but not before it destroyed sugar cane and orchid fields.

The drainage from the caldera of Kilauea which supplied the Kapoho flow caused the unsupported floor of Halemaumau to collapse about 300 additional feet.

Technically Hualalai is considered an active volcano because it has had one eruption in historic times.

Sometimes the possibility of eruptions can be predicted by earthquakes and the tilting of tiltometers that indicate inflation of the dome as lava moves in.

4. Eruptions are still occurring and continuing to build up the island of Hawaii.

Along the Chain of Craters Road one of the pit craters Mauna Ulu in 1972 became active. The lake refilled, fountaining was produced and in March of 1972 the lava overflowed.

Another pit crater one half mile from Mauna Ulu then became active by Alae Crater and started to feed a series of lava tunnels that carried the lava to the sea. The eruption continued on into 1973.

There are still many earthquakes in the region, especially offshore to the southeast. Surrounding the island are *sea mounts* or submarine volcanoes which could conceivably form still another Hawaiian Island in the future.

Haleakala National Park, Hawaii.

Geologic Column of Haleakala National Park

Erathem	System	Series	Formation
Cenozoic	Neogene	Holocene	Keoneoio Flow
			Unconsolidated Sediments
		Pleistocene	Calcareous Dunes
			Consolidated Earthy Deposits
			Kaupo Mudflow
			Hana Volcanic Series
			(Includes Kipahulu Member)
			〜〜 Great Unconformity 〜〜
			——— Kula Volcanic Series ———
		Pliocene	Honomanu Volcanic Series

Modified after Sterns and Wright.

CHAPTER 17
HALEAKALA NATIONAL PARK

Location: Island of Maui, Hawaii
Size: 27,282.78
Established: September 13, 1960

LOCAL HISTORY

According to legend a Polynesian god named Maui, who the son of Hina, once climbed to the top of a mountain and captured the suns rays and held them fast with ropes. When the sun asked for its life, Maui replied he would, only if the sun would promise to move more slowly across the sky. Maui did this for his mother, to enable her to have more daylight hours to complete her work. Thus the early Hawaiians called this great mountain HAL-A-KA-LA or "House of the Sun" (pronounced Hal-ee-ah-ka-lah).

Captain James Cook, the English explorer, named the Hawaiian Islands the Sandwich Islands after Earl of Sandwich in 1778.

The missionaries came in the 1820s to all of the islands, one of the main things they did was to devise a written form to the Hawaiian language. It was proclaimed the Hawaiian Republic in 1894, became Territory of Hawaii in 1900, was approved for statehood in 1954 and became a state on August 21, 1959.

Hawaii National Park was established in 1916 and included Haleakala section on the Island of Maui. In 1960 it became Haleakala National Park. In 1969 most of the Kipahulu Valley including the Seven Sacred Pools area was donated to the park by Lawrence S. Rockefeller and the Nature Conservancy.

The upper slopes of Kipahulu Valley is not open to the public, but is kept in a natural state for scientific study only. In 1967 a bird thought to have been extinct for 70 years was rediscovered in the Maui Nukupuu and in 1974 a new species of bird never before described was found. It is a totally new species (and genus) of the family of Hawaiian honeycreepers.

GEOLOGIC FEATURES

Haleakala is considered as having one of the world's largest extinct craters. It is 2,720 feet deep, some seven and one half miles long and two and one half miles wide, it has a circumference of 21 miles. At sea level it is 32 miles long, 24 miles wide and 10,023 feet high.

The last volcanic action occurred in the 1700s (outside the crater) so technically it is considered as being active, but for all intents and purposes is quiescent. Earthquake activity has been recorded in the vicinity of Haleakala, so it could conceivably erupt again.

Some 15 large cinder cones are scattered all over the floor, they are up to 1,000 feet high, Puu O Maui is the highest. There are also scattered numerous small ones. The cones are a mixture of spatter, pumice, cinder, dust and ash. They are multicolored, primarily because of the volcanic glass or *obsidian* that acts as prisms and reflects the sun's rays various ways. The color is dependent upon the angle of light and time of day. Some of the color is formed by gases that escaped from the cones as they were cooling, the gas either precipitated or sublimated as a crust on the surface of the cones.

Lava tubes such as Laie Cave and Bubble Cave have formed in the park. *Volcanic dikes* are intrusions of igneous material that cut across the bedding planes of the surrounding rock. These have been exposed by erosion and frequently stand out as ridges.

The so-called Sliding Sands area is made of the pyroclastic debris that was scattered around when the various cinder cones were formed.

Most of the crater is dry and barren even though it receives 150 inches of rain per year, the moisture simply soaks immediately into the bedrock. The elevation of 10,023 feet is high enough to have snow flurries in the winter, so even Hawaii can have a snowstorm. The Kipahulu Valley because of its location receives as much as 300 inches per year, thus it supports a rain forest. Waterfalls cascade down the resistant layers of lava such as Waimoku Falls. The stream then enters the ocean.

Thick clouds are frequently found around the rim of the crater, sometimes they meet in the center. Occasionally, if you are standing on the west rim, during the late afternoon it is possible to see your shadow on a cloud in the crater, surrounded by a rainbow.

GEOLOGIC HISTORY

1. **Formation of 1,600 mile fissure on ocean floor and formation of three or four submarine volcanoes.**

Sometime during the Miocene-Pliocene time about 10-25 million years ago at the southeast end of the 1,600 mile fissure five submarine volcanoes formed, Molokai, Lanai, West Maui, East Maui and Kahoolawe. Eruption after eruption occurred, until the basaltic cones built up above sea level.

2. **Wave action tried to erode the tips away, but since they are primarily lava they continued to grow.**

Inasmuch as these are shield volcanoes, the congealed lava is very difficult to erode. The pyroclastic debris, however, is easily removed by the wind and waves.

East Maui developed three rift zones, the North Rift, East Rift and Southwest Rift. A caldera developed and faults on the south side.

West Maui joined Molokai and Lanai volcanoes while erupting the Honolua lavas. Kahoolawe Volcano merely increased in size, both developed calderas. All three were still separate islands.

A good sized island formed more than twice the size of Maui of today and perhaps nearly 12,000 feet high (only 10,023 feet today). The island consisted of Molokai, Lanai, West Maui, East Maui. They were joined together by the kula lavas erupting from East Maui Volcano.

3. **Eruptions ceased temporarily and erosion was dominant, the height trapped rain, during early Pleistocene.**

The high elevation of the cones was about 2,000 feet higher than today so ocean currents were intercepted from the east, and frequent rainstorms resulted. Streams formed on the side of the

slopes and began forming canyons. Two of the largest valleys were Keanae (north end) and Kaupa (south end), on East Maui.

4. **Streams drained slopes. Two main streams through headward erosion created an amphitheaterlike basin near the summit.**

These stream drainage basins were the beginning of "Haleakala Crater," which is not a crater in the true sense of the word since it was formed by stream action as an erosional feature, and not by volcanic action which is a depositional feature.

5. **Divide between the heads of the valley broke down and the two valleys met.**

The breakdown of the divide separated the two drainage basins of Keanae and Kaupo. They continued to be deepened and enlarged by the streams.

6. **At the same time fluctuation of sea level with a last increase of sea level creating several islands from one.**

The change in sea level occurred because of the advances and retreat of the continental glaciers in the northern hemisphere. The last rise of sea level the Olowalu 250 foot rise of the sea created from a single island the islands of East Maui, West Maui, Kahoolawa and Lanai. The canyons were gradually filled with alluvium.

Maui became an island consisting of three volcanic (East Maui, West Maui and Lanai) cones during the Kahipa 300 foot drop of sea level. East Maui began erupting and filling the valleys with the Hana lavas. A lahar or mudflow occurred in Kaupo Canyon. The volcanic cone that makes up the western portion of the island is quite a bit smaller and is attached to Haleakala by a seven mile wide isthmus.

7. **Resumption of volcanic activity, valleys fill with lava, cinder cones form.**

Renewed activity occurred during the Waipio 60 foot drop of sea level and lava (Hana and Lahaina flows) poured down the valley floors almost filling them up. A series of cinder cones, lava tubes and dikes eventually built up on the newly created floor. The eruptions originated from a series of vents that extend east-west across the floor of the "crater." Pyroclastic debris of volcanic bombs, cinders or lapilli, volcanic dust and ash are scattered everywhere. Sand dunes (calcareous sands) formed on the isthmus. The Lanai Volcano became a separate island leaving the Island of Maui with the east and west volcanic cones.

8. **Last eruption occurred in 1790 changing coastline slightly no activity since then.**

An eruption in 1790 sent two minor flows at the lower elevations to the sea. They form the *Keoneoio flow,* which is visible in the southeast corner of the island above La Perouse Bay. They merely altered the coast line slightly. There has not been any signs of activity since then. Haleakala has been quiet for 200 years but occasional earthquakes testify that it is only resting and is not extinct.

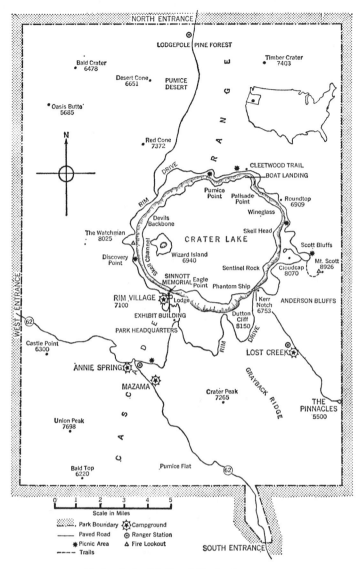

Crater Lake National Park, Oregon.

Tentative Geologic Column of Crater Lake National Park

Erathem	System	Series	Formation
Cenozoic	Neogene	Holocene	Dacite Pumice
		Pleistocene	Glacial Deposits / Mount Mazama Dacite
			Basaltic Parasite Cones
			Timber Lake Lavas / Mount Scott Lavas
			Mount Mazama Andesite / Dikes and Vent in the Wall
		Pliocene	Union Peak Lavas
			Premazama Lavas
Mesozoic	Cretaceous		Chico Series

Modified after Diller, Patton, Williams et. al.

CHAPTER 18
CRATER LAKE NATIONAL PARK

Location: S.W. Oregon
Area: 160,290.33 acres
Established: May 22, 1902

LOCAL HISTORY

According to the Klamath Indians, ancient Mount Mazama was the passageway of the below world and the above world. Llao, the chief of the below world, could be seen as a dark shadow towering over the snow on Mount Mazama. One day he and the chief of the above world, Skell (who lived on Mount Shasta), had a quarrel and Llao threw fiery rocks and volcanic ash at his enemy. His anger could be heard in the thunder. Lava spilled down the side of Mount Mazama.

The medicine men of the Klamath Indians thought the eruption, in part, was because of the evil and wicked way of the tribe. To pacify the gods they climbed to the top and threw themselves in. Skell was very much impressed by their sacrifice, so he fought even harder. After seven days Llao was driven below and the chief of the above world took his two fists and pushed the top down to imprison Llao forever.

Another legend about Llao Rock, which is a 2,000 foot high cliff, explains how many of the features developed after the lake in the caldera formed.

Llao made his home in the water of Crater Lake using Llao's Rock for his throne. In the lake with huge claws were giant crayfish that were his warriors. They would grab anyone who stood on the cliffs surrounding the lake and tear them to pieces. Skell was the god of all animals, and in a battle with Llao was captured. Llao tore his heart out and gave it to the crayfish warriors for a ball, who tossed the heart back and forth over the lake. One of Skell's servants, an eagle, managed to catch the heart in mid-air during a game and gave it to the coyote, who quickly ran with it to Skell's home. His body grew around the heart and he came back to life and was strengthened.

He fought with Llao again, managed to capture him, Skell took his body to Llao's throne and tore it into pieces, throwing the remains into the lake. The warriors ate the pieces, for they thought it was Skell's, but when Llao's head was thrown in, they realized their mistake and began to cry. Their tears helped to fill up the lake. Llao's head rested where it was thrown, we call it Wizard Island today.

On June 12, 1853, a young prospector by the name of John Wesley Hillman was letting his mule lead the way up the mountain side in search of the "Lost Cabin Mine." They went over the ridge, the mule stopped suddenly, and Hillman saw the caldera with its deep blue lake spread out before him. The location is called Discovery Point and it is located on the southwest side of the rim. Hillman named it Deep Blue Lake (or Great Blue Lake). Other members of the party wanted to call it "Mysterious Lake."

Later, in 1862, some other gold miners found it and named it Blue Lake. Two soldiers from Fort Klamath rediscovered it in 1865 and had trouble deciding between Sunken Lake, Great Sunken Lake or Lake Majesty. The latter won. In 1869 other visitors from Jacksonville, not knowing of the earlier discoveries, named it Crater Lake. This is the name it has retained.

A little boy in Kansas read about the mysterious Deep Blue Lake and vowed he would see it someday. When he was 18 his parents moved to Oregon and he started searching for it. He asked all the old timers about it, and finally after 13 years of searching, he stood on its rim looking down upon it. His name was William Gladstone Steel, the year was 1885 and by then he was a judge. It was at that time he decided it should be preserved as a park and after 17 years of effort it became a reality in 1902. In 1913 he was appointed superintendent of Crater Lake, then later the park commissioner until the year of his death in 1934.

GEOLOGIC FEATURES

Both glacial and volcanic features can be found. Crater Lake exists on the remnants of a volcano called Mount Mazama that probably was twice as high as it is today. Mount Mazama was a composite cone that was from 11,000 to 12,000 feet high. It had a very asymmetrical shape according to the direction and distribution of glacial striations with a gentle north slope that permitted snow fields and some glaciers to accumulate and a steep south slope upon which three major glaciers formed. The largest glacier flowed down Munson Valley and into Annie Creek Valley. Moraines, glacial deposits mixed with volcanic debris and other features such as striations and the broad U-shaped valley testify to this. The smaller tributary valleys today consists of V-shaped stream valleys cut in the pumice and other volcanic materials. A second large glacier flowed down from the portion of the peak that is no longer there through Sun Notch and along Sun Creek Valley, then down through Sand Creek Valley.

Glacial striations on the north side show there were probably glaciers which existed but were not as large nor extensive as those on the south side. The largest glaciers formed on the west side and joined together in the Valley of the Roque. On the east slope they flowed onto a plateau where the meltwater streams deposited outwash into the Klamath Marsh. The fact that the striations are found only on the outside of the slope and never inside also testifies to the origin of the glacier when the original peak was still there.

Bear Creek on the east slope near Cascade Springs is probably a glaciated valley. Dyer Rock (by Eagle Crags) is one of the localities with alternate volcanic and glacial deposits. There are also glacial deposits northeast of Roundtop located between two dacite flows which fill up an old valley. The same thing occurs at Steel Bay on the south side.

Mount Mazama was not the only volcano in the region. Other composite cones were Mount Bailey, Mount Theilson (north of Mazama), Mount Scott (east), and Union Peak (southwest). All that remains of Union Peak is the *volcanic neck* or congealed lava that was in the vent, the cone or *edifice* has been removed by erosion. Mount Scott's lava flowed to the east away from Mazama. It is also the highest point in the park with an elevation of 8,926 feet.

Examples of cinder cones are Timber Crater, Desert Cone, Red Cone and Bald Crater. They are all 1,000 feet or less in height, have craters a couple of hundred feet in diameter, and are 50 to 100 feet deep. Red Cone is named for the red colored scoria of which it is composed. Some of the flows were basaltic in composition. Basalt never flowed from the main Mazama crater. Red Cone shows some glacial debris mixed with its deposits and is on the same fissure system as Desert Cone.

Crater Peak is a shield volcano which has flows of three different compositions that clearly show the differentiation of the lava. The earliest are andesite, followed by basalt and last of all dacite. This pattern shows up in the entire park region. The rim of the lake and thus bulk of Mazama are dacite forming such features as Cloud Cap, Llao Rock and Rugged Crest. The basalt flowed from the cinder cones on the flank. After the collapse of the crater which formed the caldera, three andesite cinder cones developed on the floor of Mazama. They are Wizard Island, Marrian Cone, and an unnamed cone.

The caldera that developed would be close to 4,000 feet deep and five miles across if it were drained of water. The depth of the water is 1,932 feet and is presently surrounded by cliffs 500 to 2,000 feet high. Water level varies somewhat, but average depth remains about the same. Source of water is *meteoric* (mostly snow), with the average snowfall about 50 feet per year. The record snowfall was in 1932-33, it accumulated to a depth of 73 feet. Loss of water occurs through seepage between some of the flows which emerge at lower levels as springs or by evaporation. Temperature is an even 39° (once below the surface water). The water very rarely freezes, with the one exception being in 1949 when 2 to 12 inches of ice formed. The deep blue color that characterizes the lake, is a combination of absorption of all wavelengths of light except blue, and light scattering in the exceptionally pure water (20 times purer than drinking water).

Igneous dikes have formed, composed of hypersthene-andesite (the majority) or dacite. The most prominent are the set called Devils Backbone, but there are seven clustered by Llao Rock, three north of Sentinal Rock, and one each near north of Eagle Cove and another near the Watchman.

The Phantom Ship is an island that rises around 175 feet above the level of the lake. Its "sails" are a brownish black dike that rests upon a "hull" of greenish volcanic ash. It is an erosional remnant.

The Pinnacles (southeast side by Wheeler Creek) are former vents some 200 feet high. At one time a glowing avalanche of pumice barreled down the side of the cone and came to a halt some six miles from the rim. It took years for this superheated deposit to cool. Gases that were released

moved up these vents and baked them. Later running water and wind eroded this deposit and the hardened vents remained as chimneys or pinnacles. If you look carefully you will note the rapid change of the composition of the flow, for the bottom half is light colored pumice and the top half is dark colored scoria.

The Watchman was formed by a flank eruption from a vent that existed at one time beneath the lookout tower that is located on top of the Watchman.

Hillman Peak is half of a volcanic cone, the rest of the cone was destroyed with the formation of the caldera. Its feeder dike is still visible.

Llao Rock is a glacial U-shaped valley that was filled by an eruption on the valley floor, once again the feeder dike is visible. Composition of the flow is dacite. Redcloud is a triangular shaped cliff with a V on the bottom, it is an exposed crater (funnel-shaped) that formed on the flank of Mount Mazama. It also has a dacite composition.

Other erosional features are the fluted and jointed tuffs at Annie Creek and Sand Creek Canyon. The columnar jointing developed at Kerr's Notch. Springs form in the layers of glacial debris trapped between lava flows, they are usually outwash deposits.

Cleetwood Cover is a valley filled with a dacite lava flow that has a small tributary valley that is about as wide as it is high, it probably is an old lava tube that collapsed. The Cleetwood flow is an "upflow," in other words a feeder dike or pipe in which lava flowed upwards and then out onto the flanks.

Crater Castle or Cottage Rock is a pumice deposit oxidized by escaping hot gases and shaped by weathering.

GEOLOGIC HISTORY

1. About 80 million years ago the Pacific Coast Eugeosyncline existed.

This was during the Cretaceous Period and primarily graywackes and volcanic material were accumulating on the ocean floor.

2. Area uplifted above sea level, beginning of Cascade range, Western Cascade sequence.

Around 40 million years ago uplift occurred, this forced westward movement of the coastline. Volcanic eruptions during Oligocene (Paleogene period) and Miocene (Neogene period) time formed the Western Cascade sequence of andesite and basaltic-andesite flows and pyroclastic debris. This

material was deposited upon a crystalline upwarp or fold that had a trend to the northeast. They are about 20,000 feet thick and underlie the High Cascade sequence. The dip is to the east.

Because the sediments sank as fast as they accumulated, the land area was never very high. This permitted growth of Redwood forests on the wet west slopes and desert vegetation of the dry east slope (original Cascades).

3. Uplift and formation of High Cascade sequence, lava flows and volcanic cones.

Deep fissures trending north-south opened up during Pliocene and Pleistocene time, some faulting occurred. At first old valleys filled with basaltic and andesite flows, then shield volcanoes began to form. A high platform developed by the end of the Pliocene some two million years ago.

4. Composition changed to andesite, composition cones called Mounts Bailey, Mazama, Scott, Thielsen, and Union Peak formed.

The eruptions became more violent, and also the change in composition would not permit as rapid a flow, thus making a steeper cone. The composite cones developed on the "shield" platform. Mounts Bailey, Thielsen and Union Peak had *ultrabasic* or mafic intrusions in the vents and andesite on the flanks. This occurred between one and two million years ago. At first they formed as a cinder cone, then lavas with temperatures as high as 1,800°F covered the slopes. Through the years the pyroclastic debris and lava flows altered forming Mount Mazama. Towards the end eruptions occurred on the flanks instead of at the summit, e.g., Redcloud.

5. Pleistocene glaciation, glaciers forming on Mount Mazama and other Volcanoes. Severe erosion of some cones.

During the process of the formation of Mount Mazama, valley glaciers formed, especially on the southern slopes. Glacial debris alternates with dacite volcanic rocks in many areas. There is evidence of three episodes of glaciation.

Some of the erosion by glaciers was so severe that the ediface was worn away leaving only the volcanic neck standing, e.g., Union Peak and Mount Thielsen. They had just about ceased activity, so were not able to build up again.

6. Continued growth of Mount Mazama during the Pleistocene.

The Mount Mazama andesite flows which forms the core of the cone changed its composition to basalt and fed the parasite cones ever present on its flanks, such as Red and Desert cones. Bald and Timber craters were pyroclastic or cinder cones.

Mount Scott continued to erupt during the formation of the Mount Mazama andesites, with its flows moving away from Mazama (east).

The Brothers Fault Zone put a northwest trend to some of the features as they formed on the shear zone. Some dacite flows came from the subsidiary vents. Flows forming Watchman and Hillman peak occurred (andesite), including Llao Rock, Redcloud Cliff and Cleetwood Cove flows of glassy andesite. Feeder dikes can be seen on the walls of the cliffs today. From the Mazama Crater came dacite flows. In the last stages much of it was in the form of pumice.

The glaciers had grown to 1,000 feet thick and some were 10 to 17 miles long. The entire peak may have been covered with ice and snow.

7. Last stages of Mount Mazama, retreat of glaciers, eruption of glowing avalanches.

At first dacite ash developed from the crater, winds picked it up and blew it to the east covering some 350 square miles including most of Washington, Oregon and Idaho, the western portion of Montana, southern parts of British Columbia and parts of Alberta, Nevada and California. This occurred about 7,000 years ago according to Carbon-14 dating. It was even picked up by turbidity currents and transported down submarine canyons. Deposition was so widespread it is used as a marker bed to calculate the rate of accumulation of sediments for the last 7,000 years in the Cascadia Basin or Lake Washington areas.

This occurred because the dacite in the magma chamber was undergoing differentiation. The gases in the magma rose to the top making it lighter and gas rich. While the dacite at the bottom became gas deficient (or gas poor) but received all the crystals that were settling out of the lighter dacite. Thus when the lava erupted the gas rich material expanded rapidly and was blown high into the air as volcanic ash which the winds picked up and deposited everywhere.

8. Formation of glowing avalanches or nuee ardentees forming Pumice Desert pinnacles and other features.

Finally the gas poor material reached the surface, since it lacked a great deal of gas it poured over the sides as nuee ardentees clouds (at 50 miles per hour). They bowled over trees and destroyed forests and anything else in its path as far as 35 miles away before collapsing into a pile of pumice.

Some of the glowing avalanches poured down the glacial valleys such as Annie Creek and Sand Creek where the pinnacles formed as they were cooling. Different parts of the chamber were tapped accounting for the rapid change in the composition of the pinnacles, (pumice bottom, scoria top).

In other instances, because of the lack of well-defined valleys, they spread out in wide sheets forming the Pumice Desert or burying the Kalmath Marsh to the east.

Velocity was great enough that on occasions the clouds would travel uphill, and even transported 14 foot lumps of pumice some 20 miles away.

The region took many years to cool off, forming the vents for the pinnacles and other features. Everywhere was destruction. Volcanic ash has been found in Oregon caves covering some 200 Indian sandles, ropes of sagebrush and sagebrush bark mats plus tools such as scrapers. Two skulls and bones were found in another locality buried in pumice. Forests were bowled over or buried. Where the asymmetrical top stood (vent near southern end of Crater Lake) was a huge caldera.

Some ten cubic miles of material was erupted as ash and pumice. Another five cubic miles collapsed into the caldera, reducing the height to 6,000 feet. The collapse was uneven which explains why walls vary in height from 500 to 2,000 feet above the level of the lake.

9. Formation of three cinder cones on floor of caldera. Filling of lake.

The exact time when Wizard Island, Merrian Cone and the unnamed cone were formed is not known. Anywhere from 6,000 years to about 1,000 years ago. Evidence does not indicate these cones formed underwater, hence they had to form before the lake filled. However, oldest trees on Wizard Island are 800-900 years old, by allowing 200 years for soil to form, it could have been as recent as 1,000 years ago.

Merrian Cone is about one mile wide and 1,300 feet high, thus is underwater. The unnamed cone is about 1,000 feet high and 1 1/4 miles in diameter. Wizard Island is 2,600 feet high with 774 feet above water, it has a 300 foot crater on top that is 90 feet deep. All three are primarily andesite cinders but have some lava flows down their flanks.

Eventually the caldera filled with meteoric water and remains somewhat stable today but a century ago it was 13 feet higher.

Crater Lake is unique because the Indian legends actually described the events that occurred to form Crater Lake National Park.

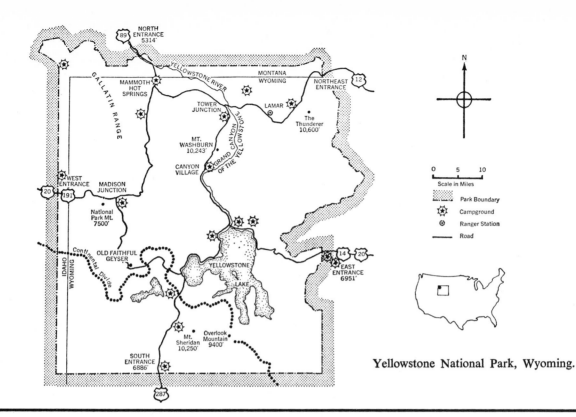

Yellowstone National Park, Wyoming.

Geologic Column of the Southern Part of Yellowstone National Park

Era	Period	Epoch	Formation
Cenozoic	Neogene	Holocene	Stream Sand and Gravel Hot Springs Deposits
		Pleistocene	Glacial Deposits and Plateau Rhyolite Supergroup
	Paleogene	Eocene ?	Absaroka Volcanic Supergroup
			Hart Lake Conglomerate
		Paleocene	Pinyon Conglomerate
Mesozoic	Cretaceous		Harebell Formation
			Bacon Ridge Sandstone
			Cody Shale
			Frontier Formation
			Mowry Shale
			Muddy Stone Member of Thermopolis
			Thermopolis Shale
			Rusty Beds Member
	Jurassic		Cloverly and Morrison Formations
			Sundance Formation
			Gypsum Spring Formation
	Triassic		Chugwater Formation (Lower Part)
			Dinwoody Formation
	Permian		Phosphoria Formation (and Related Rocks)
Paleozoic	Pennsylvanian		Tensleep Sandstone
			Amsden Formation
	Mississippian		Madison Limestone
	Devonian		Darby Formation
	Ordovician		Bighorn Dolomite
	Cambrian		Gallatin Limestone
			Gros Ventre Formation
			Flathead Sandstone
Precambrian	Precambrian		Schist and Gneiss

Modified after Ruppel et. al.

Geologic Column of the Absaroka Volcanic Supergroup in Yellowstone National Park

System	Series	Group	Formation	Member
Paleogene	Eocene	Washburn	Sepulcher	Conglomerate Facies
				Elk Creek Basalt
				Lost Creek Tuff
				Daley Creek
				Fortress Mountain
				Andesite Lavas
			Lamar River	
			Cathedral Cliffs	
		Sunlight	Mount Wallace	Stough Creek Tuff
			Crescent Hill Basalt	
			Wapiti	Jim Mountain
			Trout Peak Trachyandesite	Pacific Creek Tuff
		Thorofare	Langford	Promonatory
			Two Ocean	
			Tepee Trail	
			Wiggins	

Modified after Smodes and Plostka.

Geologic Column of the Rhyolite Plateau Flows in Yellowstone National Park

System	Cycle	Group	Rhyolite Flows			Basalt Flows		
Pleistocene	Third	Plateau Rhyolite	Roaring Mountain Member	Central Plateau Member	Shoshone Lake Tuff Member	Ospray		
						Madison River		
			Obsidian Creek Member	Upper Basin Member		Swan Lake Flat	Falls River	Basalt of Mariposa Lake
			Mallard Lake Member					
		Upper Yellowstone	Lava Creek Tuff					
						Undine Falls		
			Mount Jackson					
	Second		Does Not Occur in the Park					
	First					Narrows Basalt		
			Lewis Canyon					
		Lower Yellowstone	Huckleberry Ridge Tuff					
			Broad Creek			Junction Butte		

Modified after Christianson and Blank.

Geologic Column of Northern Part of Yellowstone National Park

Erathem	System	Series	Formation	
Cenozoic	Neogene	Holocene	Stream and Gravel	
			Hot Spring Deposit	
		Pleistocene	Glacial Deposits and Plateau Rhyolite Supergroup	
	Paleogene	Eocene	Absaroka Volcanic Supergroup	
Mesozoic	Cretaceous		Part of Landslide Formation	
			Everts Formation	
			Eagle Sandstone	
			Telegraph Creek Formation	
			Cody Shale	
			Frontier Sandstone	
			Mowry Shale	
			Thermopolis Shale	Upper Sandstone Member
				Middle Sandstone Member
				Lower Sandstone Member
			Kootenai Formation	
	Jurassic		Morrison Formation	
			Ellis Group	Swift Formation
				Rierdon Formation
				Sawtooth Formation
	Triassic		Thaynes (?) Formation	
			Woodside Formation	
			Dinwoody Formation	
Paleozoic	Permian		Shedhorn Sandstone	
	Pennsylvanian		Quadrant Sandstone	
			Amsden Formation	
	Mississippian		Madison Group	Mission Canyon Limestone
				Lodgepole Limestone
	Devonian		Three Forks Formation	
			Jefferson Formation	
	Ordovician		Bighorn Dolomite	
	Cambrian		Snowny Range Formation	Grove Creek Limestone Member
				Sage Limestone Member
				Drycreek Shale Member
			Pilgrim Limestone	
			Park Shale	
			Meagher Limestone	
			Wolsey Shale	
			Flathead Sandstone	
Precambrian	Precambrian W		Schist and Gneiss	

Modified after Ruppel et. al.

CHAPTER 19
YELLOWSTONE NATIONAL PARK

Location: N.W. Wyoming, E. Idaho, S. Montana
Area: 2,221,772.61 acres
Established: March 1, 1872

LOCAL HISTORY

First and largest National Park. First reports of the park were not believed. John Colter was a Private in the Lewis and Clark Expedition. When they went on, he requested to remain behind and trap. He entered the park from the southwest side, crossed the park and left by the northeast corner in 1807-08. Because of his descriptions of the fire and brimstone and his habit of telling tall tales people called it "Colter's Hell."

Jim Bridger, a mountain man, also visited the park and saw the things that Colter saw, but he was also known for embellishing stories so he wasn't believed either. Bridger discovered Great Salt Lake, was well informed about Yellowstone, and sketched a map in 1850 for Lieutenant J.W. Gunnison.

Other people such as W.A. Ferris (American Fur Company) in 1834; Father DeSmet (missionary) in 1852, Captain John Mullen in 1853, plus prospectors mention features of the park.

The Washburn-Langford expedition in 1870 actually brought about the concept of preserving the area for a park. Everts, a member of the expedition, managed to get lost for some 37 days. He was finally found crawling on his hands and knees. He had lost his glasses and was very nearsighted.

Many members of the party had features named after them, e.g., Mount Washburn, Sheridan Mountain, and Baronett Peak. The following year the United States Geological Survey Hayden Expedition mapped the park. Hayden had worked in the park area periodically since 1859. James Stevenson, also a geologist, was Hayden's right arm. Thomas Moran, an artist, helped to spread Yellowstone's unique beauty in the east and bring it to the attention of the public.

President Grant signed the act creating the park in 1872. Unfortunately the bill did not contain money to care for or run the park. Langford was appointed the first Superintendent of Yellowstone. He served from 1872 to 1877, Nathanial P. Langford was such a dedicated person that people change the meaning of his initials to National Park Langford. He served his five year term without pay.

Because of the lack of appropriations there was no fear of reprisal. People killed the animals for only their hides and tongues, and helped themselves to anything they wanted. This went on with civilian superintendents until 1886 when it was turned over to the United States Cavalry who protected and ran the parks until 1912.

Travel was somewhat dangerous. In one instance two tourists were killed by a party of Indians. Chief Joseph, head of the Nez Perce Indians, went through the park with his tribe in 1877 when he was being chased by the Cavalry.

In the late 1800s and early 1900s the only travel permitted was by stage coach, wagon or horses, (no automobiles were permitted.) Army posts were scattered throughout the park with the main fort at Mammoth Hot Springs—Fort Yellowstone. Because of the extreme temperatures in the winter, down to 50 degrees below zero, it is primarily a summer park.

GEOLOGIC FEATURES

Geysers are intermittent hot springs. They form because of two factors; (1) superheated water; (2) side chambers that allow the water to become trapped. A geyser consists of a twisted channel with many side chambers or pockets. Water becomes heated from a source beneath the ground, probably a magma chamber near the surface. At the elevation of Yellowstone, water ordinarily boils at 199° instead of 212°F because there is less air pressure. However, the water trapped in the side

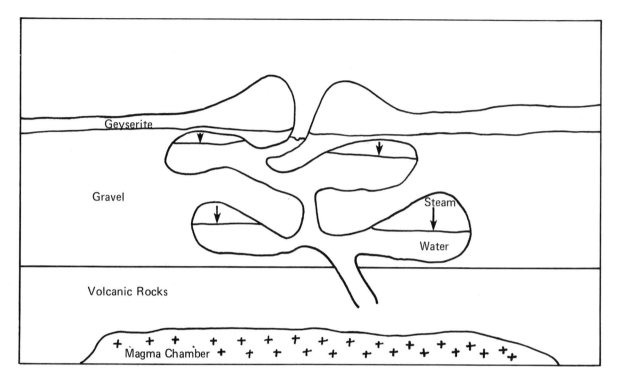

Geyser

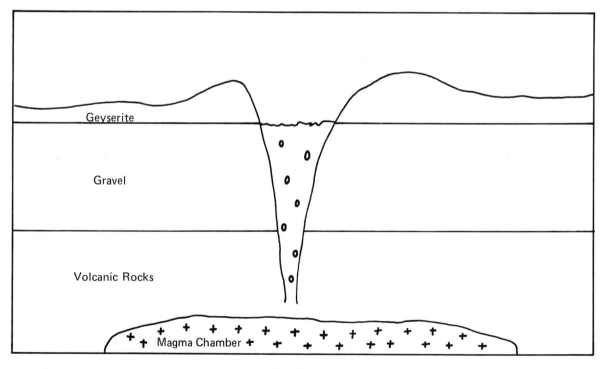

Hot Spring

pockets has pressure exerted upon it by the steam that has accumulated at the top of the pocket and is trapped. The temperature may become as high as 300°F and still not boil, this is *superheated water.* It is water with a temperature well above the boiling point but for one reason or another does not boil. Gradually the entire geyser system becomes superheated and will reach a critical point that the slightest change in temperature or pressure will trigger off the water, and it all begins to boil at once. The water near the top begins to cool off, loses gases, hence becomes denser and starts to sink. This is enough to set off the eruption. Also some of the water pours out and reduces pressure that permits the water at the bottom to boil. Because the water is in a confined system the only place there is room for expansion is at the top. The water is under tremendous pressure by the expanding gasses trapped in the water which forces the water out in eruption. Most of the water either drains back into the hole immediately because of the basin formed around the vent, or it will percolate down through the porous gravels. The amount of time it takes the system to reach the critical point again is the time interval between eruptions.

A geyser is its own worst enemy because it eventually seals off the opening by deposition of *geyserite* a material made up of silica (SiO_2). The geyserite is dissolved out of the rhyolite lava that the ground water passes through. When this occurs a new vent generally opens up nearby in another weak zone.

The deposition of geyserite around the vent will build up curious shaped mounds, thus giving many of the geysers their names. Old Faithful, Castle, Grand Geyser, Lone Star, Riverside, Clepsydra are all geysers. Most occur in geyser basins located along fault systems.

If no side channels develop the water simply boils and bubbles and is called a *hot spring.* If a body of water is 98° F or higher it qualifies as a hot spring. They also may have mounds of geyserite around them.

Scientists have been keeping careful track of the temperature of the water and have found in the last 100 years it has just about stayed the same. The water is *meteoric water* derived from rain and snow. It is stored in the Pleistocene gravels or highly fractured bedrock. Changes in the underground plumbing system by such geologic features as an earthquake can change a geyser into a hot spring, or the reverse may occur.

The variation of color in the hot spring is produced by the different species of algae that grow in the hot water. Each species can only exist in a narrow range of temperatures. A change in conditions will cause them to die. An example is the blue trumpet-shaped Morning Glory Pool. After the 1959 earthquake the temperature of the water dropped from 168°F to 158°F. This would not allow the algae at the rim of the pool where it is shallower (and therefore cooler) to exist, so the blue variety died out and the brown variety now extends to a greater depth. Hence the brown color instead of existing only at the very edge now forms down to the bowl. Examples of hot springs are the Punch Bowl, Emerald Pool, Opal Spring and Minerva Terrace.

The algae will exist up to temperatures of 170°F, with the yellow and lighter varieties living as a rule at higher temperatures than the darker greens and browns.

If the water supply is very low only steam will be ejected from the vents, usually accompanied by a hissing sound. These are called *fumaroles,* such as Steamboat Springs. If the steam contains a lot of sulfur dioxide and smells like rotten eggs, it is the so-called "brimstone" and the noxious smelling vent is called a *solfataras* such as Dragons Mouth.

When the volcanic rocks weather, the feldspar minerals are eventually broken down into clay, by the acid water impurities. The iron minerals will color the clays various shades of reds, pinks, yellows, orange, browns, black, etc. If the available ground water is limited, the clay is not removed. This forms a mixture of clay, water and gases bubbling through the thick soupy mixture to form *mudpots.* If they are highly colored by the impurities they are called *paintpots.* If the material is especially thick they form mounds called *mud volcanoes.*

One exception to the hot springs with the geyserite deposits are Mammoth Hot Springs. The ground water that supplies them flows through *limestone* formations instead of *rhyolite* (fine-grained equivalent of granite). The material held in solution is calcite ($CaCO_3$) rather than silica (SiO_2). The terraces are formed out of travertine and hot geyserite. They build up at the rim because the surface of the water is stretched as it flows over the edge. The larger area exposed permits more carbon dioxide (CO_2) gas to escape which in turn permits the deposition of travertine on the rim causing it to build even higher. The live terraces which still have water flowing over them are colored by the oxide impurities and various species of algae. The dead terraces are usually whitish in color and no longer have water flowing over them.

Another spectacular feature in the park are the *waterfalls*. Most waterfalls are formed where there is a major variation in the hardness of the bedrock.

The formation that forms the lip of the falls is usually quite resistant to weathering and has very easily eroded material beneath. The cavitation process (wearing away of rock by the collapse of microscopic-sized vapor bubbles) removes the soft beds, thus undercutting the slope. Upper and Lower falls of the Grand Canyon of the Yellowstone is a good example.

The rhyolite flows that make up the canyon have undergone various degrees of *hydrothermal* (hot water) *alteration*. This is also responsible for the yellow color of the ordinarily brown, reddish or gray color of the rhyolite. The canyon into which the 308 foot Lower Falls plunges is highly altered and thus eroded very easily. The beds that form the lip of the falls are less altered, thus more resistant. The canyon between Upper and Lower falls is carved in a rhyolite that contains more *volcanic glass* or *obsidian* and is softer than the hard, dense rhyolite that forms the lip of the 109 foot high Upper Falls.

Gibbon Falls on the other hand flows over a *fault scarp* (cliff produced by faulting) formed in the Yellowstone Tuff. The original falls was between 1/2 and 1/4 of a mile downstream by the original scarp. It took about 600,000 years to erode this far back.

Undine Falls (west of Tower Junction) is located in the Lava Creek Stream Valley. This creek is flowing through a much older stream valley that originally was filled with a basaltic lava flow and then later covered with volcanic ash. The stream of today cut through the ash fairly fast, but is having difficulty with the resistant lava, thus Undine Falls. A similar situation exists forming Wraith Falls in Lupine Creek Valley nearby.

Lewis Falls forms on the edge of a rhyolite flow. Tower Falls eroding the Absaroka Volcanics is a *hanging valley* being unable to erode as fast as the Yellowstone River Valley which it enters.

Obsidian Cliffs is made up of volcanic glass which is the Obsidian Cliff Flow within the Roaring Creek Member of the Plateau Rhyolite. It is Neogene in age which formed about 75,000 years ago during the Pleistocene. A lava flow encountered a glacier that chilled it immediately to form a volcanic glass. The glass forms because the lava cooled so quickly that its atoms are unordered and no crystals had time to grow. In other words, it is a natural glass. Flow banding can be seen that formed just before it became chilled. It contains pea sized lithophysae (spherical cavaties lined with brownish colored rock) whose origin geologists are unable to explain.

The volcanic origin of Yellowstone also shows up in the fact that two calderas exist within the park boundaries. The main caldera occupies the central portion of the park and is approximately 40 miles in diameter. The boundaries go through the Flat Mountain Arm of Yellowstone Lake, then north of Factory Hill. It cuts across Pitchstone Plateau along its divide, curves southwest of Madison Lake along the divide of Madison Plateau, then goes southeast of Mount Hayes. It follows along Gibbon River Canyon to just south of Norris and then by the edge of Solfatara Plateau. It swings north of Cascade Lake, south of Mount Washburn through the upper portion of Broad Creek, east of Pelican Cone, Turbid Lake and Lake Butte then follows along the shoreline of Yellowstone Lake and over to Plover Point.

Within this huge caldera is a much smaller younger caldera that essentially forms the West Thumb of Yellowstone Lake, formed some 125,000 to 200,000 years ago. Also there are two domes that developed after the formation of the caldera. One contains Lower, Midway and Upper Geyser basins, the other Sulphur Hills, Vermillion Springs and Sour Creek region. When they formed the bedrock fractured around them in *ring fractures*. These provided the channels for the Plateau Rhyolite that has pretty well filled in the caldera and the fissure system for the hot springs and geysers.

Another smaller (diameter 20 miles) caldera, but easier to see, exists just outside the west boundary of the park in Idaho, the Island Park caldera. Its eastern half is covered by the Yellowstone Volcanic Plateau.

With the formation of the caldera very weak nuee ardentes or *ash flows* of volcanic gases and ash came pouring out of the fissures all around the edge of the caldera (caused by sudden release of pressure). They literally blanketed the area. Winds picked up much of the debris and deposited it over an area of thousands of square miles.

Collapse occurred with the removal of material beneath, probably along a series of normal faults (footwall moves down in relation to hanging wall). The drop was several thousand feet. Through these faults the rhyolitic flows welled up, filled in most of the caldera and in some instances overflowed its borders. This is why the caldera is not obvious in Yellowstone today. The flows range in age from 600,000 years ago when the collapse occurred up to only 60,000 years age.

Not all of the fissures were within the caldera

some were outside forming such flows as the Obsidian Cliff Flow. Others were brecciated, e.g., Firehole Canyon Drive or basaltic, e.g., Tower Falls area. The basaltic flows have the tendency to form *columnar jointing*. The many sided columns are the result of the shrinkage cracks that developed as the flow cooled.

One rather unique feature can be found at Speciman Ridge. A cliff exists of volcanic ash (tuff) interbedded with lava flows plus a mixture of large fragmental rocks deposited either by streams (alluvium) or mudflows (lahars) it contains 27 fossil forests. These trees are around 50 million years old and are found in the Larmar River formation of the Washburn Group of the Absaroka Volcanic Supergroup. Nearby Mount Washburn was one of the volcanoes that supplied the material. Some of the trees were chestnut, dogwood, magnolia, maple, oak, redwood, sycamore, and walnut. Besides the stumps, the seeds, leaves, twigs, cones and needles of some 100 species (trees and shrubs) were preserved.

During a volcanic eruption the pyroclastic debris would literally bury and suffocate the surrounding forests (40 square mile area). Sometimes volcanic mudflows would pour down. This is why most of the trees are in an upright position. The trees would die, the volcanic material would weather into soil. New forests would appear only to suffer the same fate. Ground water percolating through the volcanic beds leached out the quartz and redeposited it in the trees to petrify them.

Around the hot springs area such as the Sour Lake region the drowned trees are in the process of petrification. The geyserite is gradually impregnating the trees as can be seen by the whitish appearance of the base of these trees.

Like most of the western parks in the mountains, Yellowstone did not escape glaciation. It occurred at least three times. The glacial episodes are (1) pre-Bull Lake 300,000 to about 180,000 years, (2) Bull Lake Glaciation 125,000 years to 45,000 years ago, (3) Pinedale Glaciation 25,000 years to 8,500 years ago.

The earlier stages have only a few scattered deposits here and there, such as moraines. Some of these are interbedded with the Plateau Rhyolite. The last glaciation (Pinedale) is well documented. Glaciers formed in the Absaroka Range (southeast and southwest) and Gallatin Range (north and northwest). They all converged in the park and became as much as 3,000 feet thick covering up to 90% of Yellowstone. This was a scant 15,000 years ago.

Evidence is everywhere, a 500 ton boulder (glacial erratic) was deposited near Inspiration Point, kame and kettle topography by Slough Creek, lake sediments in Hayden Valley, glacial deposits in Soda Butte Creek and Lamar valleys and Yellowstone Canyon. Around the Phantom Lake region between Geode Creek and Oxbow Creek (west of Tower Junction) is an old stream channel that used to flow along the ice edge carrying away the meltwater and debris. Once the glaciers left it dried up but its remains are obvious. Yellowstone Lake was very deeply buried under the ice. The glacial deposits are so abundant in this area that good outcroppings of bedrock are hard to find. The lake in the geologic past used to be higher and the higher former shoreline is easy to trace about 150 feet above its present level. It originally drained at its southern end (Snake and Columbia rivers) to the Pacific. Now it flows to the Atlantic via Missouri and Mississippi rivers.

Landslides both new and old can be seen in Yellowstone. Sepulcher Mountain is an extinct volcano that rests upon Cretaceous shales. These shales are weak *incompetent beds* which cannot support the weight of the overlying andesite and dacite beds. Therefore the north slope of Sepulcher Mountain is gradually moving downslope in the form of landslides.

Many of the volcanic breccias are ancient landslides of volcanic material that came cascading down the slopes for one reason or another.

The best known landslide is the one that occurred on August 17, 1959. It was triggered by an earthquake. Earthquakes are not unknown in Yellowstone. Records kept since 1871 indicate the most quake prone areas are Mammoth Hot Springs, Old Faithful, South Entrance and Yellowstone Lake. An average of five quakes per day still occur, sometimes (rarely) as many as 100.

The quake had an intensity of 7.1 on the Richter Scale (equivalent to 200 atomic bombs) and created many changes. By Rock Creek campground, the highly jointed vertical dolomite formation that was holding back the very badly weathered gneiss and schist gave way. Some 80 million tons of rock slid down one side of the Madison River Valley, dammed the Madison River creating a lake now named Earthquake Lake, then proceeded to move some 400 feet up the opposite side of the valley and then slid back to the valley floor. Nineteen people are definitely known to be buried under the debris today. There is a possibility that there are more.

The basin of Hebgen Lake was tilted, depressing and drowning the north side, and uplifting and draining the south shore. The jerking motion produced four sieches (rocking back and forth motion of the lake water). Portions of the highway slumped into the lake and were offset by faulting. Rockslides were triggered off at Firehole Falls, Golden Gate, Mount Jackson, Norris Junction, and Secret Valley. A mudflow occurred in Secret Valley, small fissure (disturbance of groundwater table) at Twin Buttes.

The water table was greatly disturbed in some places e.g., murky water in the western half of the park. The worst areas were Grizzley Lake and Straight Creek (lasted for months). The water level changed, hot springs and geysers erupted. Of the 298 eruptions in the Lower, Midway and Upper basin, some 160 of them had never erupted before. Bisquit Basin had such violent eruptions that rocks were tossed some 50 feet into the air, blocks that weighed 50 to 100 pounds.

The quake affected a total of some 600,000 square miles.

GEOLOGIC HISTORY

1. Formation of Precambrian W beds. Oldest rocks in Yellowstone some 2,700 million years.

These are a mixture of granitic gneiss, quartz-biotite schist, amphibolite and pegmatites. Some 8,000 feet of sediments lie conformably upon them. The schist was sedimentary rock that underwent regional metamorphism, was intruded by granite gneiss.

These Precambrian rocks form a part of the Beartooth Plateau (western end) that was titled and faulted when the Rockies (northern) formed. Before the blocks were tilted they had both Paleozoic and Mesozoic sediments that erosion stripped off. For the most part they are now covered by volcanics, but can be seen in Lamar Canyon.

2. Deposition of Paleozoic (3,000 feet thick) and Mesozoic (5,000 feet thick) beds in various environments ranging from the sea to beaches to floodplains.

Some 570 million years ago the region was a flat plain over which the seas advanced and retreated approximately a dozen times. The only period that is not represented is the Silurian. Some 40 separate formations were deposited with the stratigraphy of the northern end of the park varying from that of the southern. The rocks are a mixture of conglomerates, sandstone, shales, and limestones, both marine and nonmarine in origin.

They were later intruded by dikes, sills, laccoliths and stocks during the Cenozoic.

3. Laramide revolution at end of the Cretaceous. Rockies formed by folding, faulting, uplift and erosion.

About 75 million years ago, this region was uplifted into a mountain range that we now call the Rockies. At first the beds were arched upwards into *anticlines* or depressed downwards into *synclines.* Eventually enough pressure was exerted on the bedrock causing it to break or *fault,* mostly by *reverse faulting* (hanging wall moves up in relation to the footwall). Many of the folds originally created mountains, but they have since been worn down by erosion. The stripping of the highlands caused deposition of sediments in the basins (Pinyon conglomerate). The trend of folding is northwest-southeast and is independent of the Precambrian structural trend.

4. During Paleocene-Eocene time faulting and uplifting.

Around 65 million years ago the northern end of the park was broken by northwest trending reverse faults. The Gardiner Creek fault involves some 10,000 feet of displacement while the Grayling Creek fault only 1,000 feet.

The Gardiner Creek fault tilted the Beartooth block. Additional faulting by Buffalo Creek fault and the East and West Gallatin faults. The Gallatin Mountains are therefore *fault-block mountains.*

5. Eruption and deposition of the Absaroka volcanics during the Eocene.

About 54 million years ago the eruption of the Absaroka Volcanic Supergroup containing the Washburn Group (oldest) Sunlight Group and Thorofare Creek Group (youngest). They are subdivided as follows:

	Washburn Group
	Sepulcher formation
	Conglomerate Facies
	Elk Creek Basalt Member
	Lost Creek Tuff Member
	Daly Creek Member
	Fortress Mountain Member
	Andesite lavas
Early and	Lamar River formation
Middle	Cathedral Cliffs formation
Eocene	Sunlight Group
	Mount Wallace formation
	Slough Creek Tuff Member

Crescent Hill Basalt
Wapiti formation
    Jim Mountain Member
Trout Peak Trachy Andesite
    Pacific Creek Tuff Member

Thorofare Creek Group
    Langford formation
Middle to   Promontory Member
Late   Two Ocean formation
Eocene   Tepee Trail formation
    Wiggins formation

These rocks have formed most of the Absaroka Mountains, Washburn Volcano and a portion of the Gallatin Range.

The eruption was through central vents which were fairly large in size. The eruptions were mostly quiet but occasionally violent.

Streams have carved valleys in these formations. The fact that 27 fossil forests, one stacked upon the other, indicates the eruptions weren't continuous. There were very long spans of quiescence and erosion.

Intrusions of sills and dikes were common, especially around the east entrance and in the Gallatin Range. Stocks (small batholiths under 40 square miles in area) and *laccoliths* that are dome shaped intrusives with a known floor occurred. Volcanic necks also occurred in this formation.

Mount Washburn is the northern half of a volcano located on the rim of the caldera. The southern half collapsed along with the collapse of the caldera.

Approximately 40 million years ago, when the formation of the Absaroka ceased, Yellowstone was a gently undulating plateau buried under volcanics. The meandering streams had low gradients, the few remaining volcanoes were not very high. Climate was almost subtropical.

These beds can be seen by Fishing Bridge, or Dunraven Pass.

6. Moderate erosion, possible volcanic activity. No Miocene or Oliogocene beds deposited.

About 37 million years ago there may have been some volcanic activity, but these beds are either deeply buried by younger beds or have been removed by erosion.

7. Pliocene time uplift, normal faulting, erosion.

Around 10 to 12 million years ago the Rocky Mountain chain was elevated several thousand feet. This is what gave Yellowstone National Park the elevation it has today.

Normal faults produced the Gallatin Mountains uplifting a 20 mile block some 15,000 feet. This rejuvenated the streams and they once again began to downcut rapidly.

Much of the Absaroka volcanics were stripped. Around eight million years ago a basalt lava flow traveled 15 miles down the Yellowstone River Valley (between Yankee Jim Canyon and Emigrant Canyon). Cooling produced columnar jointing. It is presently a couple of hundred of feet above the road.

8. During the Pleistocene formation of the calderas and eruption of the Yellowstone Rhyloite Plateau.

About 2.5 million years ago there were three volcanic cycles, but only the first and third outcrop in the park. The second occurred with the formation of the nearby Island Park caldera. The sequence is as follows:

1st Volcanic Cycle—2.4 million years
    Rhyolite of Broad Creek (oldest)
    Junction Butte basalt (includes granite)
    Gravel of Mt. Everts
    Huckleberry Ridge Tuff of Yellowstone Group
    Lewis Canyon Rhyolite
    Sediments and basalts of The Narrows

2nd Volcanic Cycle—0.8 to 0.07 million years
    not found in park (Mesa Falls of Yellowstone Group)

3rd Volcanic Cycle—1.2 million years
    Undine Falls basalt (includes gravels)
    Lava Creek Tuff of Yellowstone Group
        Mallard Lake Member
Plateau   Obsidian Creek Member
Rhyolite   Upper Basin Member
        Roaring Mountain Member
        Central Plateau Member

Rocks of the first cycle are precaldera. Those of the third cycle are precaldera Mt. Jackson Rhyolite, Undine Falls basalt and Lava Creek Tuff of the Yellowstone Group. The rest (Plateau Rhyolites and Lava Creek basalts) are postcaldera. Sediments are mixed with some volcanics.

The first volcanic cycle produced the larger caldera some two million years ago. Many of the ash flows that developed can be seen as *welded tuffs* today, for the heat literally fused the pyroclastic debris together.

The second smaller caldera (now occupied by the West Thumb of Yellowstone Lake) developed 600,000 years ago. Later lava rose in the two vents that originally fed the volcanoes and domed them. This caused fracturing and produced some of the geyser basins. Additional flows occurred including the one that formed Obsidian Cliffs.

Paintpot Hills is a *rhyolite dome* formed like the plug dome that produced Mount Lassen (it is basic lava). Several others formed at the same time.

The 30 odd flows in some instances overflowed the calderas. The Rhyolite Plateau beds are the ones most visible in the park and are seen everywhere. They are light-colored as contrasted to the dark-colored Absaroka Super Group.

9. Glaciation, canyon cutting, thermal activity during Pleistocene and Holocene.

From about two million years ago to the present the finishing touches were put on the park. While the calderas were forming glaciation was also occurring. The ice moved across the Grand Canyon of the Yellowstone River and not down it (it buried it instead). Therefore, it is a V-shaped stream valley and not a U-shaped glacial valley.

Its development can be divided into six stages.

First, between the formation of the first and second caldera the ancestral Yellowstone River began its valley about where the Yellowstone and Lamar Rivers join today.

Secondly, after the second caldera formed and was filled in by tuff flows, it had to reescavate its valley. Thirdly lava flows dammed the drainage system south of the north rim forming a large lake. It eventually spilled over and helped to deepen the lower 15 miles.

Fourth, headward erosion cut back to lower falls (300,000 years ago) and produced a 400 to 600 foot deep channel. Fifth, glaciation (pre-Bull) buried the canyon. Afterwards, a lake formed between Inspiration Point and both falls, with little downcutting until 125,000 years ago. The canyon almost reached its present depth during the last of the two remaining glaciations. Final glacial retreat some 12,000 years ago completed the remaining development of the canyon.

Final formation of the entire park, as we see it today, occurred after the last of the glaciers retreated.

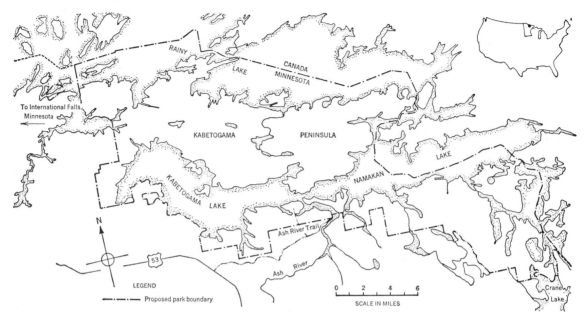

Voyageurs National Park, Minnesota.

Geologic Column of Voyageurs National Park

Erathem	System	Series	Formation
Cenozoic	Neogene	Holocene	Alluvium
		Pleistocene	Glacial Deposits
Precambrian	Precambrian Y		Keweenawan Lava Flows
	Precambrian X		Animikie Sediments
	Precambrian W		Timiskaming Sediments
			Vermillion Batholith
			Knife Lake Sediments
			Keewatin Lava Flows
			Coutchiching Sediments

Modified after Merriam et. al.

CHAPTER 20
VOYAGEURS NATIONAL PARK

Location: N.E. Minnesota
Area: 219,431 acres
Established: 1971

LOCAL HISTORY

This is canoe country, where Indians used the routes before the trappers of the various fur companies, especially Hudson Bay, plied their trade. Indian artifacts such as pottery, arrowheads and slate knives may date as far back as 5,000-1,000 B.C., (but they may also be as young as 400 B.C.). Many signs of the early Indians are found by Ash River and Kabetogama Lake.

The trappers were known as Voyageurs, and wore red sashes and red caps, usually blue cloaks and deerskin leggings. The majority were of French-Canadian ancestry. They travelled in 36 foot birch bark canoes with trade goods or furs, that could carry loads up to 1,500 pounds. As early as 1700, men such as Charles Becard de Granville were impressed by the country so wrote about it and sketched the various features and animal life.

The logging industry started in the 1880s and clear cut most of the area, so now most that is seen is second growth of white spruce, aspen, maple, and balsam in the higher areas with black spruce, tamarack, ash and white cedar in the swampy regions. There was a gold mining boom in 1893 to 1898. The boom town of Rainy Lake City (Kabetogama Peninsula) began in 1894 in spite of the fact that the closest railroad and only transportation to the area was 100 miles away. It went from a population of 500 to a ghost town in five years. The shopkeepers moved to International Falls (name today). The mines within the park are (1) Little American; (2) Lyle; (3) Big American; (4) Bushy Head; (5) Old Soldier; (6) Holman; and (7) Gold Harbor.

The main method of travel will always be by means of canoe in this park which varies from three to fifteen miles in width and about forty miles in length. The park has been created from federal, state, county, and privately owned land.

GEOLOGIC FEATURES

This area is located in the Superior Province of the Canadian Shield region and contains about the oldest beds of the North American continent, dating more than 2,700 million years ago. (Precambrian W)

From Cambrian time about 600 million years ago, to about two million years ago, this area underwent erosion. Then the large Laurentide ice sheet formed. It scraped and gouged these ancient beds four separate times, creating the lake basins that exist today and leaving its traces everywhere. There are minor moraines scattered here and there. Terraces have formed from some of the old beach lines. The glaciers finally left a scant 10,000 years ago. Their great weight depressed the crust and now *glacial rebound* is gradually occurring, elevating the land to its former elevation and thus draining many of the lakes that exist today.

This park has the most complete geologic representation of Precambrian beds of anywhere in the United States. It also contains well-developed surface glacial features, and shows the effects of Glacial Lake Agassiz.

The *greenstone belt* consists of metamorphosed lavas that are schistose (finely foliated metamorphic rock) and sheared. This often masks the variety of original rock. Most of the major islands are made up of greenstone (e.g., Dryweed, Grassy and Grindstone), also sections of Neil Point and part of lake between (South shore) Jackfish Bay and Black Bay.

The biotite and mica schist belt (Coutchiching) were originally graywackes, sandstones, mudstones and siltstones that underwent metamorphism. The majority of Kabetogama Peninsula and Black Bay area of Rainy Lake are on this belt. The contact zone can be found by State Point area (south). At the *contact region* are intrusions of granite into the

metamorphic rock. The belt is some 200 miles long and only 6 to 20 miles wide.

The Vermillion Batholith is responsible for the metamorphic zone and underlies the Sullivan Blind-Ash Bay region.

Signs of glaciation can be found at Neil Point, Grindstone Island (east end) and a small island east of Dryweed Island. Many of the deposits were used as gravel pits. These moraine deposits were left by the St. Louis sublobe (Des Moines lobe) of Mankato stage.

Pillow lavas are very common in the Keewatin greenstone formation but metamorphism has destroyed most of the structure except for the small island off of Steamboat Island. At this location they are up to three feet long.

The series of terraces at Neil Point are actually a set of fault scraps located between the greenstone belt and biotite schist. They run east-west with the greenstone block (north block) moving east relative to the biotite schist block (south block) which also moves upwards. Glaciation has scoured them and helped to accent the fault zone.

This east-west fault is important economically for it was the conduit through which the gold-bear-

ing quartz solutions travelled and crystallized. The veins have a mixture of minerals. They are tourmaline, ankerite, pyrite, plus gold and quartz.

One formation that is controversial is the Coutchiching which contains granite boulders in the conglomerate in areas such as at Neil Point indicating it is the oldest sedimentary bed. The mica-biotite schist belt in the park is tentatively assigned as being Coutchiching in age (they were originally sediments). The main problem is, did these beds form before the Keewatin beds or at the same time as the Keewatin beds? Regional metamorphism is so severe that it does not allow petrographic analysis (thin-section) that would solve this problem.

GEOLOGIC HISTORY

1. Formation of the Coutchiching sedimentary beds upon granitic basement complex, Precambrian W in age.

These beds include arkosic sandstones, graywackes, quartzites (sandstone), slate (shale), conglomerates with some iron ore and zones of

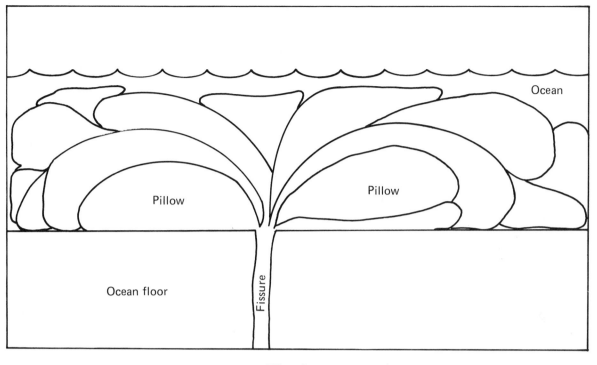

Pillow Lava

volcanic beds in an ancient ocean. These sediments were derived from the erosion of the basement rock, and include inclusions of granite which are the basement rocks.

2. Extrusion of the Keewatin lava flows during Precambrian W time.

The lava flows came from fissure eruptions, and gradually built up layer after layer of flows. They covered the Coutchiching beds in many areas. Because of the heat from the flows, some *contact metamorphism* occurred changing the sedimentary beds into metamorphic beds; e.g., shale into slate. Some were extruded under water as shown by the pillow structure. Enough time passed to have sediments mixed in with the lava flows.

3. Uplift during the Laurentian orogeny, formation of the Laurentian Mountains with the intrusion of the Laurentian granite.

The Laurentian Mountains developed by folding and faulting, and because of the deformation, a magma chamber formed which crystallized into the Laurentian granite. The intrusion caused the metamorphism of some of the Coutchiching and Keewatin beds plus some minerlization (copper and iron) e.g.: (Soudan formation). This mountain range formed in the southwest corner of the park (Saganaga Lake) and on into Canada.

4. Erosion of Laurentian Mountains, deposition of the Knife Lake series during Precambrian W time.

Like any mountain range, it was exposed to the elements and underwent the process of *weathering*, both *disintegration* or physical weathering, and *decomposition* or chemical weathering. The mountains gradually grow lower and lower with the intervening valleys filling in until there was a gently undulating sloping surface formed referred to as an erosional surface. On this surface of old age, streams heavily laden with sediments meandered back and forth, ever widening their valleys.

5. Deposition of the Knife Lake series during Precambrian W.

The streams that were eroding the Laurentian Mountains were depositing the sediments into a shallow subsiding trough called a geosyncline that formed adjacent to the mountains. At first the beaches had a composition of pieces of the granite that made the gravel for the beach. As the mountains wore down, the sediments became finer so sand and shales were deposited and the seas gradually *transgressed* or advanced over the land. The formation of *graywacke* indicated nearby volcanic activity. This caused the pyroclastic debris, (mostly dust and ash in size), to mix in with the sand and eventually broke down into clay which formed the "dirty sandstone" that graywacke is. They are at least 21,000 feet thick, perhaps as much as 50,000 feet.

6. Uplift with the Algoman orogeny forming the Algoman Mountains and the intrusion of the Vermillion Batholith. (Part of Algoman granite series)

The same processes that created the Laurentian Mountains and granite also produced the Algoman Mountains, with the intrusion of the Vermilion Batholith some 2,400 million years ago. It is 35 by 80 miles in size and has a variable composition from a light colored granite (east) to a migmatite, rich with schist (west). When the mountains were uplifted, it permitted a release of pressure, for the deeply buried sediments, with the temperatures above the melting point. The release of pressure permitted expansion and thus melting, so the magma chamber formed after the deformation of the mountains began. Remember *first* the deformation, *then* the magma chamber! The granite intruded the distorted Knife Lake sediments metamorphosing them into a biotite-mica schist (to the north). And formed a gneiss belt adjacent to it, in which the gold bearing quartz veins were emplaced.

All these events occurred between two to four billion years ago. About 1.7 billion years ago (Precambrian Y) a mafic (dark minerals) dike intruded near State Point. These are the youngest rocks in the park.

7. Erosion of the Algoman Mountains.

Once again the process of weathering took over and reduced the mountain range after many millions of years to a erosional surface, producing a major *unconformity,* or a break in the sequence of deposition. The geologic history in this region seems to end here until the Pleistocene. If additional beds were deposited in this area, they were later removed without a trace.

8. Formation of Laurentide Ice Sheet and glaciation during the Pleistocene.

Somewhere around 1.5 million years ago, the climate grew colder and in Canada a large ice sheet called the *Laurentide Ice Sheet* developed over most of Canada and northern United States. This ice sheet advanced four times, scraping and gouging the ancient Precambrian rocks and carrying much of the granite and metamorphic rock further south as *glacial erratics.*

9. **Ice action produced the lake basins that exist today. Last retreat was approximately 10,000 years ago.**

Wherever the beds were softer because of composition or structure, the ice sheet scraped and gouged out basins which filled with water as the glaciers retreated for the last time. Terraces formed (Neil Point). Also glacial erratics, chatter marks, gravel deposits, glacial striations and glacial pavements are scattered throughout. The more resistant areas were left as highlands such as the Kabetogama Peninsula that makes up the bulk of the land area of Voyageurs National Park.

10. **Formation of glacial Lake Agassiz.**

When the Mankato lobe retreated uncovering many of the buried valleys, it allowed the rivers to reclaim their valleys, only to find there was no outlet. Hence, Lake Agassiz (named after the Father of Glacial Geology—Louis Agassiz) formed. The lake eventually covered up to two million square miles and covered the park with water between 150 to 700 feet deep. The depth was dependent upon the location of the ice lobe. About 9,500 years ago it began to retreat north, along with the lobe from the park area, leaving the landscape buried under the clays from the lake bottom. Most of the lakes in Voyageurs National Park are remnants of Lake Agassiz.

11. **Glacial rebound has uplifted the region around 200-300 feet in elevation and drained some of the lakes.**

The weight of over 10,000 feet of ice caused the bedrock to become compressed. Now that this weight is removed, the land is gradually returning to its original elevation, this isostatic readjustment is called *glacial rebound.* Some of the larger lakes with uneven basins have drained partly to form a series of smaller lakes. Uplift is not complete and probably won't cease for another 12 to 15 thousand years. This means eventually other lakes will be drained.

12. **Formation of present landscape by erosion and deposition of streams.**

The retreat of the glaciers permitted vegetation to grow upon the debris that was left, covering the region with forests that were logged in the late 1800s. The streams have further modified the park through the processes of erosion and deposition.

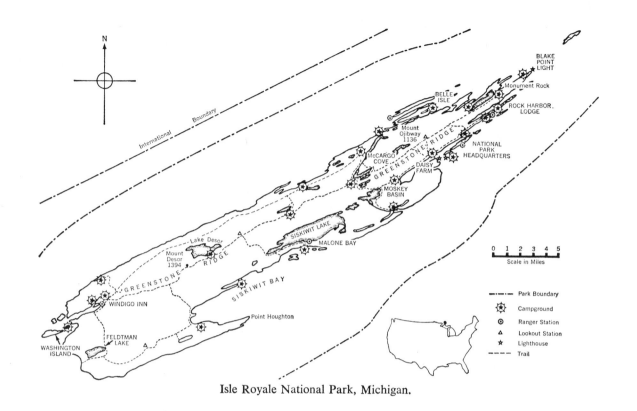

Isle Royale National Park, Michigan.

Geologic Column of Isle Royale National Park

Era	Period	Epoch	Formation	Member
Cenozoic	Neogene	Holocene	Alluvium	
		Pleistocene	Glacial Deposits	
Precambrian	Precambrian Y		Copper Harbor Cong.	
			Portage Lake Volcanics	Scoville Point Flow
				Edwards Island Flow
				Middle Point Flow
				Long Island Flow
				Tobin Harbor Flow
				Washington Island Flow
				Greenstone Flow
				Grace Island Flow
				Minong Flow
				Huginnin Flow
				Hill Point Flow
				Amygdaloidal Flow

Modified after Books, White, Wolff, Huber et. al.

CHAPTER 21
ISLE ROYALE NATIONAL PARK

Location: Michigan (An Island in Lake Superior)
Area: 539,341.01 acres (Land area 133,844 acres)
Established: April 3, 1940 (Authorized March 3, 1931)

LOCAL HISTORY

The Indians called it Minong which meant "the good place to get copper." As long ago as 4,500 years the Indians (Algonquin, Sioux, or Iroquois) mined the *native copper* (pure copper) in shallow mining pits. To extract the copper they built a fire on the location that they wanted to work. After the rocks became very hot they dashed cold water on them. The sudden cooling caused them to shatter. They then used stone hammers to beat out the copper. Since the type of rock found in the stone hammers are not found on the island they must have come from the mainland. More than 1,000 of these pits are scattered all over the island. Some of them are actually 10 to 20 feet deep.

French explorers looking for areas to trap furs found the island in 1669 and named it after Louis XIV, Isle Royale or "Royal Island." They claimed it in 1671. At that time Chippewa Indians were living on it. It became British property in 1763 after the French lost the French and British Indian Wars. The Peace of Paris Treaty gave it officially to the United States in 1783, but the British really controlled it until after the 1812 War. Six fur-trading posts were on Isle Royale at one time, one Hudson Bay, the other five the American Fur Company. As far as the Chippewas were concerned it was still their territory until they gave up their last claim in 1843. When it was ceded to the United States in 1843 prospectors overran the island. They burned the trees to find the copper ore or cut them down to use for building. Mining on a commercial basis was tried in the 1870s and finally in 1899 all attempts were abandoned since there is not enough ore for this type of mining. The companies occasionally found chunks of native copper that weighed several thousand pounds.

One chunk was exhibited in 1876 in Philadelphia during the Centennial Exposition. It weighed 5,720 pounds.

People built summer homes on Isle Royale in the early 1900s. Albert Stoll Jr., a newspaper man from Detroit, was the prime mover in establishing the National Park in 1940.

GEOLOGIC FEATURES

The copper occurs as *amygdules* in the amygdaloidal basalts. These amygdules are the bubble holes or vesicles left in the lava flows that as hot water solutions passed through they deposited material (in this case native copper). In other words—filled in vesicles.

Another mineral that occurs in the amygdules in the gemstone *chlorastrolite* or *pumpellyite*. It is either a variety of zeolite or prehnite, the origin is uncertain. It sometimes forms with *thompsonite* (variety of prehnite). Its formula is $Ca_2 MgAl_2 (SiO_4) (Si_2 O_7) (OH)_2 \cdot H_2 O$. It was first named and described by Dr. Charles T. Jackson in 1849 in a geological report. The name chlorastrolite means "green star." The crystals are very rarely any larger than 1/2 an inch although some up to two inches has been found. Sunlight will fade them. The crystals can be found at Scoville Point, between Rock Harbor and Suskowit Lake, at Old Light and Todd Harbor. Unfortunately since this is now a National Park it is illegal to collect them within the park boundaries.

Isle Royale is technically an *archipelago* (island groups in a body of water) since it is surrounded by some 200 tiny islands. Total area for the park is 810 square miles, 600 of it is water since the boundaries extend 4 1/2 miles in every direction.

The bedrock is primarily Precambrian W basaltic lava flows. A paleomagnetic study (magnetic field of the geologic past) from 44 samples on Isle Royale were taken from the Middle Keweenawan lava beds. (None were collected from the Lower Keweenawan beds.) These studies are being used

for correlation purposes. The Lower and Middle Keweenawan rocks are subdivided into three groups, oldest normal, reversed and youngest normal according to the direction of the flow of the magnetic field or *polarity*.

The structure of the beds that form Isle Royale is a gentle limb of an asymmetrical syncline dipping about 15° to 20° on the north side of the fold. The syncline formed when the Keweenawan lava flows were extruded onto the land, thus removing support from the basin which caused it to subside forming the syncline. Since this is an island, features produced by wave action can be seen. Most of them formed during the Pleistocene when the level of the lake was much higher than today. The two best exposed abandoned shorelines are those of Glacial Lake Minong about 450 feet above sea level (140 feet lower than Superior is today) which formed before *glacial rebound* occurred. Glacial rebound has elevated the Minong Beaches some 80 feet. When an area is buried by several miles of ice it undergoes compression. If the weight is removed by the ice melting or retreating the bedrock can once again expand, this is called *glacial rebound*. The other lake level that is well displayed is that of Lake Nipissing that occurred after glacial rebound. It was 605 feet above sea level some three feet above the present level of Lake Superior today (602 feet). The Beach line today is at a 650 foot level which means glacial rebound has elevated it almost 50 feet. Average rebound today 1.33 feet every hundred years.

Monument Rock on the eastern end of the island is a *sea stack* that was eroded during the Lake Minong stage. On Amygdaloid Island (northwest side of east end) a *sea arch* or wave cut arch can be found formed during the Nipissing stage. At the end of Feldman Ridge (west end) is a barrier beach that is presently closing off a swamp about 200 feet above the present sea level, some ten feet above it is a wave-cut cliff with a sea cave formed by the enlargement of the joint system. There is a possibility that another set of beaches occurs some 300 feet up, but glacial debris and other material plus the lack of an outcropping of rock obscure the necessary proof.

The drainage system is, in part, controlled by joint system and in part by tilted beds of varying resistance. The joints were the result of stress and strain as the synclinal structure developed. At the

west end they trend almost due north, as they approach the east end they shift 30° to the east. McCargoe Cove has formed along it.

Some faulting has occurred along some of these joints producing closely spaced openings that weather more rapidly than the areas where the joints are far apart. The weathered area has a tendency to form ravines.

Glaciation has played a major role in carving the landscape. However it is more noticeable at the western end than it is the eastern. The reason is the direction of ice flow. At the eastern end the glacier flowed parallel to the strike of the beds. It gouged the bedrock, deepened basins, created many lakes and left few deposits. During the last stage the glaciers covered up the eastern end and *ice margin deposits* (possible recessional moraines, kames, outwash, etc.) built up a natural barrier. The glaciers that covered the western end flowed almost due west instead of southwest as for the eastern half. Since it flowed across the ridges *roche moutonnees* formed with a gentle smooth side and steep rugged side produced by glacial plucking.

Also formed by the glacier as a depositional feature are the *drumlins*. These inverted teaspoon-shaped, asymmetrical hills, made up of till, formed when the glaciers plasted debris behind some knob or obstacle. Occasionally roche moutonnees and drumlins are confused because of the shapes, but the chart below shows the differences.

Roche moutonnees	*Drumlin*
Feature of erosion	Feature of deposition
Solid bedrock	Till
Profile—direction of ice movement	Profile direction of ice movement
May occur alone	Occurs in swarms

Drumlins can be found by Lily Lake west of Siskiwit Bay. The direction of the *glacial striations* or scratches found scattered all over the island agree with the direction of ice movement according to the drumlins and roche moutonnees.

The highly jointed bedrock and alternately tilted layers of sedimentary and igneous rocks was very conducive to *glacial plucking* or *quarrying.* Water seeped into the pores of the bedrock and froze, when the glacier moved, the section of rock (separated from the rest by bedding planes and joint systems) would be pulled or plucked out. This is how the uneven jagged side of the roche moutonnees formed. Also, the asymmetrical ridge on the western side of the island. The blocks that are pulled out are deposited elsewhere as *glacial erratics*.

Glacial abrasion has removed most of the deeply weathered rock, but in a few protected localities some of the basaltic lava flows show *spheroidal weathering*. The fracture system broke up the lava flows into a series of blocks. The corners of the blocks would weather first since more area was exposed and thus became rounded. Then the entire rock would weather at an even rate, spherical layer by spherical layer. Usually a core is left surrounded by the weathered debris.

GEOLOGIC HISTORY

1. **Formation of Portage Lake series (Middle Keweenawan) of Upper Precambrian time or Precambrian Y.**

On Island Royale about 1.1 billion years ago a series of *flood basalts* (lava flows from fissures instead of a central vent) poured out from a fissure system trending northeast-southwest. It built up to a thickness of over 10,000 feet in a series of twelve separate flows that usually have layers of sedimentary, pyroclastic debris or mottled ophitic (crystals of pyroxene that cover crystals of plagioclase feldspars) volcanic rocks.

The flows from youngest to oldest are as follows:

12. Scoville Point Flow—100-200 feet thick, porphyritic.
11. Edwards Island Flow—50 feet thick, fine-grained dark rock.
10. Middle Point Flow—50 feet thick, porphyritic.
 9. Long Island Flow—50 feet or less, fine-grained dark rock.
 8. Tobin Harbor Flow—50-100 feet thick, porphyritic.
 7. Washington Island Flow—80-200 feet thick, ophitic.
 6. Greenstone Flow—100 feet ophitic zone, 75 feet pegmatie, 175 feet ophitic, 50 foot columnar joints.
 5. Grace Island Flow—50 feet thick, *lath-shaped euhedral* plagioclase feldspar phenocrysts.
 4. Minong Flow—77 feet thick, fine-grained dark rock, amygdules of agate.

3. Huginnin Flow—67 feet thick, lath-shaped, euhedral, plagioclase feldspar phenocrysts.
2. Hill Point Flow—158 feet thick ophitic with large augite crystals.
1. Amygdaloid Island Flow—50 feet or less, andesite with agate amygdules.

2. **Metamorphism and mineralization of the Portage Lake volcanics.**

As each layer was extruded and weathered, they formed clastic sedimentary rocks (less than 10 percent), heated pressure built up metamorphosing the lower beds slightly. This is when the thompsonite and pumpellyite formed being deposited by the metamorphic solutions. Also, agate, copper, calcite and other materials occur as amygdules.

3. **As lava poured out, the area sagged and formed an asymmetrical syncline with a gentle north limb (15°) and steeper south limb (35°).**

The removal of all the lava caused the elongated basin or syncline to subside at approximately 50,000 feet along the axis. The settling was uneven so the syncline is asymmetrical. The rocks that make up Isle Royale are from the gently dipping limb.

4. **Deposition of the Copper Harbor conglomerates (Middle Keweenawan) during Upper Precambrian or Precambrian Y.**

A clastic wedge of sediments was deposited on alluvial fans, as flood plain deposits or in shallow water. Thickness ranges from 1,500 feet to over 6,000 feet. The thickness increases eastward changing from a sand to a coarse gravel. This indicates the source area was to the west probably along the northern shore of the lake in Minnesota. The boulders are a mixture of andesite, basalt, latite, rhyolite and trachyte which are igneous. Claystone and sandstone (arkosic) are of sedimentary origin and the quartzite metamorphic. Most of the boulders are igneous but were not derived from the volcanic rocks on the island but from older Keweenawan beds. They probably came from the North Shore Volcanic Group. The beds contain shallow water features such as ripple marks, mud cracks and graded bedding.

5. **Continued subsidence, further tilting of sediments, erosion of softer sediments.**

When the syncline first formed the streams deposited the Copper Harbor sediments. Further subsidence increased the gradient and much of the newly deposited material was stripped off forming the ridge and valley topography. The Copper Harbor sediments are seen only on the southern third of the island surrounding Siskiwit Bay and making up Feldtman and Houghton ridges.

6. Formation of Isle Royale fault and other related minor faults.

The Isle Royale fault is a high angle thrust fault. Smaller north-south trending faults have formed along the northwest side of the island. East-west faults are predominate on the eastern end of the island. A possible stream system formed in the Lake Superior syncline draining to the northeast.

Faulting occurred on both sides of the syncline (Keweenawan fault of the peninsula).

The main faulting created a *horst* or the elevation of a block in relation to the two adjacent blocks. This horst was in the center of the syncline forming Isle Royale and the Keweenawan Peninsula. The transverse faults developed later.

7. Deposition of Jacobsville sandstone and subsequent removal.

Somewhere around Precambrian Z and Cambrian time a continental sandstone (a blotchy red and white deposit) formed. It forms the floor of Lake Superior today. Either erosion has removed them or it has been buried by Paleozoic sediments. It is not found on Isle Royale.

8. Glaciation during the Pleistocene several times.

The glaciers scraped away most of the softer sediments, deepened all the valleys and further emphasized the ridge and valley topography. The eastern end of the island has the glacial basins filled with water because the glacier flowed parallel to the ridges. The western half has the roche moutonnees topography since the glaciers flowed against the grain of the ridges.

The last glacial lake, Superior Lake, left 11,000 years ago leaving the glacial deposits seen on the island today.

9. Retreat of ice drained Lake Superior to the southwest, later to south, drop in lake level

(Minong Stage). Another level dropped it even more.

When the glaciers filled the Lake Superior basin they dammed up the drainage system forming large glacial lakes. As the glaciers finally retreated they uncovered lower and lower outlets. Thus the lake drained in different directions as these outlets were uncovered. Sometimes the levels would remain stable for a long time and beaches would form. Rapid drainage and then stabilization would occur and a beach at a lower level would develop once again.

The original Lake Duluth drained to the southwest into the ancestor of the Mississippi River, with an elevation of 1,085 feet above sea level. Later it drained south into Lake Michigan and formed the so-called Minong Stage around 10,500 years ago and dropped down to 450 feet above sea level. When another outlet formed at Sault Saint Marie about 9,800 years ago the level dropped to 375 feet above sea level and was called the Houghton Stage.

10. Glacial rebound has been elevating entire regions at the rate of 1.33 feet every 100 years.

The rebound began about 10,000 years ago, rapidly in the beginning and today very slowly.

After 5,000 years the rise in lake level created the Nipissing Stage now located 650+ feet above sea level. The Minong Stage is presently 680 feet above sea level. It took a total of 10,000 years to elevate the Minong beach some 80 feet and while only 5,000 years to elevate the Nipissing 50 feet. (Superior is 602.4 feet above sea level). Since the lake and present outlet are now rising at the same rate, the level of the lake will stay stable.

11. Modification of present landscape.

Alluvial fans are filling up valleys, deltas, and lakes. Lakes are silting in and becoming swamps. The glacial debris is being converted into soil and exposed bedrock, on the ridges is undergoing weathering to form the landscape that makes up Isle Royale.

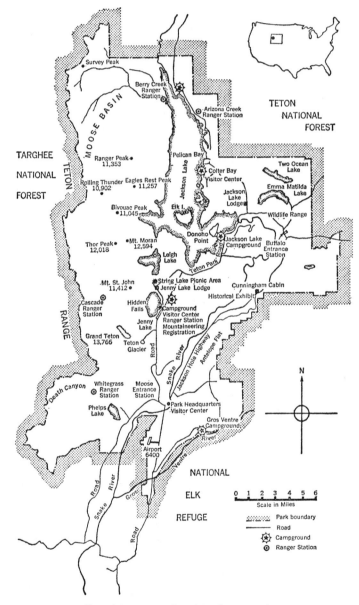

Grand Teton National Park, Wyoming.

Geologic Column of Grand Teton National Park

Erathem	System	Series	Formation	Igneous Rock
Cenozoic	Neogene	Holocene	Alluvium	
		Pleistocene	Pinedale Glacial Deposits	Volcanic Ash
			Loess	
			Bull Lake Glacial Deposits	
			Oldest Glacial	
			Upper Lake Sequence	
			Lower Lake Sequence	Andesite and Basalt
			— Bivouac Formation —	— Rhyolite —
		Pliocene	Teewindt Formation	Basalt and Rhyolite
			Camp Davis Formation	Rhyolite
		Miocene	Colter Formation	Andesite and Basalt
	Paleogene	Oligocene	Wiggens Formation	Andesite and Basalt
		Eocene	Tepee Trail Formation	Andesite and Basalt
			Aycross Formation	Tuff
			Wind River and Indian Meadows Formation	Tuff
		Paleocene	Pinyon Conglomerate	
Mesozoic	Cretaceous		Harebell Formation	
			Meeteetse Formation	
			Mesa Verde Formation	
			Unnamed Sandstone, Coal and Shale	
			Bacon Ridge Sandstone	
			Cody Shale	
			Frontier Formation	
			Mowry Shale	
			Thermopolis Shale	
			— Cloverly and Morrison (?) —	
	Jurassic		Sundance Formation	
			Gypsum Spring Formation	
	— ? —		— Nugget Sandstone —	
	Triassic		Chugwater Formation	
			Dinwoody Formation	
Paleozoic	Permian		Phosphoria Formation	
	Pennsylvanian		Tensleep and Amsden Formation	
	Mississippian		Madison Limestone	
	Devonian		Darby Formation	
	Ordovician		Bighorn Dolomite	
	Cambrian		Gallatin Limestone	
			Gros Ventre Formation	
			Flathead Sandstone	
Precambrian	Precambrian		Gneiss and Schist	

Modified after Love et. al.

PART IV
FORMED BY MOUNTAIN BUILDING AND UPLIFT

CHAPTER 22
GRAND TETON NATIONAL PARK

Location: N.W. Wyoming
Area: 310,442.52 acres
Established: February 26, 1929

LOCAL HISTORY

Until the 1800s this was Indian Country. They called them Teewinot or "Pinnacles." Various tribes would summer there; the Blackfeet, Crow, Flathead, Gros Ventre, Nez Perce, Shoshone and Utes. Sometimes peacefully, other times at war with each other.

Artifacts indicate that the Indians first came into the region around 5,000 B.C. to 2,500 B.C. when it was relatively dry and the glaciers had retreated.

On the same trip in 1808-09 that John Colter discovered Yellowstone he crossed over the Tetons, thus making him the first white man to see these mountains.

A group led by William Price on their way to Astoria, Oregon (usually referred to as the Astorian Party) crossed the Tetons in 1811 on their way to the Columbia River. They called them "Pilot Knobs."

In 1819 French-Canadian trappers came to trap beaver and other game and named the three peaks (Grand, Middle, and South Teton) "Les Trois Tetons" or "The Three Breasts."

The three partners Jedediah Smith, David E. Jackson (Jackson Hole was named after him in 1829), and Captain William Sublette lived and trapped there. Jim Bridger, Thomas Fitzpatrick and Joseph Meek, all mountain men plied their trade here.

In 1832, at nearby Pierre's Hole (mountain man term for valley) located to the west of Tetons, the mountain men held their Fur Rendezvous. A yearly event that no mountain man or trapper, if he was alive and able to move, would miss.

The golden years of trapping in Teton country were from 1825 to 1840. In 1845 styles changed in both North America and Europe, and furs, especially beaver for hats, were no longer desired. This ended the trapping industry.

In 1836 Washington Irving wrote the story about the Astoria Party and used the name "Teton." Owen Wister wrote the book called "The Virginian" in 1902 and had Jackson Hole as the locality.

Main scientific parties were The Hayden Geological Surveys. They were in 1871, 1872, and 1878. They assigned such names as Bradley, Jenny, Leigh, Phelps, and Taggart lakes, also Mount Saint John. From these trips photographs (1872) by William H. Jackson and paintings by Thomas Moran (1879) helped to make this area popular.

In the 1880s settlers came, settled near Teton

Pass, and founded the towns of Driggs, Jackson, Tetonia and Victor. Cattle were brought into the region, first by rustlers hiding the cattle, then by European promoters and blizzards almost wiped them out. Wars between the cattlemen and sheepmen occurred in 1892.

Two homesteads, Menors Ferry and the Cunningham place have been partially restored to show what the country was like.

First attempt to climb the Tetons in !870 failed. Supposedly N.P. Langford of Crater Lake fame and James Stevenson climbed it in 1872. It took until 1930 for all of the peaks to be climbed.

Grand Teton National Park was established in 1929. Jackson Hole Monument in 1943 and the upper valley in 1950. Much of the land (33,562 acres), with the intention of donating it as a National Park, was acquired by John D. Rockefeller. After some legal battles, they were all incorporated (a compromise) into Grand Teton National Park on September 14, 1950. It is now almost 500 square miles and about 24 miles by 38 miles.

GEOLOGIC FEATURES

The Tetons were created by *block faulting,* the one west block was uplifted to create the Teton Mountains, the east block dropped to form Jackson Hole. Approximately 7,000 feet can still be seen today forming the spectacular face that has been modified by erosion and glaciation. Total displacement may have been up to 30,000 feet.

Jenny Lake formed 9,000 years ago when moraine dammed up the stream. The Cascade Canyon glacier came down Cascade Canyon changing the V-shaped stream valley to the U-shaped glacial valley. (The valley is named for the many cascades and falls it contains.) The glacier then dumped its moraine. Evidence in this region indicates the area has been uplifted some 2,000 feet since the creation of Jenny Lake.

According to some geophysical evidence the discovery of another parallel (normal) fault to the Teton fault (along Wyoming-Idaho state line) may mean that the Tetons are a horst block. A *horst* is created when one fault block moves up in relation to the two adjacent blocks.

The bedrock of the mountains is Precambrian W gneiss, schist, some granite and diorite. However, the western slopes of many of the mountains such as Mount Moran, Mount Hung, Rendezvous Peak, and the Wall at the head of Avalanche Creek have sedimentary beds deposited 550 million years ago (Cambrian). They can be found forming most of the slopes north of Webb Canyon (northern end of park) and south of Granite Canyon (southern end

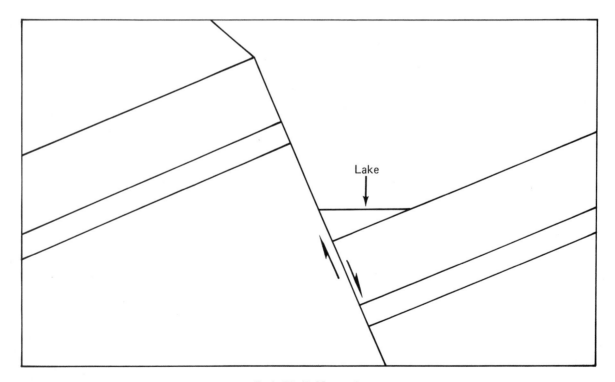

Fault Block Mountain

of the park). They rest uncomfortably on the Precambrian beds.

On Mount Moran a vertical basalt dike can be seen on its eastern face, which has been truncated by the Cambrian bed and must therefore be Precambrian. This is a good example of how the law of cross-cutting relationships (a bed is younger than the bed it cuts across) and unconformities can be used to determine relative dating. Some of these dikes are very resistant and stand out as ridges, others are weak and form depressions called chimneys for mountain climbers to inch their way up.

Mount Leidy and other areas have been folded and eroded. Cretaceous beds were deposited on their crests. These sediments were deposited between 60 and 120 million years ago. They underwent deformation and erosion, then somewhere between 60 and 70 million years ago gravels were deposited on top. Around 64 million years ago they were uplifted and now form the top of Mount Leidy, Pinyon Peak, and Gravel Mountain.

Some 25 million years ago by Two Ocean Lake existed a volcanic vent which spewed out a tremendous amount of material. Some 7,000 feet of water deposited debris accumulated.

The *continental divide* is the dividing point of drainage to the Atlantic or Pacific, it is gradually shifting. The reason for the shift is the faster flowing, higher gradient streams (because of the steeper slope) on the east slope are cutting back by headward erosion and eventually steal the headwaters of the less favorable situated streams. Extension to the west from the original divide is three to five miles.

The core of the Tetons consists of metamorphic and igneous rocks. Found in these rocks are small pods of metamorphosed talc commonly called *soapstone.* Because of its green color it is referred to as "Teton Jade." Its hardness of one on Moh's scale readily distinguishes it from Jade with a hardness of around five to eight (dependent on variety).

Signal Mountain (near Jackson Lake Dam) has ledges on the north and east faces that are sheets of volcanic rock formed from avalanches of volcanic ash that came from Yellowstone about two million years ago.

Blacktail Butte (south of Antelope Flats) is a remnant of an old arch (Teton-Gros Ventre) that was destroyed by the Teton fault.

A wedge between Buck Mountain fault and Teton fault was uplifted higher than the rest of the Teton block forming the highest peaks such as Grand, Middle, and South Teton. Some of the smaller faults had streams flowing along them (since they are weaker) that have carved canyons such as Open and Phillips canyons.

The Tetons were modified by glaciers and the evidence is everywhere. Taggart, Bradley, Jenny, Leigh and Jackson lakes basins were all once occupied by glaciers.

The Potholes were formed when chunks of ice from Jackson Lake Glacier were buried in the outwash. They eventually melted leaving depressions. Antelope Flats and Baseline is composed of glacial outwash from one of the earlier glaciations. The mounds and ridges around Jackson Lake, now covered with trees, are glacial moraines, i.e., Burned Ridge. Glacial hanging valleys on Moran, St. John and Teewinot mountains, large troughs of Cascade, Death, Leigh, and Moran canyons all indicate glaciation.

Many of the small lakes by the Snake River are *oxbow lakes* formed when a cutoff occurs and the Snake abandoned its former meander.

GEOLOGIC HISTORY

1. Deposition of sediments and volcanics along an island arc system during Precambrian W time, metamorphism.

Soon after deposition of these sediments (probably sandstone, shale, limestone and lava flows with pyroclastic debris) the metamorphic process began and the rocks that form the *layered gneiss* underwent folding.

About 2.8 billion years ago *regional metamorphism* due to folding and faulting occurred and four Precambrian units underwent deformation. The *Layered gneiss* was folded a second time, the *Hornblende quartz monzonite gneiss* which is not layered, the *Biotite granodiorite gneiss* that is layered and *Hornblende–plagioclase gneiss,* a metamorphosed diorite were deformed. All four formations have sharp contacts with one another.

2. Intrusion of granodiorite and pegmatites about 2.4 billion years ago (Precambrian X).

The granites and pegmatites (formed from hydrothermal solutions) clearly intruded after the earlier beds were folded and faulted. Any opening such as fracture, faults, foliation planes were invaded by the molten magma or its hot water solutions some 2.4 billion years ago. Their resistance to weathering causes them to form jagged ridges spires, and pinnacles.

3. Intrusion of diabase dikes at boundary between Precambrian X and Precambrian Y, uplift and erosion.

The diabase dikes range from 5 feet to 150 feet thick. The largest are on Mount Moran and are traceable for some seven miles. Resistance to weathering varies, sometimes forming ridges (eg., Mount Moran) or depressions (eg. Middle and Grand Teton).

Uplift and erosion formed an unconformity between Precambrian and Cambrian beds.

The shape of the mountains are fairly well controlled by their structure. Lineation of the metamorphic rocks are responsible for the smooth tilted eastern slope and the almost vertical western slopes.

The joint systems that developed during uplift explains many of the tops that are leveled. Also, the East-West trending canyons formed by streams and modified by glaciers are the result of the jointing.

4.   Deposition of Paleozoic beds between 225 to 270 million years ago.

The Cambrian beds are as follows:

> (1) Flathead Sandstone—marine, reddish brown, 50-60 meters thick.
> (2) Gros Ventre formation—marine, green shale and limestone 180-240 meters thick.
> (3) Gallatin Limestone—marine, blue-gray, hard, 55-90 meters thick.
>
>   The Ordovician beds are as follows:
>
> (4) Big Horn dolomite—marine, light colored dense 90-150 meters thick.

There are no beds Silurian in age in the Tetons. The Devonian beds are:

> (5) Darby formation—marine, dolomite and shale, 60-150 meters thick.

The Mississippian beds are:

> (6) Madison Limestone—marine, blue-gray, some red shale, 300-370 meters thick.

The Pennsylvanian beds are:

(7) Amsden formation—marine, light gray sandstone.   } 200-500
(8) Tensleep formation—marine, dolomite, sandstone, shale.   } meters thick

The Permian beds are:

(9) Phosphoria—marine dolomite, shale, phosphate, 50-75 meters thick.

All these Paleozoic beds can be found on the north and west flanks of the Tetons. They form the peaks of such ranges as Mounts Bannon, Peak, Fossil, and Red. They were shallow water deposition in an advancing sea over the eroded Precambrian surface. They are found mostly west of Webb Canyon and south of Granite Canyon.

5. Deposition of Mesozoic beds around north end of Tetons and parts of Jackson Hole. After Nugget Sandstone tilting southward, after Meeteetse, formation Teton-Gros Ventre uplift.

Triassic

(10) Dinwoody formation—marine, siltstone, 60-120 meters thick.
(11) Chugwater formation—partly marine, siltstone, shale, marine limestone 300-500 meters thick.
(12) Nugget Sandstone—continental, red sandstone, 0-11 meters thick.

After the Nugget was deposited the area was tilted and the Nugget was removed by erosion from the northern portion of Jackson Hole.

Jurassic

(13) Gypsum Spring formation—partly continental, partly marine, gypsum, shale, dolomite 240-300 meters thick.
(14) Sundance formation—marine, sandstone, shales, limestone, 150-200 meters thick.
(15) Morrison formation—sandstone and claystone, 200 meters thick.
(16) Cloverly formation (Morrison included)—continental, sandstone, claystone, 200 meters thick.
(17) Thermopolis shale—marine, black shale, sandstone on top, 50-60 meters thick.
(18) Mowry shale—marine, shale, has fish scales and bentonite, 200 meters thick.
(19) Frontier formation—marine, sandstone, shale, bentonite, 300 meters thick.
(20) Cody shale—marine, shale, green sandstone, bentonite, 400-760 meters thick.
(21) Bacon Ridge Sandstone—marine and continental, sandstone, coal, shale, 280-370 meters thick.
(22) Lenticular Sandstone, Shale and Coal Segments—continental, sandstone, shale, coal, 1-100± meters thick.
(23) Mesaverde formation—continental, sandstone, coal, shale, quartzite, sandstones, 0-300 meters thick.
(24) Meeteetse formation—continental, sandstone, siltstone, coal, bentonite, 0-200 meters thick.
(25) Harebell formation—continental, sandstone, siltstone, shale, quartzite, pebble conglomerate, 0-1500 meters thick.

While the Harebell was being deposited about 80 million years ago the ancestral Teton Gros Ventre Arch was formed. It trended southeast and formed during the Laramide Revolution. The Targhee uplift occurred at the northwest end of the arch.

6. Folding of Precambrian, Paleozoic and Mesozoic rocks into a southeast trending arch that was continuous with the Gros Ventre range between 65 and 80 million years ago (Laramide Revolution).

As the uplift reached its maximum elevation, quartzite was stripped and deposited as in Pinyon conglomerate during the Eocene. Another fold was the beginning of the Wind River Range (to the southeast). This is the conglomerate found on top of Mount Leidy. It is also found at Pinyon Peak Highlands (type area), the east section of Jackson Hole, and the northern part of the Tetons.

Paleocene $\left\{\begin{array}{l}\text{(26) Sandstone and Claystone—continental, sandstone, claystone, quartzite conglomerate,} \\ \qquad \text{coal at base, 0-1,200 meters thick.} \\ \text{(27) Pinyon conglomerate—continental, sandstone, claystone, quartzite conglomerate, coal} \\ \qquad \text{at base, 0-1,200 meters thick.}\end{array}\right.$

Other areas also contributed material to form this conglomerate; (1) Northwest of the Gros Ventre uplift another asymmetrical fold formed, the southwest side underwent thrust faulting. This is also when the Washakie Range just northeast of the park boundaries was uplifted and thrust faulted to the southwest; (2) The Targhee uplift (northwest of Tetons); (3) Ancestral Gros Ventre uplift was raised and folded asymmetrically with the southwest side being steep.

Eocene $\left\{\begin{array}{l}\text{(28) Indian Meadows—claystone, siltstone, sandstone, varigated, coal 30-90 meters thick in} \\ \qquad \text{the center, tuff-hornblende, andesite composition, 0-1,000 meters thick.} \\ \text{(29) Wind River—claystone, siltstone, sandstone, varigated, coal 30-90 meters thick in the} \\ \qquad \text{center, tuff—hornblende, andesite composition, 0-1,000 meters thick.} \\ \text{(30) Aycross formation—rhyolite and andesite came from Absorka Mountains, claystone,} \\ \qquad \text{siltstone, sandstone, bentonite quartzite conglomerate, 0-150 meters thick.} \\ \text{(31) Tepee Trail formation—andesite and basalt came from vents at east edge of Jackson} \\ \qquad \text{Hole, volcanic conglomerate tuff, claystone, 0-300 meters thick.}\end{array}\right.$

During the Eocene the uplifting was produced by unrest of the crust, faulting weakened areas and volcanic activity began. Material came from the nearby Absarokas that form one side (east) of Yellowstone National Park, some came from the Jackson Hole area.

A series of thrust faults occurred southwest of the park shoving the beds over the newly formed conglomerates. This thrusting produced a series of mountain ranges that all trend northwest-southeast. They are the Hogback Range, the Wyoming Range, Salt River Range, and Snake River Range.

East of the park, Tripod Peak was elevated 2,600 meters by a northwest-southeast trending thrust fault that was shoved to the southwest. It was right after this that the volcanic activity began.

While volcanic activity was occurring in the Absarokas, a portion of Jackson Hole (south central) was raised and the newly deposited Pinyon conglomerate was stripped and redeposited on the other side (south) of the Gros Ventre Range as part of the Aycross formation. All these events occurred during the Eocene, some 38 to 54 million years ago.

7. **Intermittent volcanic activity up to 25 million years ago. Oligocene—Miocene time. More uplift and thrust faulting.**

Oligocene $\left\{\begin{array}{l}\text{(32) Wiggins formation—andesite and basalt vents along NE side of Jackson Hole, mostly} \\ \qquad \text{agglomerate and ash, volcanic conglomerate, tuff layers, 0-1,000 meters thick.}\end{array}\right.$

Miocene $\left\{\begin{array}{l}\text{(33) Colter formation—pyroxene andesite basalt vents in north and east parts of Jackson} \\ \qquad \text{Hole, volcanic conglomerate tuff, sandstone, some basalt and andesite frag-} \\ \qquad \text{ments—0-2,100 meters thick.}\end{array}\right.$

The volcanic activity continued but most of the material this time came from the Jackson Hole region, primarily along the northern and eastern sides. Conglomerates, tuffs accumulated along with andesite and basaltic flows during the Oligocene (26 to 38 million years ago). These beds are called the Wiggins formation. All this time the processes of weathering were occurring, streams formed and transported sediments for we find these fragments mixed in with the overlying Miocene age Colter formation. The Colter also has volcanic material and can be seen along Pilgrim and Ditch creeks.

The Teton-Gros Ventre region elevated, with the eastern section being pretty well stripped of Cenozoic rocks. With all the volcanic activity occurring along the north side of Jackson Hole the region sagged several thousand meters.

Near the Two Ocean area a large lake had formed and some 7,000 feet of volcanic sediments deposited by water, had accumulated.

8. **Folding and thrust faulting (Cache Thrust). Fresh water lake formed in Jackson Hole region between eight to ten (Pliocene) million years ago, accompanied by volcanic activity.**

Pliocene
{
(34) Camp Davis formation—conglomerate, tuff, diatomite, limestone, claystone. 0-1,700 meters thick.

Rhyolite Tuff—Yellowstone.

(35) Teewinot formation—limestone, tuff, claystone conglomerate. Rhyolite flows Jackson Hole 0-1,800 meters thick.
}

At the north end of the Tetons, folds developed trending north. Gros Ventre (southwest end) thrust–faulted over lower Pliocene beds and other thrust blocks. This fault called the Cache Thrust slid to the northeast through Jackson, Teton Pass and into Idaho.

The Camp David formation with its conglomerates (exposed at south tip of Jackson Hole) tuff and diatomite reflects some of the unrest. The diatomite accumulated in the Teewinot Lake produced by faulting (normal) near the Gros Ventre Buttes. Above these beds and resting on Blacktail Butte, the hills south of Kelly, Antelope Flat (east margin) is the Teewinot formation deposited in this lake. Flows from both the Yellowstone area and Jackson Hole reflect the restless situation.

9. Arch broken forming N-S fault blocks (Teton fault) during Pliocene (eight million years ago) west block uplifted, east block depressed, both tilted westward.

During the Pliocene the Teton-Gros Ventre arch was broken by the north-south trending Teton fault in a ten mile wide belt that completely shattered the area. Blacktail Butte with its lake sediments on top and the East and West Gros Ventre Buttes are sections of this zone. The east block moved down and were promptly overridden by limestones (Cambrian) which covered the middle Pliocene beds (southwest side of Snake River Range). The west block moved up and both tilted west and the total displacement to this day in the eight or nine million years that have passed is around 9,000 feet (3,000 meters). The movement has been unequal. Sometimes a few feet in a matter of minutes, other times an average of 1/10th inch per year.

Basaltic flows extruded on the eastern side of Jackson Hole, probably moving along the new fault surface.

Pliocene or (?)
Pleistocene
{
(37) Bivouac formation—conglomerate with pebbles of Precambrian rock, rhyolite tuff. 0-300 meters thick.

Rhyolite Welded Tuff.
}

The age of the Bivouac formation is uncertain, it is either Pliocene or Pleistocene and is a conglomerate of Precambrian pebbles eroded from the newly uplifted beds. These beds outcrop on Signal Mountain and the West Gros Ventre Butte.

Rhyolite tuffs possibly from Yellowstone blanketed the area forming the welded tuffs today. They are found all around the Tetons (west, north and northeast) also on Signal Mountain and eastern side of Antelope Mountain.

Some intrusive rocks occurred as plugs along with both acidic and basic lava flows. They are scattered north and west of the Tetons, by Teton pass on the Gros Ventre Buttes and at Flat Creek (northside). Thickness is variable.

10. Uplift and displacement approximately 3,000 meters and is still occurring.

At the southern end of Jackson Hole a hogback developed by normal faulting dropping another 3,000 meters (9,000 feet). The central section continued to tilt west along the fault.

11. Erosion of Paleozoic and Mesozoic sediments from main peaks exposing Precambrian igneous and metamorphic rocks. Uplift created new lake.

With the uplifting creating peaks and the sagging in the Jackson Hole region a lake formed and in it were deposited during the Pleistocene was the Lower Lake sequence that now make up the low hills by Kelly (south).

Pleistocene
{
(38) Lower Lake sequence—shale, claystone, siltstone.

Some (lower) deposited in deep water.

0-60 meters thick.

Pyroclastic debris, Andesite intrusions.
}

Volcanic activity occurred in adjacent areas.

12. **At same time Buck Mountain fault thrust a wedge, of east front of range, above the level of the Teton block and forming peaks (three Tetons and Mt. Owen).**

The Jackson Hole area continued to sag and tilt. The region was raised another 300 meters (1,900 feet) tilting the Lower Lake sequence. The Buck Mountain fault thrust raised a wedge of rocks.

A second lake was created and the Upper Lake sequence occurred from 0-150 meters (0-500 feet). They are a mixture of shale, sandstone and conglomerate. An outcropping occurs in the valley of the Gros Ventre River.

13. **Glaciation several times, some nearby volcanic activity.**

The oldest glacial deposits vary in thickness, composed mostly of sand, gravel, and silt, that make the moraines difficult to detect. The best are exposed on Timber Island and on slopes below 8,000-9,000 feet elevation.

The Bull Lake deposits and outwash, range up to 120 meters (350-400 feet) and are mostly moraines. They are mostly found in the southern portion of Jackson Hole.

Loess deposits, the angular silt-sized particles picked up by the wind from the surface of the glaciers, range from 10 to 60 feet. About this time some faulting in the National Elk Refuse occurred offsetting some features.

14. **Pinedale glaciation and nuee ardentees around north and east ends of park forming hard volcanic rock. Additional tilting.**

Much of this came from the Yellowstone Park area, but some are from unidentified sources. Some flowed around the north end to the west of the Tetons and south at least 40 miles and others 25 miles down the east side. Depositing volcanic ash that fused into hard volcanic rock that closely resembles lava.

During the Pinedale glaciation the outwash plains were tilted westward because of additional movement along the faults.

15. **Erosion, landslides, shifting of continental divide.**

When the glaciers finally left 8,000 years ago the streams once again reclaimed the area and are in the process of stripping the area, landslides have occurred, talus is accumulating at the base of the slopes. Minor faulting is offsetting alluvial fans, especially in the National Elk Refuge. The glaciers present today appeared about 4,000 years ago. The geologic forces are still shaping Teton National Park.

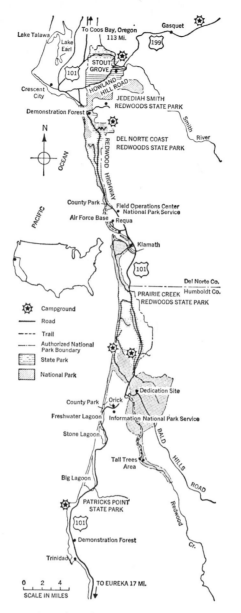

Redwood National Park, California.

Tentative Geologic Column of Redwoods National Park

Erathem	System	Coast Range Section		Klamath Section	
Mesozoic	Cretaceous to Jurassic	South Fork Mountain Schist			
		Dothan — ? — Galice	South Fork Mountain Franciscan		
					— ? —
— ? —	Triassic to Paleozoic			Western Triassic and Paleozoic Belt	Apple Gate Group (?) Stuart Fork Formation (?) Chancellulla Formation (?)
Paleozoic					

Modified after Bailey, Irwin, Jones et. al.

CHAPTER 23
REDWOOD NATIONAL PARK

Location: California
Area: 56,201.38 acres
Established: October 2, 1968

LOCAL HISTORY

This is the region of the Klamath Indians whose name means "the encamped." They used dugout canoes, mainly to gather the seed pods of the yellow water lily that was their staple food item. They never fought the white man, but were known to raid other tribes for females to be used as slaves.

According to legend, the Klamath River once had its outlet about six miles north (Wilson Creek). It was kept confined by high parallel bluffs. When Po-lick-a-quare-ick or "the Wise One" finished all his work on earth he got ready to leave. He collected his belongings and placed them in a canoe and began to paddle towards the sea. His worshippers (for he was the Indian equivalent of Christ) began to follow him. To prevent this when he reached the high bluffs he ordered them to part and the Klamath River flowed through the new opening into the ocean. By the time his people reached the sea he had disappeared into the setting sun. So they lost their Wise One, but the Klamath River has kept its new course ever since.

The Yurok, Tolawa and Chilula of more modern times inhabited the area. They used dugout canoes, and their main diet consisted of fish, especially salmon. They also ate elk and acorns. They made dwellings of redwood planks and also sweat houses. To down a tree for the acorns they would build a fire around the base.

Portola's Expedition of 1769 (Spanish) referred to these trees as "Palo Colorado" or "red trees." Juan Crespi (Franciscan Missionary) on this expedition was the first to write about them. During the 1794 expedition, Archibald Menzies realized they were a new species of tree. The name wasn't actually given until 1823 when A.B. Lam-bert called them *Taxodium sempervirens,* or evergreen sequoia. A German, Steven Endlicher felt it did not belong to the bald cypress group so renamed it *Sequoia sempervirens.*

Jedediah Smith was probably the first white man to venture into this area in 1828. He camped near Crescent City, his group averaged about a mile a day through this maze. The region was settled in the 1860s. Travel across the Kalamath River was by ferry until 1926.

Traces of an old split log stage coach road can still be seen by the Howland Hills Road in the Jedediah Smith Redwoods State Park section. Also Indian trails and mule trails from the miners are scattered throughout all the parks.

In 1918 the Save the Redwoods League was established with its main goal being the creation of Redwood National Park.

Studies were made in the 1920s and again in the 1930s, but nothing came of them. A bill in 1946, the Douglas Bill, tried to establish a memorial National Forest, but it did not pass.

When the National Park Service made a survey in 1963 and 1964, the frightening facts came out. Only 15 percent of the original two million acres of redwood trees still existed. Only 50,000 acres or 2.5 percent were preserved in state parks. At the present rate of logging, all virgin forests of redwoods would disappear (except the ones in protected parks) by the year 2,000. The tallest of these trees were also on private land and not protected.

Finally in 1968, by linking three state parks together (from north to south), Jedediah Smith, Del Norte and Prairie Creek, at a total of 27,770.99 acres plus 28,430.39 additional acres of private land, the park was created. The proclamation was signed by Lyndon B. Johnson. It is a mixture of residences and small business inside its boundaries, cutover forest, forest, beaches and bluffs. It is stretched out along the coastline for 46 miles with a maximum width of seven miles.

The largest Coast redwood is 367.8' high, with a diameter of 14 feet. It was discovered in 1963 by the National Geographic Society. It is found near Redwood Creek (southern end) on an alluvial flat in Tall Trees Grove.

These trees lived about 2,000 years ago and are called the coast redwoods, or *Sequoia sempervirens*. They grow today in a rather restricted belt some 15 miles wide and 450 miles long, along the Pacific Coast. Fossil forms have been found as far east as North and South Dakota. They also occur in Montana and Yellowstone National Park in Wyoming. These trees grow only in the fog belt where the temperatures are moderate, rainfall around 100 inches per year mostly during the winter and have dry foggy summers.

The term sequoia is derived from a misspelling of Sequoyah, a Cherokee Indian (1770-1843) who invented a phonetic alphabet for his language during a 12 year period. Once it was finished, the Cherokee nation learned to read and write in six months (after they were convinced of the importance of being able to read).

Redwoods grow rapidly, a 20 year old tree is about eight inches in diameter and around 51 feet high. They reproduce three ways, by seed, which is 1/16 of an inch in diameter (123,000 seeds per pound), by sprouts from cut stumps, and from roots and burls. The cone is the size of an olive and leaves are two ranked needles.

The seedlings sprout around the parent in circles forming groves. They have a shallow root system that only extends down four to six feet and forms a 80-100 foot circle, hence they are susceptible to being blown down. Sometimes their own weight is enough to cause them to fall over. The dull chocolate brown bark and reddish brown wood is very resistant to decay and fire. They can be hit by lightning, burned in forest fires, have tunnels drilled through them and survive. They are relatively immune to insects, their enemy is man. It takes 1,000 years for the tree to mature and only minutes to cut it down.

Jedediah Smith State Park was established in 1929 with a total acreage of 9,526 acres. It has a mixture of Coast redwoods and Ponderosa pine. Stout Memorial Grove has the tallest tress in the park. A popular tree in this portion of the park is the Boy Scout Tree. The Tall Trees Unit is under private ownership (lumber companies) but eventually will become a portion of the park. The average rainfall here is about 100 inches per year.

Del Norte Coast Redwoods State Park was founded in 1925, its acreage is 5,897 acres. It contains four memorial redwood groves (most of the memorial groves have been purchased by matching funds from the state and Save the Redwoods League). The steep slopes have trees growing almost to the shoreline, in a natural transition from redwood to shoreline plants. Part of it was cut over in 1927 so both virgin trees and second growth trees exist.

Prairie Creek Redwoods State Park was created in 1923, and contains 9,629 acres. The Coast redwoods grow with Douglas fir, Sitka spruce, Western hemlock, and broach leaved trees. This area approaches a rain forest and is also a sanctuary for the Roosevelt elk that is native to this region (herd of 200). Elevation varies from 1,500 feet to sea level. The average rainfall here is only 70 inches, hence a greater proportion of other types of trees.

Gold Bluffs is named because of the gold dust that is still mixed in the sand today. This is where the elk herds roam. A canyon called Fern Canyon, because of their abundance, is also found here. People have been panning since the 1850s, unfortunately there isn't enough to make it a worthwhile venture. The state parks are still under jurisdiction of the State of California, and not the United States Government, but will eventually become part of the Park Service.

GEOLOGIC FEATURES

The marine beaches are typical, steep cliffs and headlands, sand beaches, e.g., Gold Bluff in some of the protected areas. Also sea stacks, arches, tidal pools, and sand dunes.

Part of the park is located in the Coast Ranges that consists of Upper Jurassic and Cretaceous Franciscan rocks that are the eugeosynclinal facies of the Pacific Coast geosyncline. These beds are a mixture of graywacke, conglomerate, shale, minor limestone, chert, metamorphic rock and mafic volcanic rocks.

The rest of the park is located in the western portion of the Klamath Mountains, an area that has not been studied very well. The rocks are a complex mixture of both igneous or metamorphic rocks that are either Precambrian in age or Paleozoic. The lack of fossils make dating difficult.

Uplift is still occurring today which accounts for the frequent earthquakes in this region. This region is crisscrossed with faults that trend northwest-southeast. These faults, to a large degree, determine the shape of the drainage basins and directions of flow of the streams. Because these stream valleys lie parallel to the areas where most storms occur, flooding is common.

GEOLOGIC HISTORY

1. Subsidence and deposition during the Cambrian.

In the seas of the Cordilleran geosyncline which was forming; sands, muds and limestones were being deposited.

2. Subsidence, deposition and extrusion during Ordovician.

By this time the Cordilleran geosyncline had developed into its two components—the *miogeosyncline* located in Nevada and *eugeosyncline* in California. Extrusion of lavas from the island arc system occurred by the border of California and Nevada.

3. Deposition and extrusion during Silurian and Devonian.

During the Silurian, mostly subsidence occurred, with sand and mud being deposited, then uplift to the east in Nevada. By Devonian time the Antler orogeny began affecting western Nevada and southern California with the Roberts Mountain thrust fault forming. Extrusion occurred in the vicinity of the Coast Ranges, but deposition occurred around the Redwoods Park region.

4. Antler orogeny to the east, continued subsidence and deposition during Mississippian.

The Antler orogeny formed a mountain range that ran north-south through central Nevada and into southern California. This area acted as the source area for the sediments that were being deposited in the Redwoods National Park region, primarily sands and muds because of the uplifting. Some extrusion of lavas took place during Meramac time in the Coast Range area.

5. Mountain building and erosion during the Pennsylvanian.

The effects of the Antler orogeny was dying out but not before a mountain range formed in northern California. It was stripped of its Paleozoic sediments about as fast as it rose, depositing sediments both east and west of the range. Some extrusive activity occurred in the region of the Cascade Mountains.

6. Subsidence and deposition occurred during the Permian plus the Sonoma-Cassiar orogeny along the coast and in southern California. Extrusive flows in park area.

By the time the Permian arrived the mountains of the Antler orogeny had been completely worn

down. The area was depressed as the mountains from the Sonoma-Cassiar orogeny formed. Again these were stripped very rapidly so, even though total uplift was great, the mountains themselves were never very high and were eroded away by the end of the Permian. But not before some of the sediments were buried by lava flows along the coast in the area of the Redwoods National Park and in the Cascade Range.

7. **Nevadian orogeny just starting, reached a climax at the end of the Jurassic. Formation of the Southfork Mountain Franciscan group with igneous and extrusive beds.**

Over 135 million years ago the area was once again low. Deposition occurred forming a mixture of metamorphosed rocks, conglomerates, sandstones (graywacke), shales, and cherts. These in turn have been intruded by ultrabasic rocks. Many unconformaties have developed and facies changes are present. With the Nevadian orogeny the Mesocordilleran geanticline was beginning to form in Nevada during the Jurassic period.

During middle and late Jurassic time the Galice and Dothan formations were deposited. Because of lack of fossils, there is some disagreement as to which formed first, most authors place the Dothan before the Galice. The Dothan is primarily graywacke and shale, while the Galice is schist, in other areas it is a mixture of graywacke, slate and phyllite.

These sediments were deformed and metamorphosed locally into a mica schist and green schist (Southfork Mountain schist). They then underwent intrusion (granitic) before some of the late Jurassic-early Cretaceous beds were deposited above them (Myrtle formation or possibly the Franciscan).

Because of its complex history the Klamath Mountains have been divided into four belts they are: (1) Western Jurassic belt; (2) Western Paleozoic and Triassic belt; (3) Central Metamorphic belt; (4) Eastern Paleozoic belt. Only the western Jurassic belt occurs in Redwoods National Park.

8. **Laramide orogeny, continued deposition of Southfork Mountain Franciscan group.**

During early Cretaceous the Nevadian orogeny ended, the Mesocordilleran geanticline developed fully and underwent erosion. Dinosaurs wandered on this surface. In the area of Redwoods National Park the area was underwater receiving sediments from the geanticline as the area continuously subsided. Since it is an eugeosyncline the volcanics are not unexpected.

By the end of the Cretaceous the Laramide revolution began and the Rockies were born.

9. Formation of Klamath Mountains and Coastal Ranges after uplift during early Cenozoic. Appearance of redwoods during the Oligocene. Cascadian orogeny during the Pleistocene.

During early Cenozoic the area began to be gradually raised as the Laramide orogeny ended and the Cascadian orogeny began. The Klamath Mountains and Coast Ranges were formed by the uplifting and faulting of the Mesozoic intrusive beds which were both granitic and ultrabasic and are mixed in with metamorphosed rocks and metasediments. To the east, lava plateaus (Modoc) and volcanoes formed. About 30 million years ago during the Oligocene the Coast Redwood evolved and grew in great forests all along the coast of California and further east.

The bedrock of the park is therefore either the Franciscan group of the Coast Ranges or the Western belt of the Klamath Mountains. Some continental sedimentary deposits of sand and gravel occur. They are stream deposits and alluvial fans probably Pleistocene in age. Also some landslide deposits are found around Crescent City.

Once the redwood forests began the environment was changed into three areas because of the tremendous size of the trees. They are as follows:

a. Crown Area—Continuous canopy where certain insects and other forms of life never touch the ground. Mixed in with Douglas fir.
b. Understory Area—The open regions that contain the young redwood, spruce, hemlock, fir and cedar.
c. Duff Area—The material that covers the ground such as ferns and shrubs, e.g., huckleberry, rhododendron and ferns.

The park itself is divided into four units according to the environment. The Redwood Forest, the Marine and Shore, North Coastal Scrub (narrow strip of cliffs with poor soil, poor drainage and constant salty winds), and Cutover Forest. Combined together they create a unique park.

Sequoia-Kings Canyon
National Parks, California.

Generalized Geologic Column of Sequoia-Kings Canyon National Park

Erathem	System	Series	Formation
Cenozoic	Neogene	Holocene	Alluvium
		Pleistocene	Terrace Gravel
		Miocene (?)	Olivine Basalt Capping Friant
			Friant Formation
	Paleogene	Eocene	Ione Formation
Mesozoic	Jurassic	Late	Muscovite Granite
			Pyroxene Quartz Diorite
			Hornblende Biotite Quartz Diorite
—— ? ——	—— ? ——	Early	Hornblende Gabbro and Diorite
Paleozoic			Meta-Gabbro
			Serpentine
		Late	Metavolcanics
			Metasediments

Modified after Bateman, Clark, Huber, Kingsmore, Rinehart et. al.

CHAPTER 24
SEQUOIA NATIONAL PARK

Location: S.E. California
Area: 386,862.97 acres
Established: September 25, 1890

LOCAL HISTORY

Early prehistoric Indians hunted in this region. A rock called Hospital Rock has many petroglyphs on them. The Yokut Indians that lived there in the 1800s knew of the pictographs but didn't know what they meant. The Indian camp nearby has mortar holes in the rock where the Indians ground acorns.

The trees were first discovered in 1833 when a party led by Joseph R. Walker crossed the mountains. The clerk from the group, Zenas Leonard, made notations of the sizes of some of these trees. But their find was not publicized.

In 1845 a section of the third Fremont Expedition traveled along Pu-sun-co-la (Indian name) in its explorations. It is now known as Kern River (named after the topographer).

In 1852 A.T. Dowd, a miner, was following a bear that he wounded and rediscovered the trees. He brought his friends to see them and so received credit for discovering them. It may have been this tree that was later cut down and had a dance hall on its stump. This was the Calaveras Grove.

Hale D. Tharp was looking for a ranch site in 1856. He was from nearby Three Rivers and got along quite well with the Indians. In 1858 Chief Chappo of the Potwishas (Yokut ?) and other Indians took him along a trail that went under Moro Rock and showed him the trees. He was the first to see the Giant Forest. For some 30 years he used the area as a summer range. He built a cabin out of a tree in which he had carved his name and date of discovery. The room was 56 feet long and eight feet high. The front of the log had a door, window and stone fireplace fitted into it. It has been preserved and can be visited today. Actually John Muir gave the name Giant Forest in 1875 to the grove which Tharp discovered.

The Indians suffered with the arrival of the white man who brought with them measles, small pox and scarlet fever killing most of the Indians off. They even requested their friend Tharp to ask the settlers to leave. When Tharp told Chief Chappo and his warriors that the settlers refused to go, they actually sat down and cried. This was in 1862, three years later they retreated to the mountains for survival.

William H. Brewer of the United States Geological Survey led a team which mapped the area in 1861. In 1864 a Mr. Clarence who was on the mapping team crossed the Kings—Kern divide and climbed Mount Tyndal (14,018 feet).

In 1873 A.H. Johnson, John Lucus and C.D. Begole were the first to climb Mount Whitney, the highest mountain in the United States (14,495 feet).

James Wolverton, in 1875, a trapper was the discoverer of the tallest tree. He named it after his commander from the War of the States—the General Sherman Tree. The tree is 272.4 feet high with a circumference of 101.6 feet giving it a diameter of 36.5 feet (would have been taller if it hadn't been struck by lightning).

Atwell Mill was built in 1879 to cut the trees into lumber. Remains of the mill and the home of the mill operator still can be seen today.

In 1885 members of a socialistic group filed 55 claims of 160 acres each, to the land above Marble Fork of the Kaweah River. This was to be their Utopia. They formed the Tulare Valley and Giant Forest Railroad Company. The plan was to lumber the sugar pine and yellow pine and to develop the land. After visiting giant grove they decided not to cut the sequoia trees but still planned to cut other varieties. They named many of the trees after their heroes. The government refused to grant a title to them until an investigation was held. Law suits were finally enacted so they could have title to their land.

In 1886 they reorganized forming the Kaweah Cooperative Commonwealth Colony. They built a

sawmill and a railroad and called their community Colony Mill. They began to cut down trees (which didn't belong to them) and put them through the saw mill.

George W. Stewart (Visalia editor) also known as "The Father of Sequoia National Park" and Guslav Eisen (California Academy of Sciences) found out about the mill and promptly went to work to save the trees. Their efforts culminated in the establishment of Sequoia National Park in 1890. The cavalry was sent to expel both the colonists and the settlers. The colonists left without too much effort on the part of the cavalry. But the ones to give the army trouble were the settlers, who for the last 30 years have been using this land as if they owned it. It was used for grazing of cattle and sheep and hunting.

The other assignment of the cavalry was to administer the park. Unfortunately they had no legal way to punish any offenders. Captain Dodst was the first administrator and also had nearby General Grant National Park (now part of King National Park) to care for. Just as he was getting things to work smoothly the Spanish-American War began and they were withdrawn. No sooner than the

troops left when the settlers started grazing and hunting again. So troops were sent out again (First Utah Volunteer Company for one month until the Regulars could arrive). It stayed under military rule until 1914 when civilian administrators took over.

GEOLOGIC FEATURES

Sequoiadendron giganteum (Sequoia gigantea) has only two living relatives the *Sequoia sempervirens* or Coast Redwood and the dawn redwood *Metasequoia glyptostroboides* discovered in China in 1944. The sequoia trees are evergreens while the metasequoia sheds its leaves.

The trees grow in a belt about 250 miles long between 4,000 and 8,000 feet in elevation on the western slopes of the Sierra Nevadas only. They also grow on the oldest soils and do not grow on Pleistocene soils. They are scattered in groves rather than growing as a single forest.

These trees reproduce by seeds only. The cones are larger than the redwoods (3x) between 1.5 and 2.5 inches. A dark gritty substance can be found in the cones and lower part of the trunk which, when dissolved in water, becomes a purple ink.

The trees bloom in the winter (February or March) with a yellow flower. It takes the cones 1 1/2 years to mature. They have 36 overlapping scales, that protect two to six seeds (about 1/4 × 1/4 inch in size). The bark is reddish brown, the wood is coarser, texture is different and narrower rings form than the redwood. There is no tap root only a broad root system that extends all around for about 200-300 feet. This is one of their weakest points, these roots are on and near the surface so are slowly being killed by tourists walking on them since they cannot stand the bruising.

Man in his desire to protect the trees has not permitted forest fires to occur and this is detrimental since it has allowed thickets of young pines, incense cedar and white fir plus bushes and shurbs to cover all available space. The seeds have to fall on a clear space with sufficient moisture or they can't grow. Fires set off by lightning are actually beneficial to the trees. The Indians used to burn this debris off also. This does not harm the trees as the spongy bark that is up to two feet thick pretty well fireproofs the tree. Controlled burns are being tried in some regions to permit the young trees to have a place to grow.

For the first 75 years growth rate is an inch every three to five years, after that it takes 20 years to add an inch. Young trees are cone shaped but when they become 250 to 300 feet high they shed the lower branches and develop large limbs. As long as the narrow yellow band just beneath the bark called the *sapwood* remains the tree can survive. This is why it can actually grow over burned areas. The large amounts of tannin in the tree is its secret of immunity against insects, fungus and other diseases. No tree has yet been found dying of old age and some are close to 4,000 years old.

Volume wise they are the largest trees. General Grant is the largest known sequoia tree, it is 272.4 feet high, 101.6 feet around at the base and has a volume of 49,600 cubic feet. It is located in the Giant Forest Grove in Sequoia National Park.

An unexpected feature of the park are the *caves* formed by the solution of the marble. Crystal Cave is open to the public. The temperature is a constant 50 degrees, that is four degrees colder than Mammoth Cave in Kentucky. The reason for this, is a cave reflects the average annual temperature of the region, and Mammoth Cave is located further south. It wasn't discovered until 1918 (28 years after the park was created) and was not opened to the public until 1940. Crystal Cave contains the usual cave features, stalactites, stalagmites, columns, solution pits, draperies and pools. Collapse blocks can also be found.

Moro Rock and Alla Peak, are exfoliation domes as are Beetle Rock, Sunset Rock and The Watchtower (a granite pinnacle, 2,000 feet high). The Table Meadows are an exfoliated tableland. The lakes are mostly tarn lakes, an exception to this is Moraine Lake (Chagoopa Plateau) for it lies entirely within a moraine and is impounded by a moraine loop. Franklin Pass (near Mineral King) has a sand made up from fragments of granite. Kern Valley follows an ancient fault that hasn't moved in the last 3.5 million years. Many of the small gulches follow the master joint system.

There are both glacial U-shaped valleys, e.g., Kaweah Canyon and also V-shaped stream valleys such as some of the tributaries to the Kern River. The hanging valleys in Kern Canyon are deeply trenched back into the rim in an attempt to cut down to the floor of the Canyon, e.g., Wallace Creek and Whitney Creek. Crescent Meadow and Bearpaw Meadow are alpine meadows created by glacial action. Sand Meadows was deposited by meltwater streams.

Diamond Mesa and The Boreal Plateaus, Cirque Peak and Mount Whitney were never glaciated because the high winds would not permit enough snow to accumulate on them.

Avalanche chutes are a common feature in Kern Canyon, Bearpaw Meadow and on Mount Whitney. Some cirques (and canyon walls) have so many of these chutes that the sides appear scalloped or fluted.

A *Rock Glacier* can be seen in Kaweah Basin, it is occupying a cirque of a former glacier. The moraines from the glacier can still be seen on the sides of the valley.

The drainage system of Sequoia is different than that of Kings Canyon even though they are adjacent to one another. The reason is the secondary mountain system called The Great Western Divide that runs from north to south and literally splits the park in half. The Kaweah River on the west drains to the southwest (same as Kings River) while Kern River on the east side of the divide drains due south. This parallels the crest and it does not turn southwest as it should for 75 miles. Part of the reason for its southerly flow is that the Kern River flows along a fault system and is thus confined by it.

The Great Western Divide is a true mountain range and is sometimes mistaken for the crest of the Sierra Nevadas. Table Mountain, a portion of this divide, has volcanic flows capping it, thus the reason for its name. It is part of the Whitney erosional surface that formed before the Sierra Nevadas of today. Remnants from the first and second glacial periods are such features as badly weathered erratics, e.g., Bighorn Plateau, Little Cottonwood Creek. Outwash can be found along Generals Highway.

Granular disintegration and *spheroidal weathering* is obvious at the Siberian Outpost where the granite is falling to pieces grain by grain, and weathering layer by layer inward because the granite has been split by frost action into blocks. Granular disintegration and chemical weathering are very evident along the Generals Highway because the solid looking rock is so soft it can be shoveled away.

GEOLOGIC HISTORY

See King's Canyon.

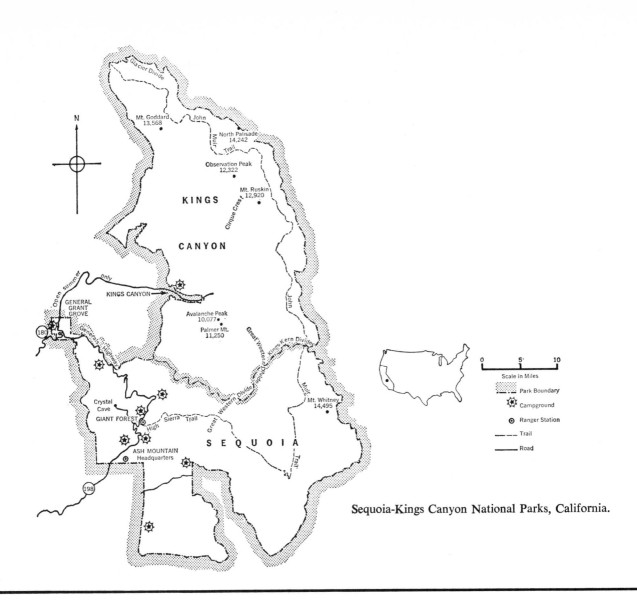

Sequoia-Kings Canyon National Parks, California.

A List of Specific Rock Formations in Sequoia-Kings Canyon National Park
The Age Relationship is Uncertain

Igneous Intrusions	Metamorphic Rocks
Alaskite	Amphibolite and Amphibole Schist
Ash Mountain Complex	Marble
Big Baldy Group	Metavolcanics
Big Meadow Pluton	Mica-Feldspar-Quartz Schist
Cactus Point Granite	Quartzite
Clover Creek Granodiorite	
Cow Creek Granodiorite	
Elk Creek Gabbro	
Giant Forest Pluton	
Lodgepole Granite	
Pear Lake Quartz Monzonite	
Potwish Quartz Diorite	
Tokopan Porphyritic Granodiorite	
Weaver Lake Group	

Modified after Ross et. al.

CHAPTER 25

KINGS CANYON NATIONAL PARK

Location: S.E.California
Area: 460,330.90 acres
Established: March 4, 1940

LOCAL HISTORY

Spanish explorers in 1805 crossed King's River on a Sunday and named it "Rio de los Santos Rexes" or River of the Holy Kings which is now shortened to Kings River.

The Indians did not live in Kings Canyon but only hunted for the simple reason that few oak trees grow there and acorns were a staple in their diets.

A few settlers used this area for grazing sheep and cattle. Miners found copper in 1888 but not enough for commercial mining. This was also one of the favorite areas of John Muir. The John Muir Trail follows the crest of the Sierra Nevadas. The original park was called General Grant National Park and was created on October 1, 1890 when the proclamation was signed by President Harrison. General Grant National Park was abolished and became part of King's Canyon National Park in 1940. Its name was changed to General Grant Grove and is a detached section of the park.

The sequoia trees in General Grant Grove were discovered in 1862 by Joseph Thomas. In 1867 Mrs. L.P. Baker of Visaliz, California visited the Big Grove areas of Kings Canyon and discovered one tree bigger than the rest. She named it The General Grant Tree and even sent some small branches of it to General Grant when she informed him of this. It is still the tallest tree in Kings Canyon National Park being 267.4 feet high, 107.6 feet in circumference around the base and a volume of 43,038 cubic feet. It is also the Nation's Christmas tree (set aside in 1925) and a memorial to veterans. It is the second tallest tree in the world.

In 1870 settlers came in the area. One fallen log called the *Fallen Monarch* started out as a home, then a saloon, then became a restaurant run by a widow, and finally a stable for the United States Cavalry (it held 32 horses). In 1872 the Gamlin brothers built a log cabin that is still preserved today.

A tree was cut down in 1875 to ship the base to Philadelphia for the 1876 exposition. However, everyone thought it was a fake. The stump is called Centennial Stump.

GEOLOGIC FEATURES

Most of them are glacial—horns, aretes, cirques, tarn lakes, cols, glacial U-shaped valleys. Avalanche chutes are common. They usually develop in the massive granite that is not fractured, but will also develop in the granite if it is evenly fractured.

Rock glaciers form in the higher altitudes where there is a lack of water, if it occurs it is more likely to be frozen. Since there is no running water, the talus and other material loosened by mass wasting, accumulates on the floor of the valleys. Water becomes trapped in the spaces between the rocks and freezes. When water freezes it expands, when it expands it pushes the adjacent rock a little further down the slope. So large masses of mostly rock and some water gradually creeps down the slope like a glacier. It even has an outward appearance of one and follows the same path down the slope.

There are also caves formed in the marble which has the same composition as limestone, for it is metamorphosed limestone. The caves contain the usual stalactites, stalagmites, columns, dripstone, etc.

Most of the streams trend southwest. They flow down the west slope, as is to be expected, unlike those in Sequoia National Park. The streams in Sequoia National Park flow in several directions because it has more than one divide.

Exfoliation is seen here more than it is seen in Sequoia National Park because of the type of granite in that portion of the park. Good examples are Tehipite Dome, Kettle Dome and North Dome.

Glaciation was more extensive in the geologic past for example, Kings Glacier (Pleistocene), in spite of its low altitude, was 44 miles long and flowed to elevations as low as 2,500 feet above sea level. This indicates the tremendous volume of snow that had to accumulate. It has left its marks everywhere.

Kings Canyon, of course, is a U-shaped glacial valley, while the Middle Fork of Kings River is V-shaped. The canyon has two beautiful hanging valleys with waterfalls (Roaring River Falls and Mist Falls).

GEOLOGIC HISTORY

1. Deposit of sediments in Cordilleran geosyncline during the Paleozoic.

The Cordilleran geosyncline formed sometime around Cambrian time. It existed until the end of the Mesozoic. Conglomerates, sandstones, shales and limestones were deposited in the miogeosynclinal portion and graywackes and volcanics in the eugeosynclinal. Remnants can be seen today as roof pendants within the batholith. The most complete section in the eastern Sierra Nevada is the Mount Morrison roof pendent, an area about 40 miles north of Kings Canyon National Park. It contains some 32,000 feet of Ordovician to Permian (?) beds.

2. Uplifting and folding of the first set of Sierra Nevadas at the end of the Permian and beginning of the Mesozoic.

About 225 million years ago these beds were uplifted and folded into a mountain range. They were predominately of the miogeosynclinal facies. Towards the end or during the folding volcanic activity occurred.

3. Erosion and wearing away of mountains, then subsidence.

As soon as the mountains began to rise the erosional forces took over and began to erode the mountains, gradually they developed an erosional surface. The seas gradually transgressed and buried this folded set of mountains that were eroded down to their roots.

4. Formation of the Pacific Coast eugeosyncline and Rocky Mountain miogeosyncline with the intervening Mesocordilleran geanticline.

There was deposition of sand, silt and limy ooze in the depression. Because of the depression

of this region the adjacent region was uplifted and acted as a source area. The crustal unrest produced all types of volcanic activity. Volcanic debris is mixed in with the detrital material. There is an unconformity between the Paleozoic and Cenozoic beds. In some regions Triassic beds are completely missing.

5. Uplift and folding near end of Jurassic-Cretaceous time forming second set of Sierra Nevadas.

When the geosyncline filled, the beds were uplifted a second time (Nevadian orogeny) and the second set were formed. They were folded into a snyclinorium which was complexly faulted. The structure of these new beds had a northwest trend. Volcanic activity increased with andesites, tuffs, pillow lavas, basalts and volcanic breccia forming. Several things happened: first, *regional metamorphism;* the sandstone became quartzites, the shales—phyllites and slates, limestone—marble, the volcanics—metavolcanics. Where the deformation was the greatest mica-feldspar-quartz schists, amphibolite schists and amphibolites formed. Secondly the location of the axis of the synclinorium became a weak zone, it lies where the Sierra batho-

lith now stands. The eugeosynclinal sediments are to the west, with the miogeosynclinal to the east. Third, both strike faulting and thrust faulting have totally disrupted the sequence. The original axis probably trended N 40° W. The series of faults that form our present day range cut across these structures. The majority of deformation occurred during the Nevadian orogeny but additional deformation occurred both before and after. Major deformation happened before the intrusion of the batholith.

6. Intrusion (repeatedly) forming the Sierra Nevada batholith.

Apparently the first intrusions were basic in composition for they are preserved in the pendants which is all that is left of the original bedrock. Compositionwise, these intrusions were quartz diorite, diorite and hornblende gabbro.

The batholith occurred as a series of *plutons* (general term for any igneous intrusion). Sharp contacts can be found between these plutons or they are separated by a thin wall of metamorphic rocks. Most of them are parallel to the long axis of the batholith. Radioactive dating indicates the intrusion occurred during three main periods. Late

Triassic-Early Jurassic (between 183-210 million years; Late Jurassic (124-136 million years) and early Late Cretaceous (80-90 million years). The first set are concentrated on the east side of the batholith, the second along the west side and the last (a belt 25 miles wide) along the crest.

Evidence indicates this material crystallized out from a melt and being lighter material was squeezed gradually upwards. The composition is variable being alaskite (light colored granite), granite, quartz monzonite, granodiorite, and quartz diorite.

The inclusions are of two types, *xenoliths* that are sections of country rock not completely assimilated or *autoliths* which are fragments of the chamber that have crystallized earlier and are incorporated within the later material. This is the category to which the basic inclusions that are found belong. Many of them are stretched out indicating they were soft and were shaped by the flowing and intruding magma.

Lineation occurs, zoning and other features associated with igneous intrusion can be found. They vary from intrusion to intrusion. The joints however, cut across the individual plutons indicating they formed after cooling. Two large pendants which are remnants of the original bedrock are traceable, they are the Bishop Creek and Mount Goddard.

It was at this time that the quartz veins containing gold were injected into the bedrock forming the Mother Lode deposit which was responsible for the gold rush of 1849. Copper, lead, and zinc also formed.

7. Erosion stripped sediments from the batholith and reduced the area to a lowland with northwest trending hills.

During the Cretaceous and Cenozoic the mountain system under went a long period of erosion which stripped between 7 and 12 miles of sediments off the batholith at the rate of 1/2 to 1 1/2 feet per thousand years. All that is left are remnants of the metamorphosed Precambrian, Paleozoic and Mesozoic sediments as roof pendants and occasionally as xenoliths.

The stability of the land, the long span of erosion reduced the mountains down to an erosional surface that had low hills that trended to the northwest. In some areas deposition of the Chico (?) (Cretaceous) formation followed by the Ione (Middle Eocene) occurred. If they existed, they have since been stripped by erosion.

8. During the Paleocene a series of four major uplifts, elevating the range (50 mile by 400 mile block) several thousand feet. Volcanic eruptions.

The four erosional surfaces that can be seen in Sequoia-Kings area are:

a. Cirque Peak Erosional Surface 13,480 feet 5-10 million years ago.
b. Boreal Platform Erosional Surface 11,500 feet   20 million years ago.
c. Chagoopa Erosional Surface 9,000 feet 10-15 million years ago.
d. Canyon-Cutting Stage 7,000 feet   1 million years ago.

The stepped topography in canyons which form benches testify to the periodic uplift. Changes in gradient of the streams indicate the rate of tilting (some 5,000 feet in the last 9.5 million years). Volcanic centers spread basalts, rhyolites and ash flows on nearby area, starting with the middle Oligocene (33 million years ago) and ending with the late Pliocene (five million years ago).

After the second uplift a great deal of material was stripped away and added to the sediments on the valley floors.

9. Between each uplift, erosion produced deep broad valleys in the western flank.

The Kern valley was cut to almost its present depth some 3.5 million years ago. The Chagoopa Plateau is slightly over 2,000 feet above the present floor. The Boreal Plateau is the level upon which Funston Lake lies. Kern Canyon is straight because it lies along a fault surface. During the Pliocene the region (western margin) was tilted 1°30″ to 5° west (dip decreases eastward). It was at this time that King, Kern, and San Joazuin rivers cut their inner gorges. Most of the cutting occurred between 25 and 30 million years ago. The mountains were between 3,000 to 5,000 feet high at their crests. Climate was warm and tropical. This is also when the previously mentioned volcanic activity occurred.

10. Vertical faults along eastern front of range with depression of adjacent block. e.g., Owens Valley.

Around middle Pliocene the Sierra Nevadas underwent major tilting. A series of vertical faults occurred, that parallel the eastern side of the Sierra Nevada Mountains. A block 40-80 miles and more than 400 miles long was tilted to form an impressive mountain range. This, of course, resulted in the adjacent block being depressed, thus creating structures such as in the Owens Valley area that is presently being filled with alluvium. This is responsible for Mount Whitney being elevated some 11,000 feet above the floor of Owens Valley. The faulting lagged behind the tilting, which is to be expected. In all probability a large monoclinal fold (single flexure with beds above and below parallel) formed, when the stress and strain became too great a series of vertical faults occurred. Most of the movement ended about two million years ago. The oldest glaciation occurred before the faulting, the youngest after.

This block faulting created the set of mountains that we see today, so they are actually the third set of Sierra Nevadas. Before and after the final uplift, many of the quartz veins containing gold and other precious minerals were exposed to weathering and the material weathered out, was transported by running water and deposited as placer deposits. This process is still continuing today.

11. Glaciation during the Pleistocene, part occurred before faulting, rest after faulting.

When the Pleistocene occurred continental glaciers came down from Canada but did not reach this far south. However, the altitude was high enough that mountain glaciers formed.

Two glacial stages occurred on the west side of the Sierra Nevadas the Point Glacier Stage and Portal Glacial prior to the faulting. This was followed by the various Wisconsin stages and substages. On the east side occurred the Sherwin and McGee Glacial Stages prior to faulting and again the Wisconsin after.

Main glaciers occurred in Kern Valley and Keweah River Valley (Sequoia National Park) and Kings Canyon (Kings Canyon National Park). The original V-shaped stream valleys were modified and deepened by glacial action and changed to the typical U-shaped glacial valleys. The rapid erosion left many of the adjacent valleys (both stream and

glacial) high above the floor and thus became hanging valleys. Maximum extent of the glaciers was 10,500 to 12,000 feet.

The first two stages on both sides are difficult to detect since they have been covered by the younger deposits, they have undergone weathering and have been disturbed by faulting. Evidence for the Wisconsin stage is easily seen.

Differences in type of bedrock, joint systems, degree of weathering and location have effected the glacial process. Scattered all over the park are glacial grooves and striation, erratics, signs of glacial plucking and quarrying, cirques, tarn lakes, U-shaped valleys and moraines. Some of the higher areas because of lack of sufficient precipitation were not glaciated, e.g., Mount Whitney.

12. Development of modern landscape and growth of sequoia trees.

The Pleistocene glaciers left about 9,500 years ago when the climate became too warm. Finally about 2,000 to 3,000 years ago the climate once again became cool enough for the small mountain glaciers that exist today to develop.

Metamorphic rock still exists either as roof pendants or on the flanks of the composite batholith. Hence in the *marble* formations, that are metamorphosed limestones, several caves have formed. The best known is Crystal Cave.

Sequoia trees, of course, have been around since the Mesozoic, but the trees we see today are no older than 4,000 years to the best of our knowledge. The trees that are growing today do not grow on recently glaciated ground (20,000 years ago) although some of the glaciers were within five miles of the groves. The former presence of nearby glaciers and glacial action may help to explain the scattered distribution of the trees today.

When the glaciers left for the last time they modified the landscape to some degree. Stream erosion and mass wasting has furthered modified it to its present topography which translated means "snow, sawtoothed mountain range" which is a very apt description.

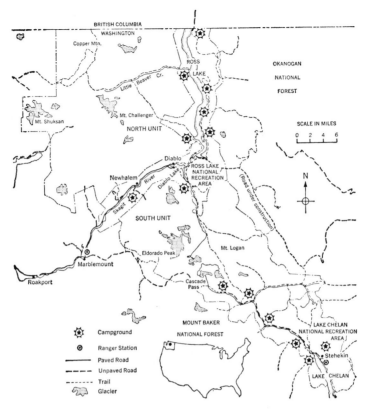

North Cascades National Park, Washington.

Geologic Column of East and West Belts (Geosynclinal Rocks) of North Cascades National Park

Erathem	System	Series	West Side	East Side	
Cenozoic		Oligocene	Hannegan (Skagit)		
		Paleocene	Chuckanut = (Swank)		
Mesozoic	Cretaceous	Early	Nooksack Group		
	Jurassic	Late		Jack Mountain Phyllite	
	Pre-Jurassic to Post-Permian	Late		North Creek Volcanics	Elijah Ridge Schist
Paleozoic	Permian to Devonian	Middle Middle	Hozomeen Group	Chilliwack Group	
	?	?	Marblemount	Eldorado	

Modified after Mish.

Geologic Column of Shuksan Metamorphic Suite
North Cascades National Park

Erathem	System	Complex	Formation	
Mesozoic (?) or Paleozoic (?)	Pre-Triassic Post-Mid-Devonian	Shuksan Metamorphic Suite	Shuksan Greenschist	
			Darrington Phyllite	
	Pre-Mid-Devonian	Yellow Aster Complex	Marblemont	Eldorado

Modified after Kendall, Miller, Mish and Bryant.

Geologic Column Methow Valley North Cascades National Park (Graben)

Erathem	System	Series	Formation	Canadian Equivalent
Mesozoic			Pipestone Canyon Formation	
	Cretaceous	Upper	Midnight Peak Formation	Pasayten Group
			Ventura Formation	
			Winthrop Formation	
	Cretaceous	Lower	Virginia Ridge Formation	Jackass Mountain Group
			Harts Pass Formation	
			Panther Creek Formation	
			Goat Creek Formation	
			Newby (Upper)	Dewdney Creek
	Jurassic	Upper	Newby (Lower)	

Modified after Barksdale.

Geologic Column of Main Core of North Cascades National Park

Erathem	System	Series	Chilliwack Composite Batholith Western Intrusive Zone	Eastern Intrusive Zone
Cenozoic	Paleogene	Upper Oligocene to Lower Oligocene	Late (Perry Creek) Phase	Perry Creek Intrusive
		or Late Eocene	Main Phase	Golden Horn Batholith
Mesozoic	Cretaceous (?)	Lower ?	Early Phase	Black Peak Batholith
			Undifferentiated	Ruby Creek Hetrogeneous Plutonic Belt
				Diorite Stocks

Modified after Kendall, Miller, Mish and Bryant.

Geologic Column Skagit Metamorphic Suite North Cascade National Park

Erathem	System	Complex	Formation	Time Sequence
Paleozoic (?)	Post-Mid-Devonian		Gabriel Peak Orthogneiss	Postskagit suite
			Alma Creek Intrusive	Latest Metamorphic Intrusion
			Haystack Creek Intrusive	Late Metamorphic Intrusion
			Skagit Gneiss	High Metamorphism
		Cascade River Schist	Isochemical Remnants in Skagit Low Grade Metamorphosed Zone Medium Grade Metamorphosed Zone	Medium Metamorphism
			Marble Creek Orthogneiss Intrusion	Premetamorphic Intrusion
	Pre-Mid-Devonian	Yellow Aster Complex	Marblemount Eldorado Meta Quartz Diorite Orthogneiss	Preshuksan Suite Pre-Mid-Devonian

Modified after Mish.

CHAPTER 26

NORTH CASCADES NATIONAL PARK

Location: North Central Washington
Area: 505,000 acres
Established: October 2, 1968

LOCAL HISTORY

According to one of the Indian legends, the region where the Cascades presently stand was a flat region when the world was young. At that time rain did not exist, instead moisture came up through the ground to feed the plants and trees. However, for some reason it stopped coming up in the Cascade area. So the Indians sent a delegation to the god Ocean (in the west) to plead with him for water. In response, Ocean sent Cloud and Rain, his children, to help them. With their help the land flourished, however the greedy Indians would not allow them to return and kept digging holes for Rain and Cloud to fill. Even after their father Ocean promised the Indians all the water they needed they refused. Therefore, Ocean prayed for punishment of the people, so the Great Spirit scooped up a large pile of dirt from the Earth and created the Cascade Mountains from it. Ocean filled the hole that was made and we call it Puget Sound. The east side of the Cascades, where the Indians were, dried up since Ocean refused to send Rain over the mountains. The only water they had left is the pits they had dug. The largest is Lake Chelan.

A different tribe has another legend about the origin of Lake Chelan. Long ago the land was a flat prairie with many animals and a happy way of life for the Indians. A terrible monster came and eliminated all the animals so the people starved. The Indians beseeched the Great Spirit for help and he answered their prayers by killing the monster. However, it kept coming back to life, after he killed it for the third time the Great Spirit drew out a large knife and struck the ground. The ground trembled, a huge cloud formed and hid the land. When all cleared, a high mountain range

stood on the former plain, the dirt being obtained from the deep canyons that now existed. Into the deepest canyon the Great Spirit threw the monster, then filled it with water, this became Lake Chelan. But its tail never died, and keeps thrashing around creating huge waves so it is unsafe for the Indians to paddle their canoes on it. Behind these legends is a nugget of truth, as we shall see a little later.

Around 150 years ago the first white men into the area were hunters and trappers. Before the 1850s, gold, copper, molybdenum, and zinc were mined on a small scale basis. As a matter of fact, the Ruby Arm of Ross Lake (now flooded) had *placer* (gold mixed in with stream gravels) deposits that were discovered. Areas outside the park boundaries were large enough to be developed as commercial enterprise, but not inside of what is now North Cascades National Park. Today mining is not important.

The main industry presently is timber and now logging is king. The park was created to prevent the remaining trees from being removed by industry. A study team was established in 1963, a report submitted in 1965, and the park established in 1968.

The park is divided into four units: (1) North Unit, which includes the former Picket Range Wilderness Area, (2) Ross Lake National Recreation Area, containing Diablo, George and Ross lakes formed by damming the Skagit River (for electrical power for Seattle), (3) South Unit, which includes the former Eldorado Peaks Wilderness area, (4) Lake Chelan National Recreation Area.

GEOLOGIC FEATURES

North Cascades National Park contains a grand total of 756 glaciers which cover a total of 103 square miles. This is not a surprising number when one realizes because of the geographic location four or five feet of snow can accumulate in a single storm, and a total of 30 feet during the winter. The Indian legend of the dry east side and moist west side is correct. The wind coming from the moist Pacific is cooled as it is elevated over the Olympics, thus causing as much as 110 inches of rainfall per year.

Glacial valleys are such valleys as the Lake Chelan Valley that the glaciers scoured so deeply it now holds a lake. This lake is about 1.5 miles wide and 55 miles long, therefore, winds coming from certain directions quickly build up high waves, making it unsafe for small boats. Hence the "monster's tail" legend.

All the major valleys have a broad U-shaped trough that have been carved to base level. They have braided streams in them, e.g., Ruth Creek and the upper part of North Fork of Nooksack Valley. Others, such as Baker and Chilliwack rivers and Big and Little Beaver creeks are in equilibrium, and have *graded streams* where the rate of erosion is equal to the rate of deposition. The one exception is Skagit River Valley. Above Beaver Creek and below Diobsud it is the typical glacial valley. But that 25 mile stretch between is a narrow V-shaped valley. Apparently what happened was that region was a divide and the glaciers flowed away from it (north and west) forming glacial valleys in that direction (e.g., Rose Lake Basin).

The tributary valleys never had much of a chance to downcut rapidly, so are frequently hanging valleys with cirques at their heads. The most spectacular is from Green Lake (Bacon Peak) with a 1,500 foot high waterfall.

The majority of glaciers are on Bacon and Eldorado Peaks and Mounts Challenger and Shuksan. Moraines can be seen by Luna Creek, Challenger and East Nooksack Glaciers. The usual glacial features can be seen such as horns (Mount Despair) and aretes (Indian Creek). East of Ross Lake the glaciers did not extend as far so some indications of earlier glaciations can be seen. West of Ross Lake the last episode of glaciation was by far the most extensive and hence obliterated the traces of earlier advances and retreats.

A special type of rock that can be seen here is a *migmatite.* It is a combination of a foliated (layered) metamorphic that has been injected in very thin alternate layers, so it is neither igneous nor metamorphic but a combination of each.

Also evident in this area are two types of faults. One is the *thrust* fault, the other the *graben.* There is some disagreement as to how a thrust fault develops. It is also quite possible that they can form two ways. One method is the result of grav-

ity on a gentle slope. Sections or plates literally slide one upon another. The movement is *downwards* and slightly *outwards* and up. A gap is left (stratigraphically) at the head of the slide (or fault). The second is a compressional force where the layers of beds are thrust *upwards,* slightly *outwards* and *down.* The last block overrides the earlier beds and this may be the only exposure of the marker bed.

The graben fault is created when one block drops down in relation to the two adjacent blocks. Two grabens in North Cascades National Park are the Methow Graben and the Chiwaukum Graben.

To complicate matters, the bedrock can actually be divided into four zones from west to east. Each has its own stratigraphic sequence all resting on the basement complex of gneiss.

The first zone which is bounded on the east by Shuksan fault is sedimentary and volcanic rocks, Paleozoic in age (Chilliwack Group). It also includes Mount Baker, a volcano, formed during the Pleistocene. It is called the Western foothills region.

The second zone is located between the Shuksan fault and Ross Lake fault. These are gneiss and schist intruded by granite. It is the core of the Chilliwack batholith. The metamorphic beds are highly deformed.

The third zone is between the Ross Lake fault and the Chewack Pasayten fault. It is primarily the Methow Graben consisting of Upper Mesozoic beds that in turn have been intruded by Cenozoic granites. The Jack Mountain thrust is contained within this zone, it is a wedge that forms the tops of several peaks. As far as the park boundaries, they end shortly after the Jack Mountain Thrust.

The fourth zone, although not in the park but east of it, is from the Chewack Pasayten fault on the Eastern Highlands, a mixture of gneiss, schist, and granite.

It is for this reason that it is almost impossible to have a single geologic column, since the stratigraphy is completely different from one section to another.

As mentioned earlier, metamorphic facies indicates the conditions of metamorphism according to the environment. The rate of progression from low to high is (1) greenschist facies; (2) epidote-amphibolite facies; (3) amphibolite facies. These are the result of regional metamorphism. The hornfel facies occurs at temperatures of 7,000° C and higher is the result of contact metamorphism and the rock is baked by the intrusion.

GEOLOGIC HISTORY

1. Formation of the crystalline basement, the Yellow Aster Complex.

The term complex gives a clue as to the origin of this type of rock. It has a long history of igneous and metamorphic activity. It was formed before Middle Devonian time and also before the Cascade metamorphism occurred. The age could be anywhere from Precambrian to early Paleozoic. There is a major unconformity separating it and the overlying beds.

Two members of the Yellow Aster Complex are the Marblemount Metaquartz diorite in the west belt where it appears in tectonic slices, and the Eldorado Orthogneiss in the east belt. Within the core itself the beds are folded into anticline.

2. Cascade metamorphism forming the Cascade metamorphic suite that has three subdivisions and beds are divided by faults that occurred after metamorphism.

These sediments were derived from geosynclinal deposits and thus vary in stratigraphy, underwent different degrees of metamorphism at slightly different times. They are as follows:

(a) Shuksan Metamorphic Suite—west side of core, underlies Shuksan structural belt with the early to mid-Eocene Straight Creek fault (east) and Shuksan Thrust (west) of mid-Cretaceous time, that forms a thrust plate. The members of the suite are the Darrington Phyllite that is overlain by Shuksan greenschist. The metamorphic facies is within the greenschist facies. They are Paleozoic in age, probably before the Devonian. The metamorphism occurred either during Late Permian or Early Triassic time.

(b) Skagit Metamorphic Suite—this lies beneath most of the core. It has two members, the Cascade River schist, the older of the two which was originally sediments and the Skagit Gneiss, a migmatite formed from the Cascade River schist. The metamorphic facies is greenschist to amphibolite. The original rock was probably Paleozoic in age, metamorphism probably occurred before the Jurassic, perhaps at the same time as the Skagit metamorphic suite formed.

Intrusives also occurred both before and after the suite formed e.g.: Gabriel Peak Orthogneiss (after suite formed). On the east side, the Ross Lake fault separates the Jack Mountain Phyllite (lower Cretaceous?) which is underlain by the Elijah Ridge schist from the Skagit metamorphic

Suite. While on the west are Paleozoic and Meso-zoic eugeosynclinal sediments occur.

3. Deposition of eugeosynclinal sediments during middle and late Paleozoic time. Deformation.

The typical eugeosynclinal beds occurred with a mixture of volcanics and sediments. To the west is the Chilliwack Group (middle Devonian to middle Permian), to the south is the Stillguamish Group (late Permian). On the east is the Hozomeen Group (Pennsylvanian?–Permian) and North Creek Volcanics (late Paleozoic? pre-Jurassic). Some deformation occurred late Permian–early Triassic.

4. Deposition of eugeosynclinal sediments during the Mesozoic. Thrust faulting on both sides of core the Mesozoic.

Deposition on the west side of the Cascades Mountains is Cultus-Bald Mountain (late Triassic–early Jurassic) sequence. It forms an unconformity (or disconformity) with the overlying Wells Creek Volcanics (middle to late Jurassic). Another disconformity occurs and formation of Nooksack Group (upper Jurassic and lower Cretaceous) follows. Thrust faulting occurred during mid-Cretaceous time.

On the east side is the equivalent of the Dewdney Creek Group, it is early Cretaceous in age. These are located in Methlow Valley Graben. They are the Newby (late Jurassic-early Cretaceous) separated by an unconformity with the overlying early Cretaceous beds of the Goat Creek, Panther Creek, Harts Pass and Virginian Ridge formations. Thrust faulting occurred mid-late Cretaceous time on the east side.

5. Deposition on land and in the oceans during end of Cretaceous and beginning of Paleocene.

On the western side of the core were deposited the Chuckanut-Swauk formation (Paleocene to Eocene) and the Nanaimo Group (late Cretaceous to Paleocene). These are both marine and continental deposits and contain some coal beds. To the east in the Methow Graben are the late Cretaceous Winthrop, Ventura, Midnight Peak formation that are all nonmarine. An unconformity separates them from the Paleocene (?) Pipestone Canyon nonmarine beds.

6. Folding during early Eocene. Additional deposition.

The beds were simply folded, no faulting or

any major complex structures developed. Deposition of the Huntingdon (middle to late Eocene) that rests unconformable on the beds below. Plus the Hannegan-Skagit Volcanics (Oligocene). Both a slightly deformed. Above these are Pleistocene deposits and modern alluvium.

7. Formation of intrusions including the Chilliwack Composite batholith that makes up the core.

Paralleling the Ross Lake Thrust fault but on its east side is the so-called eastern intrusive belt, it consists of first the Ruby Creek Heterogeneous Plutonic belt which is primarily a quartz diorite to granodiorite composition. It may be the same age as the Black Peak batholith that has the same composition. If so they are both late Cretaceous in age. Second came the Golden Horn batholith (late Eocene-early Oligocene) followed by the late Oligocene Perry Creek intrusive (occurred after the Skagit Volcanics).

Forming the main core is the Chilliwack Composite batholith. It began with the intrusion of stocks with a diorite or gabbro composition, followed by the main intrusion of quartz diorite that gradually changes into granodiorite (late Eocene) which in turn intruded (late Eocene-early Oligocene) Leucogranitic (light colored granite) stocks and a late intrusion that occurred at the same time as the Perry Creek intrusion (on east side of Ross Lake fault). Two other intrusions occurred, the Hidden Lake Stock (late Cretaceous early Paleogene) which is concordant with the rest of the beds and the discordant Cascade Pass quartz diorite (Miocene).

8. Folding and thrust faulting during the Cretaceous.

In the park the Shuksan fault occurred on the west and it was at this time that the ultramafic (very basic minerals) rocks were formed, possibly by the stress of the shearing. Rocks were sheared, shattered, metamorphosed producing complex structures.

On the east side of the core the Ross Lake fault occurred and the Jack Mountain Thrust (brought Paleozoic rocks over lower Cretaceous beds and Methow Graben). Intrusions have destroyed part of this fault zone.

9. Additional faulting and folding in the Cenozoic. Glaciation.

Most of the events occurred during the Eocene forming such features as the Boulder Creek fault, Straight Creek fault and folding of Chuckanut formation.

Later uplift and erosion stripped these beds. Additional uplift and faulting about ten million years ago produced the mountains that were scoured and shaped during the Pleistocene to create the rugged North Cascades of today.

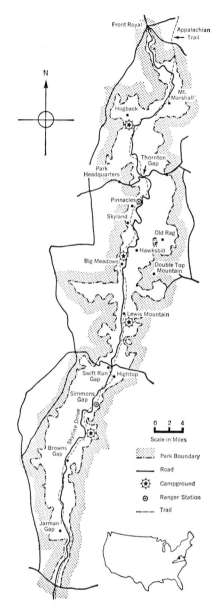

Shenandoah National Park, Virginia.

Geologic Column of Shenandoah National Park

Erathem	System	Formation		
Cenozoic	Neogene	Alluvium		
Paleozoic	Cambrian	Chilhowee Group		
Precambrian	Precambrian Z	Catoctin Formation		
	Precambrian Y	Old Rag Granite	Pedlar (?) Granodiorite	

Modified after Reed and Hack.

CHAPTER 27

SHENANDOAH NATIONAL PARK

Location: N.W. Virginia
Area: 193,536.91 acres
Established: December 26, 1935

LOCAL HISTORY

Shenandoah means "Daughter of the Stars" and some of the mountains that are more than 3,000 feet high appear to reach for the stars.

In the early 1600s Captain John Smith was the first known white man to see these mountains. A German scholar, John Lederer was the first real explorer in 1669. The Colonial governor of Virginia (in 1716), Alexander Spotswood, led his group, the Knights of the Golden Horseshoe, with their high boots and plumed hats through what is now called Swift Run Gap (lower 2/3rds of the park). Their purpose was to blaze a trail through the mountains for those settlers who wished to travel further west. It must have been somewhat of a joyous time as the records show an occasion where they drank to the King's health, fired a volley, drank to the Princess's health, fired a volley, and so on down the royal family's line.

The Indians in the region raised corn, squash, and pumpkin, ate fish and game. They built dugout canoes by setting one side of the log on fire and scraping out the burned and charred wood.

The area was settled in the mid 1700s (18th century) by what we call mountain people. Some were colonial settlers, others were British soldiers that deserted the army during the Revolutionary War. They were a simple people that spoke an Elizabethan English. They raised cows, chickens, and hogs and grew most of their own food (e.g., corn, cabbage, beans, and apples). A rather unique culture was developed and they became very self sufficient people, developing their own methods of doing things. Unfortunately, their method of farming was to set fire to the land to clear it, this of course, destroyed most of the original vegetation. Most of it is now second growth with the excep-

tion of a few isolated spots such as White Oak Canyon where hemlock trees (about 100 of them) are up to 500 years old.

History is everywhere from the game and Indian trails that still can be seen, to the earthworks in Brown's Gap built by Confederate forces during the War of the States. At this time it was called the "bread basket of the confederacy" and was the "back door to the government." Sheridan and Jackson fought here. During the French and Indian War George Washington had his headquarters in Winchester. This was the homestead of Abraham Lincoln's father. People such as Sam Houston (first president of Texas), Cyrus McCormick (inventor of the reaper), and Admiral Byrd (Arctic explorer) were born here.

In 1886 George Freeman Pollock fell so in love with the place that he built a cabin and later a resort (Skyland) that still exists today on Stony Man Mountain. This was literally the birth place of this park for Pollock and his friends were the prime movers of establishing the park.

A national committee was created in 1925 to find suitable areas in the East for parks. Pollock and some of his friends proposed Shenandoah to them and managed to finish the required questionnaire just hours before the deadline.

The task of acquiring the land was given to the State of Virginia. Therefore, Harry Flood Byrd Sr., the governor, (who supported the proposal) appointed a Commissioner of Conservation and Development. The approval was in 1926 but it took the Commissioner some 10 years to collect $1,200,000.00 from the people. It was donated in pennies, nickels, and dimes. An additional $1,000,000.00 was finally appropriated by the State at a time that money was very difficult to obtain.

This didn't end the task, for once the land was purchased some 400 families had to be relocated. This was a problem, for these people, even though the land was worn out by improper farming methods, stripped by soil erosion, and most of the timber gone, did not want to give it up. They had a tremendous pride of ownership and love of the land of their forefathers and did not care to leave for better land elsewhere. The entire park was created from private land and not public. Eventually all was accomplished. Even Herbert Hoover who owned a summer camp near Rapidan, gave it to the state of Virginia (when he left the White House) to become part of the park.

Part of the Blue Ridge Parkway runs through Shenandoah, some 105 miles of it. It took nine years to build and has a total of 75 overlooks. It also includes 94 miles of the Appalachian Trail (runs from Georgia to Maine) besides the other 270 plus miles of trails. The park is some 75 miles long and varies 2 to 17 miles wide and follows the ridge top of the southern Appalachians.

GEOLOGIC FEATURES

Waterfalls and *Rapids* are very conspicuous as there are more than 200 of them in the park (e.g., Jones Run, Park Hollow, Cedar Run, White Oak, and Hazel Falls). Waterfalls occur where there is a difference in the resistance of the beds to weathering. The weaker beds are removed more quickly while the resistant beds erode more slowly and stand out as ledges over which the water falls. The process of *cavitation* (where the vapor bubbles col-

lapse and strike the walls and on a microscopic level) chip away pieces of rock undercutting the bedrock. Eventually the ledge no longer supported, collapses and the falls recede back. The cavitation process frequently forms caves behind the falls. The falls eventually disintegrates into rapids. Given enough time it will form a steep slope and gradually becomes gentler.

Springs also are very abundant. Water seeps along some plane of separation such as bedding planes, joint systems, etc., or flows in a permeable bed between two impermeable beds. Regardless of its origin eventually the layers intersect the hillside and water comes out as a spring.

Wind gaps are created when a *superimposed* or *antecedent stream* abandons its *water gap.* In the case of a superimposed stream, the stream was flowing over a deformed surface where the structure was buried or at the time the stream was flowing the surface was smooth and level. If the area was gradually raised and the stream did not change its course it would be let down onto the structure and eventually cut a water gap.

An antecedent stream is one that flows on a surface that is gradually being deformed (e.g., raised into an arch), it is able to erode as rapidly as it forms, again forming a water gap. If for some reason either type of stream is diverted by *stream piracy* (stream loses its head waters to another stream) or for some other reason, the gap no longer

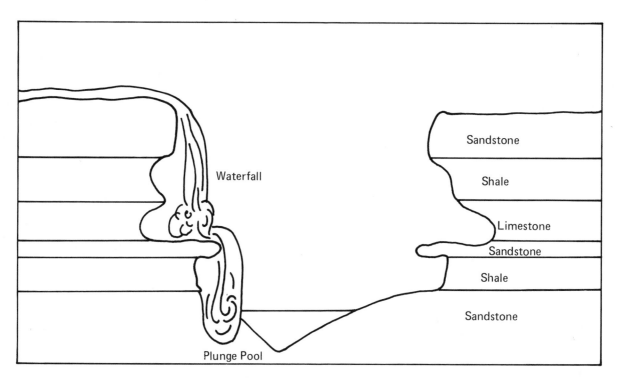

Resistant and Weak Beds

has a stream flowing through so is called a wind gap. Some of the gaps are Hawksbill, Swift Run, Hughes River or Simmons Gap.

Some of the lava flows have the *columnar jointing* developed as the lava cooled and shrank producing five or six sided columns that in this case are up to 150 feet thick. These can be seen near Little Stony Man Mountain parking lot (200 feet south).

In this same lava are *vesicles* or holes in the lava left by the trapped gases. They frequently become slightly stretched if the lava hasn't completely cooled, giving them an almond shape. If they become filled with another mineral at a later time, they resemble almonds and are called amygdules (Greek for almond). The greenstone in the park is amygdaloidal.

GEOLOGIC HISTORY

1. Formation of the Precambrian Old Rag granite and a fine-grained granodiorite.

About 1.1 billion years ago (Precambrian Y) a magma chamber formed (magma is the parent material of igneous rocks) deep within the earth some 30,000 to 40,000 feet deep. As it cooled two types of rocks formed dependent upon the amount of time involved.

As can be seen by the chart, the main difference between granite and granodiorite is that granite usually has muscovite and orthoclase (potassium feldspar) and is more abundant than the plagioclase feldspar (sodic-calcic feldspar). While the granodiorite usually lacks muscovite but has a pyroxene (usually augite) and the plagioclase feldspars are more abundant than the orthoclase. Both contain quartz and have a phaneritic texture (most of the crystals can be seen with the naked eye).

The hypersthene granodiorite (Pedlar formation) probably formed first using up the iron, calcium, sodium, and magnesium thus forcing the granite to form. The granodiorite can be seen on the western slope and on the crest of the Blue Ridge (Mary Rock Tunnel) while the granite is on the east side of the ridge and on Old Rag Mountain (area after which the Old Rag granite was named). The contact between the two is not sharp but gradual over a distance of several miles.

These two beds underwent metamorphism in some areas causing the crystals to recrystallize in response to the stress and strain (long axis perpendicular to area of greatest pressure ($\rightarrow$ [] $\leftarrow$) causing a layered appearance forming a *gneissic texture.*

As these deeply buried beds were slowly crystallizing and undergoing metamorphism, the overlying sediments were being stripped by erosion and after millions of years were exposed on the surface.

The Old Rag granite and Pedlar formations belong to a group of beds called the Virginia Blue Ridge Complex (Precambrian) that are as follows from youngest to oldest:

Lovingston formation (granite, gneiss, quartz monzonite).
Old Rag formation (granite).
Roseland anorthosite (plagioclase).
Marshall formation (granite, gneiss and quartz monzonite).
Pedlar formation (granite, granodiorite, syenite, quartz diorite, anorthosite and unakite).

They are sometimes so difficult to separate that they are mapped as the Virginia Blue Ridge Complex instead of the individual formations such as in the northern end of the park near Front Royal.

These were then carved into mountains and valleys with relief as great as 1,000 feet. As erosion continued the mountains were lowered and the sediments derived from them were deposited into the valleys, gradually filling them. This continued until 600 million years ago.

Type of Rock		
	Granite	Granodiorite
Main Minerals	Muscovite, Biotite, Hornblende	Biotite, Hornblende Pyroxene
Quartz	Present	Present
Feldspar Ratio	Orthoclase-Plagioclase	Plagioclase-Orthoclase
Texture	Phaneritic	Phaneritic

2. Volcanic activity, formation of Catoctin formation a series of lava flows from fissure eruptions. Formation of columnar jointing, erosion.

Some 600 or so million years ago (Precambrian Z) the scene changed. Unrest occurred and huge fissures opened up and poured out a series of lava flows that is now known as the Catoctin formation. It is primarily basic lava flows, some of which is amygdaloidal. It also contains schist and gneiss (composition of amphibole, chlorite, epidote and plagioclase), plus arkose, conglomerate and phyllite.

These flows poured out onto an uneven surface first filling up the valleys before they began to bury the rest of the landscape. The average thickness was about 150 feet. Some of the individual flows were more than seven miles long. They cooled forming columnar jointing (e.g., Crescent Rocks, Franklin Cliffs).

Over a period of thousands of years the flows came one after another and gradually buried some of the highest peaks.

Between these flows were long periods of erosion bringing in sediments such as gravel, arkose (sand with 25% or more feldspar), and clay. These flows were later changed to greenstones by metamorphism.

We do not know the total thickness of flows for the upper beds have been removed by erosion. However a clue can be found in the Big Meadows–Stony Mountain area where some 1,800 feet of 12 separate flows can be seen.

The erosion changed the area to a flat plain with very little relief, only an occasional mountain peak projected from its former surface that erosion unburied. About the only form of plant life at that time was algae.

3. Formation of the Appalachian geosyncline during the Cambrian. Over a 300 million year period, deposition of 30,000 feet of marine sediments.

Upon the lava flows was deposited the Chilhowee group resting unconformably upon the Virginia Blue Ridge Complex. Lower beds are a mixture of sedimentary rocks and volcanics. They were derived from the craton and were deposited eastward (source area in the west) as a wedge-shaped deposit of coastal plain sediments.

These sediments accumulated in the miogeo-

synclinal portion with the source area being somewhere to the southeast (probably ancestral Europe). The water was warm, shallow, and very rarely over 400-600 feet deep. Life forms that were abundant are unfamiliar to us today. At first there were only primitive brachiopods (a bivalved animal) and trilobites (extinct form of life related to crabs and lobsters of today). A little later came the corals, cephalopods (squid family) and gastropods (snail family) followed by the crinoids (sea lilies) and echinoids (sea urchins).

Almost as rapidly as the geosyncline filled, it sank so the water remained the same depth but the sediments were able to accumulate to a great depth. With this depression the region to the southeast was gradually being elevated thus an ever-rising source of sediments. The weight of the sediments above (around 30,000 feet) helped to change the sediments below into sedimentary rocks.

4. **Appalachian Revolution some 200 million years age. Folding, metamorphism, and erosion of beds.**

By the end of the Permian period the Appalachian geosyncline was destroyed by uplifting, folding, and faulting. The main force came from the southeast, so the *anticlinorium* (series of anticlines and synclines arched upwards) trends to the northwest. Reconstructing according to the angle of the beds, the mountains may have been as high as 20,000 feet. What is more likely, is that they were eroded almost as fast as they were elevated, so were probably not much higher than they are now.

The elevation caused the seas to drain from the interior and a low swampy, gentle landscape evolved, a perfect environment for the generation of reptiles that were to take over—the dinosaurs.

The uplift and folding caused metamorphism changing the igneous beds into gneiss and schist. The Catoctin volcanics were metamorphosed into greenstones, gneiss and schist and some of the interbedded sediments into phyllites.

5. **Continued erosion during the Mesozoic, stripping away the sediments and exposing the metamorphic and igneous core.**

During the following 100 million years these mountains underwent erosion and formed an erosional surface. Remnants of it can still be seen today. On this surface alluvium was spread burying the underlying structure. Streams meandered back and forth over this low surface broken only by an occasional monadnock.

6. Rejuvenation occurred elevating area some 2,000 feet. Granite gneiss and greenstone became ridges, softer sediments and metasediments the valleys. Superimposition of streams.

A series of uplifts occurred with intervening periods of quiet. Gradually the more resistant rocks, especially the greenstone, began to form ridges. Because the streams were originally flowing on surfaces covered with alluvium they were able to downcut without changing their courses and continued to do so when they reached the deformed beds. This made them superimposed streams and they formed water gaps. Later stream piracy caused streams to lose their headwaters, abandon the water gaps changing them to wind gaps.

The beds on either side of the Blue Ridge Mountains are soft limestones and shales that are easily eroded, thus forming the intervening wide broad valleys with meandering streams such as the South Shenandoah River. They are also very susceptible to formation of caves and springs thus explaining the abundance of them in adjacent areas.

Even though Shenandoah National Park is part of the Appalachian Mountain system its geologic history is completely different than that of the Great Smoky Mountains with its softer sedimentary beds, instead of the older, harder, metamorphic and igneous beds of the Blue Ridge region.

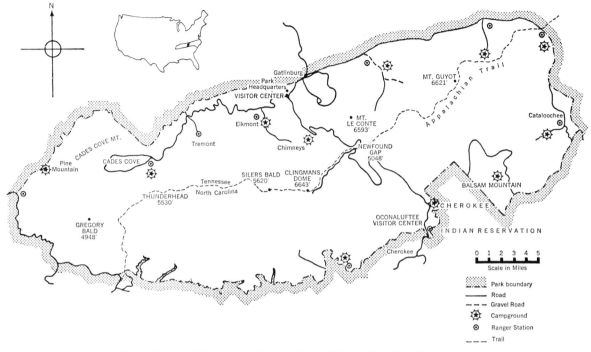

Great Smoky Mountains National Park, North Carolina-Tennessee.

Geologic Column of Precambrian Stratigraphy Below and North of Greenbrier Fault of Great Smoky National Park

System	Group	Formation			
Cambrian (?) and Cambrian	Chilhowee	Cochran Formation			
Precambrian Z (?)	Walden Creek Group	Sandsuck Formation			
		Wilhite Formation			
		Shields Formation			
		Licklog Formation		Fault	
		Western Section		**Eastern Section**	
	Unclassified Formations	Cades Sandstone		Webb Mountain and Blue Ridge Rocks	Rich Butt Sandstone
	Snowbird Group	Metcalf Phyllite		Pigeon Siltstone	
				Roaring Fork Sandstone	
				Longarm Quartzite	
				Wading Branch Formation	
Precambrian Y		Basement Complex			

Modified after King, Newman, Hadley.

Geologic Column of Paleozoic Beds of Great Smoky Mountain National Park

Erathem	System	Series	Formation	
Paleozoic	Mississippian	Upper	Greasy Cove Formation	
		Lower	Grainger Formation	
	Devonian	Upper	Chattanooga Shale	
	Ordovician	Middle	Bays Formation	
			Sevier Formation	
			Chota Formation	
			Tellico Formation	
			Blockhouse Shale	
			Lenior Limestone	
		Lower	Knox Group	
	Cambrian	Upper	Knox Group	
		Middle	Conasauga Group	
		Lower	Rome Formation	
			Shady Dolomite	
			Chilhowee Group	Helenmode Formation
				Hess Quartzite
				Murray Shale
	Cambrian (?)	Lower (?)		Nebo Quartzite
				Nichols Shale
				Cochran Formation
Precambrian	Precambrian	Ocoee	Sandsuck Formation of Walden Creek Group	

Modified after King, Newman and Hadley.

Geologic Column of Precambrian Stratigraphy Above and South of Greenbriar Fault of Great Smoky National Park

System	Series	Group	Formation
Precambrian Z (?)		Rocks of Murphy Marble Belt	Nantahala Slate Early Paleozoic (?)
	Ocoee	Great Smoky Group	Unnamed Sandstone
			Anakeesta Formation
			Thunderhead Sandstone
			Elkmont Sandstone
		Snowbird Group	Roaring Fork Sandstone
			Longarm Quartzite
			Wading Branch Formation

Modified after King, Newman and Hadley.

CHAPTER 28
GREAT SMOKY MOUNTAIN NATIONAL PARK

Location: North Carolina—Tennessee
Area: 516,626.02 acres
Established: February 6, 1930 Officially dedicated
    by Franklin D. Roosevelt on September 2,
    1940

LOCAL HISTORY

According to the Cherokee Indians, when the world began there was only sky and water and all of the animals lived in the sky. When they needed additional room they looked upon the water and wondered what was below it. Finally a water beetle agreed to find out and went down to the water.

He skated over the surface looking for some place to rest but found none. Finally he dived and brought back a small amount of mud that grew into the planet Earth. They fastened it with four ropes to hold it to the sky.

Since the new earth was very flat, soft, and wet, the birds kept checking to see if it was dry enough. But each time the answer was the same, it was still too wet. When it was the buzzard's turn, he flew down and also had no success but kept trying. As he grew tired his wings kept striking the ground forming valleys. When he raised them he created mountains. So the animals called him back before he changed the entire world into mountains and that is how the Smokies were formed.

This was Cherokee country and they named the mountains "Great Smoky" because of the haze that perpetually seems to be over them. They had many legends, according to them the rivers were a path to life after death. Thunder was simply a noisy man wandering around the hills. The animals were people and spoke (the Uncle Remus stories originated from here). They lived in log cabins and were among the best educated because Sequoyah or George Guess (Indian mother, white father), after he was crippled by an accident, invented an alphabet (85 characters) over a 12 year period.

Hernando de Soto, the Spanish explorer, in his quest for the "Spring of Eternal Youth" was prob-

ably the first white man in this area in 1540. He also named the Appalachian Mountains.

William Bartram in 1778 visited the Smokies and wrote of the wide, unique variety of plants. This area was never glaciated and literally acted as a sanctuary for plants that ordinarily grew further north. For example, in all of Europe there are only 85 different varieties of trees, in the Smokies alone there are more than 140 and of these 129 are native. It is natural, rich, botanical garden.

Gold was discovered in 1815 and this was the downfall of the Cherokee. The prospectors found little gold but instead found land that they wanted for themselves and not for the Indians. President Jackson, in 1830, signed the bill called the Indian Removal Bill. In the winter of 1838-1839 the Indians were forced to walk all the way to Oklahoma on the "Trail of Tears." About 1/4 of them died during the trip. The few that managed to escape formed the "Eastern Band of the Cherokee" and their descendants live at the adjacent Qualia Indian Reservation.

The people that settled the Smokies were mainly English or Scotch-Irish. They have resisted changing their ways. They still speak "Elizabethan English," the English spoken when "Elizabeth the Virgin" was queen. They were self-sufficient people who preferred a simple way of life and found no need to leave their valley for any reason. Some of the names reflect their way of thinking, e.g., Chunky Gal Mountain was named after a fat girl. A very popular place today, Charlie's Bunion was named after a guide's sore feet. Fighting Creek Gap was named as the result of an argument over the location of a school building. Huggins Hell was named after Irving Huggins, who got lost in a "slick" and took a few days to find his way out. Some Indian names are preserved, e.g., Oconaluftee and Cataloochee streams, but they are not abundant.

In 1780 these mountain people beat the British forces at King's Mountain under Colonel Ferguson, they also defeated Cornwallis at Yorktown. They were excellent shots and almost eliminated all the game before it became a park. Poaching and the manufacture of corn liquor is still a problem today.

Some cabins are still preserved, such as the John Oliver Cabin. He was an 1812 War veteran and became the first permanent resident in 1818. Also other homes, such as the Becky Cable House (post-Civil War), the first frame dwelling, are preserved. At Cades Cove and Roaring Fork the old ways are being carried on for future generations to see.

The park was first suggested in 1899, it was authorized by Congress in 1926. The task of acquiring the land was given to North Carolina and Tennessee. It took them 20 years to gain possession of 6,600 parcels of land. A total of five million dollars was collected mostly by donations but it wasn't enough (only half the amount needed). John D. Rockefeller, in memory of his mother, set up the Laura Spelman Rockefeller Memorial and matched the other half. Later, federal funds acquired the rest. The formal dedication was in 1940, some 14 years after it was authorized.

GEOLOGIC FEATURES

The smoke is actually caused by the trees. The trees give off hydrocarbon molecules called terpenes. They were derived from the oils of the trees. The action of sunlight causes them to break down into fragments and then recombine into big molecules. These molecules are large enough to bend or refract light rays forming the haze.

The "smoke" becomes very thick in the fall because of the decaying leaves adding more material. The molecules react with each other and other pollutants, to create the haze. That is why it is heavier sometimes than others.

There are also treeless areas called *balds* or *slicks* that are regions of almost a solid mass of branches of intertwining shrubs 8 to 12 feet high. They have even been known to grow over and hide small valleys. The vegetation is a mixture of rhododendron, mountain laurel, blueberry shrubs and sand myrtle. The *slicks* are produced by excessive evaporation due to altitude and high winds, by fire, landslides, or clearing by man. They never last a long time geologically speaking, as they are eventually claimed by the forest.

The *grass balds,* on the other hand, have thus far defied an explanation of how and why they form. They are meadows with herbs, grass and sedges. They form mostly on the southwest part of the park around the 5,000 foot level. Size wise they range from 100 to 300 feet wide and are usually several hundred yards long, but they may by up to two miles. They are frequently ringed by azaleas. Exposure to prevailing winds helps to keep them in existence as does grazing animals.

Basement complex is what the name implies, these are igneous and metamorphic rocks that have undergone several periods of deformation and are the base upon which the rest of the rocks lie. It is usually impossible to determine the original rock type making up the complex.

A *fenster* or window is a break created by erosion through a thrust fault or recumbent fold exposing the beds underneath the structure, (e.g., Cades Cove, Tuckaleechee and Wear Cove). While a *klippe* is an erosional remnant of a sheet that has been overthrust such as the hills around Cades Cove which are Klippen of the Greenbrier Thrust fault. Both of these features are very common in the Appalachians.

Inliers (older surrounded by younger) and *outliers* (younger surrounded by older) e.g., Chilhowee Mountain, are also erosional features but faulting does not have to be involved, the beds may simply be folded. These also can be found. For example, the hills northwest of Cades Cove are all outliers of the Ocoee series.

Many of the straight line parallel valleys are the result of high angle faulting such as the valleys around Gatlinburg. These are along the Oconaluftee and Gatlinburg faults.

Metamorphic zones indicate the degree of metamorphism. As the metamorphism increases different minerals will appear in the rocks creating zones within the bedrock. The progression is as follows: chlorite zone → biotite zone → garnet (almandite) zone → staurolite zone → kyanite zone → sillimanite zone. These can be seen in the Great Smoky Mountain. The chlorite zone near Mount LeConte (upper park Great Smoky conglomerate) as it approaches the range going north changes to the biotite zone. North of the crest the garnet zone is found, some three miles southeast it changes to staurolite and then kyanite in the schist layers. These zones have a northeast trend, and end with this sequence because conditions did not permit the formation of the higher sillimanite zone. The sillimanite zone or higher does not occur anywhere in the Great Smoky Mountain.

Another form of metamorphism can also be found, and that is *granitization* (production of an igneous rock bypassing the magmatic stage). If the metamorphism is great enough, most rocks can be changed into a rock that resembles granite but never crystallized from a liquid. Some of the Ocoee

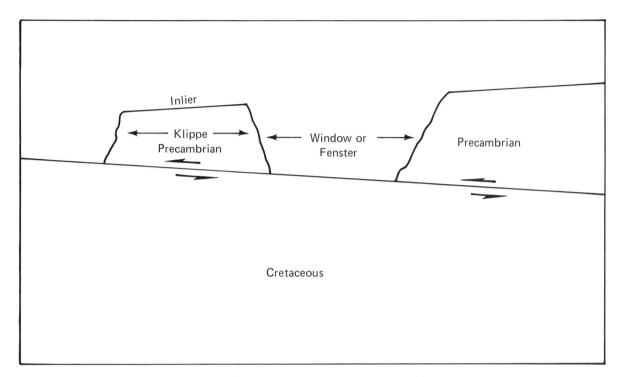

Inlier (Older Beds Surrounded by Younger)

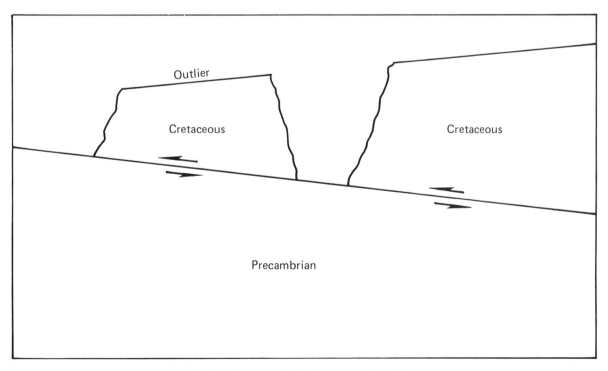

Outlier (Younger Beds Surrounded by Older)

near Bryson City appears to have formed from the granitization of the older Carolina gneiss (and associated rocks).

An area known as The Sinks (near Cove Mountain), is not a sinkhole, but is the result of a river cutting through an abandoned river channel called a meander scar. The name is derived from the fact that there is a very rapid drop of the river. The original channel can be traced around the knobs to the northeast (bedrock probably Great Smoky conglomerate).

By the Chimneys, *crossbedding* is well exposed as is *graded bedding* where the sediments have been sorted with the coarse sediments gradually grading into the fine sediments and then the sequence begins all over again. Also in a few locations *scour and fill* can be seen. In this type of bed, erosion has created channels on the surface forming small troughs, later when the area was under water the channels would fill first being an uneven surface, hence the term scour and fill.

GEOLOGIC HISTORY

1. Formation of basement complex during the Precambrian.

These beds are a mixture of gneiss and schist that were derived from sedimentary and volcanic rocks (layered variety) or intrusive bodies (nonlayered). They have undergone folding, faulting and metamorphism several times, so its almost impossible to determine the original rock.

Radioactive methods can only date the time they were deformed, the oldest being around one billion years ago (Precambrian Y). However, they were disturbed each time the younger layered Precambrian beds were deformed. The basement complex also contributed material to the younger beds, as fragments of the older material are found incorporated in the younger.

2. Deposition of the Ocoee series during Precambrian Y time, accumulating to a thickness of 20,000 feet. Erosion.

The Ocoee series even though it is Precambrian in age is younger than most of the rocks in the Blue Ridge area. The bedrock that makes up the Blue Ridge are older than one billion years, while those of the Smokies are about half that age.

The Ocoee series is divided into three groups, from oldest to youngest they are, the Snowbird Group, Great Smoky Group and Walden Creek Group. There is also a sequence that has been broken by faults and has been so disturbed, that so far, it has been impossible to assign them to any group. They have been identified as to the locality where they occurred and placed between the Snowbird and Great Smoky group stratigraphically.

The Ocoee beds are for the most part cliff formers and make up the majority of the Great Smoky Mountains. They form a band 175 miles and up to 30 miles wide and have been thrust faulted over Paleozoic rocks by the Great Smoky fault. These beds, of course, have undergone metamorphism and range from the chlorite to the kyanite zone (northeast to southeast). When a rock undergoes metamorphism it can be changed into many possible types dependent upon the degree of metamorphism for example, the sedimentary rock shale can be changed as follows:

$$\text{shale} \rightarrow \text{slate} \rightarrow \text{phyllite} \rightarrow \text{schist} \rightarrow \text{gneiss}$$

The metamorphic zones occur within these. Phyllite will usually contain the chlorite zone; schist the biotite, garnet, and staurolite zones; gneiss the kyanite and sillimanite. They are all interelated to one another.

The age of the Ocoee beds are late Precambrian according to evidence thus far and they do not contain any fossils. The stratigraphy of Ocoee is as follows:

The Snowbird Group is a mixture of sandy shale, siltstone, feldspathic sandstone, arkose and phyllite and lies unconformably on the basement complex. It was derived in part from the weathered surface of the granitic rocks and varies in thickness from 2,000 to 17,000′+.

Snowbird Group lies on the basement complex and is over-lain by either Unclassified formation and Walden Creek or Great Smoky Group.

The individual formations within the group are:

youngest	Pigeon Siltstone	
		Metcalf
	Roaring Fork Sandstone	Phyllite
	Longarm Quartzite	
oldest	Wading Branch formation	

The Snowbird Group is a basal and marginal formation deposited unconformably on the basement complex. They are exposed primarily on the north side of the Smokies and also southeastern portion.

The Unclassified Group are transitional, both vertically and laterally, and have features of both the Snowbird and Great Smoky Groups. There are some differences such as the lack of blue-tinted quartz grains. They are mostly sands but some phyllites and are 3,000-4,000 feet thick. The entire sequence has the tops faulted off, plus they are thrust faulted within the sequence.

These beds are named after the localities in which they occur, hence the Cades Sandstone, Webb Mountain and Big Ridge rocks, and Rich Butt Sandstone.

The Great Smoky Group is a thick monotonous mass of pebble conglomerate, coarse to fine sandstone and silty or clayey rocks. The conglomerates were derived from the weathering of granite so the pebbles are either quartz or feldspars. These granites formed the mountains which were the source area for these sediments. Because of their origin they somewhat resemble granite and are as thick as 25,000 feet. They accumulated in a deep sea trench where the sediments were transported by turbidity currents.

This group forms most of the Great Smoky Mountains and are the beds that most people see. The subdivisions are as follows:

youngest Unnamed sandstone
Anakeesta formation
Thunderhead Sandstone — Great Smoky Group, undivided
oldest Elkmont Sandstone

Walden Creek Group is mostly shale and siltstone but has lenses of conglomerate, sandstone, quartzite, limestone and dolomite. Conglomerate pebbles are a mixture of milky quartz, black quartzites, granite, limestone and other rock. Total thickness is around 8,000 feet.

The Walden Group was formed on an unstable shelf that fringed the continental interior towards the northwest. Some of its deposits were laid down in shallow water, others slumped from the shallow water into somewhat deeper water at the edge of the shelf.

The Walden Creek Group is subdivided as follows:

youngest Sandsuck formation
Wilhite formation
Shields formation
oldest Licklog formation

Erosion at the end of the Precambrian has produced an unconformity between the Precambrian and Paleozoic beds.

3. Deposition of Paleozoic beds.

Most of these beds are better exposed northwest of the park. Within the park they can be found in three separate situations. They are with the Great Smoky thrust sheets (Chilhowee Mountain and Miller Cove, English Mountain and Green Mountain) and form knife shaped ridges. Next they are found below the Great Smoky fault (northwest of its main edge which forms low rolling country with knobby hills). Last of all they are found as windows in the Great Smoky fault (Wear Cove, Tuckaleechee Cove, Cades Coves). They appear as low cleared areas surrounded by wooded hills.

4. Deposition of the Cambrian Chilhowee Group which contains worm tubes as fossils, and other Cambrian beds.

These beds are a mixture of conglomerates, quartzites, sandstones both glauconitic and arkosic plus siltstones. The cumulative thickness is about 3,250 feet, and they lie unconformably upon the Precambrian Sandsuck formation.

youngest Helenmode formation
Hesse quartzite
Murray shale
Nebo quartzite
Nichols shale
oldest Cochran formation

These formations contain the oldest fossil remains in the Great Smoky Mountain region. All formations above the Cochran formation contain *Scolithus* (burrows of primitive sea worms) and their tracks and traits. In the Murray Shale the ostracod *Indiana,* and in the Helemode the Trilobite *Olenellus* and other fossil shells.

The remaining Cambrian beds are primarily limestones and dolomites with some shale and sandstones layers. They have a total thickness of some 5,200 feet.

youngest Knox Group (U. Cambrian–
L. Ordovician
Copper Ridge dolomite (Lower Knox)
Conasauga Group
Rome formation
oldest Shady dolomite

This sequence indicates long quiet deposition in the Appalachian geosyncline where source areas were very distant and low in elevation.

5. Deposition of Ordovician beds, mainly limestones which contain brachiopod, trilobite and gastropod fossils.

After the deposition of the Knox Group that is partly Ordovician, erosion occurred, producing another unconformity.

These beds are a sequence of mostly clastic rocks ending with a red mudstone on top, indicating a rise of land adjacent to the geosyncline which acted as the source of sediments. They were somewhere near where the present day Smokies exist. The thickness is between 9,000 to 10,000 feet.

youngest	Bays formation
	Sevier formation
	Chota formation
	Tellica formation
	Blockhouse formation
oldest	Lenair limestone

Again erosion occurred for there are no beds until upper Devonian time. So upper Ordovician, all of the Silurian and lower and middle Devonian are missing.

6. Deposition of Devonian and Mississippian beds.

These beds are preserved in a narrow syncline caught under the Great Smoky thrust sheet. They are conglomerates, sandstones, siltstones, shales and limestones. There is a total thickness of 1,425 feet.

Mississippian	Greasy Cove formation
	Grainger formation
Devonian	Chattanoga shale

7. Appalachian Revolution 200 million years ago folding mountains into anticlines and synclines and thrust faulting Precambrian Ocoee beds over Cambrian Chilhowee beds 35 miles north and west. (Great Smoky overthrust). Metamorphism.

These forces folded this region into a huge anticlinorium, but the pressures were so great that many of the folds were overturned and finally gave way and became thrust faulted.

Three major faults formed during the Appalachian Revolution.

(1) *The Greenbrier fault,* together with its related faults, occurs within the Ocoee series between the Snowbird and Great Smoky group. It is found in the eastern half of the Great Smoky Mountain and has been broken by younger faults. The fault is wrapped around the Cataloochee anticlinorium. Its age is early Paleozoic.

(2) *The Great Smoky fault,* together with its related faults, occurs between the Ocoee series and the Paleozoic rocks. The location is at the border between the foothills and the Appalachian Valley in the north part of the park. The age is later Paleozoic. It carried older rocks over younger and is responsible for many of windows or fensters in the park such as Wear, Tuckaleechee, and Cades coves. In some areas it branches.

(3) *The Gatlinburg fault,* and its related faults, occurs within Ocoee series (near the northwest foot of the Great Smoky Mountains). It is as young as or younger than Great Smoky fault (late Paleozoic).

The folds range from broad arches (Mr. LeConte) to small sharp folds (foothills to northwest). They trend east and northeast and are asymmetrical (steep dip northwest), some are overturned.

Farther south (metamorphism is greater) a complex relationship occurs because folds of several ages intersect. Folding began about the time of Greenbriar thrusting and continued after Great Smoky thrusting.

Regional metamorphism affected mainly the rocks to the southeast and barely those to the northwest. The latered Precambrian and Paleozoic sediments were metamorphosed after the Greenbrier faulting. This means the basement complex underwent metamorphism once again.

Additional faulting occurred (Great Smoky and Gatlinburg) which moved beds even further and placed metamorphosed beds over unmetamorphosed beds. Total movement of these faults ranged form 12 to 25 miles.

8. Uplift and folding caused intrusion of diorite dikes, that contained copper, formation of milky quartz veins by hydrothermal solution and formation of Precambrian gneissic granite.

Within the Ocoee Group are blocks of milky white quartz formed as the result of secondary filling of fractures, locally called "Flint Rock." Also thin bodies of diorite intrude these beds, especially in Hazel and Eagle creek. These same intrusions emplaced copper in the country rock in the southwest part of the Smokies. In these portions is gneissic granite (Raven Fork area) formed both by intrusion and granitization. The sills are believed to have been intruded after the initial metamorphism of the Ocoee series and before the final metamorphism during some part of the Paleozoic.

9. Erosion of older Ocoee beds to form "windows" or "fensters" exposing the younger Chilhowee beds underneath.

During the Mesozoic and Cenozoic erosion cut through the overthrust sheet in the cove area and created "windows" through which the overridden beds are exposed (Ordovician). Because limestone is soluble, the windows have been enlarged into level-floored coves or valleys.

10. **Continued erosion during Mesozoic and Cenozoic with periodic uplifts, causing hard rocks to form ridges and soft rock to form valleys.**

From Paleozoic time to the present this land area has been continuously eroded, although erosion may have been faster from time to time by renewed uplift (without accompanying deformation) of the rocks. *The modern Great Smoky Mountains are entirely a product of this long erosional period and sporadic uplift. They are not, themselves, a direct product of large deformation which is visible in the rocks that they are made of.* The Smokies are characterized by valleys cut between the ridges out of the same rock formations that project above them.

11. **About 1/2 million years ago up to 20,000 years glaciation in the north intensified physical weathering and frost action down south. It split up boulders, and mass wasting carried them to the valley floors (some are 30+ inches in diameter).**

The mountains were well south of the region of the ice sheets, but they caused the climate to become colder. It is possible that the highest ridges of the Smokies were bare of trees and were covered with *snow.*

Some of the steep mountain valleys are covered by angular boulders of huge size which have broken off from the ledges. The fact that no boulders are breaking off today, the existing boulders are covered with vegetation (e.g., lichen), and breaking down into soil, indicate this was a process that occurred in the past but is not occurring today. We can see this type of boulder accumulation (which is angular) occurring in alpine and subarctic regions. Today the process forming them is *frost action.* Also the floors of many of the valleys are studded with these boulders for which it is impossible for the present or past streams to have moved them. Once again this was a process of *frost action* with its alternate freezing and thawing, slowly moving talus downslope.

12. **Development of modern landscape last 20,000 years.**

Erosion has continued, but has made only minor changes. The major changes have been in the plant life. As the ice sheets retreated northward, the forests grew higher on the mountains to stay within their climatic zones. They eventually covered the summits. Many of the plant species also migrated upward onto the higher slopes of the mountains instead of following the ice sheet northward. The same conditions that sheltered these mountains also protected this late Cenozoic vegetation. Because more of these plants have survived in the Smokies than anywhere else there is a greater variety of vegetation here than in most other places in the world. Hence spruce and fir forests which belong to the Canadian zone are found where there should be warmer type vegetation. Between the geology and the vegetation the Great Smoky Mountain it is a unique situation, which is one of the reasons it is preserved as a National Park.

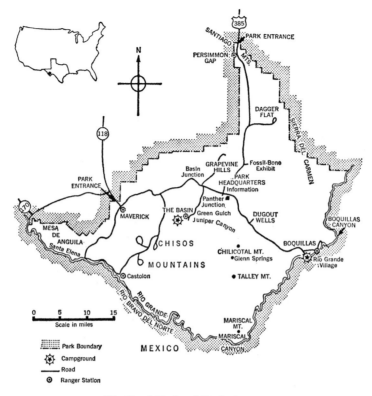

Big Bend National Park, Texas.

Geologic Column of Big Bend National Park

Erathem	System	Series	Formation
Cenozoic	Neogene	Holocene	Alluvium
	Paleogene	Pliocene to Eocene	Chisos Volcanics
			Sandstone, Clay and Conglomerate
Mesozoic	Cretaceous	Gulf	Javelina Formation
			Aguja—Continental Phase
			Aguja—Marine Phase
		Comanchean	Pen Formation
			Boquillas Flagstone
			Santa Elena Limestone
			Del Carmen Formation
			Glen Rose Formation
Paleozoic	Pennsylvanian		Tesnus Formation
	Devonian		Caballos Formation
	Ordovician		Maravillas Formation

Modified after Maxwell et. al.

CHAPTER 29
BIG BEND NATIONAL PARK

Location: West Texas
Area: 708,221.2 acres
Established: June 20, 1935 June 12, 1944

LOCAL HISTORY

First inhabited by the West Texas Cave Dwellers, who lived in dry cave shelters. They used the atl-atl (notched stick with hand grip that threw a short spear) for hunting. The yucca plant was utilized for rope, twine and basketry, it was chewed and pounded into fibers to make a material, which could be woven into cloth. The dead were buried in basketry wrappings. These people also left pictographs on the surrounding rocks.

This region contains more than 200 sites from the ancient Indians. They are such things as hearths, mounds, burial sites, sotol pits (for baking this variety of cactus), irrigation ditches, mortar holes, even an ancient battlefield. From one burial site some 56 items that were used daily were excavated.

The West Texas Cave Dwellers eventually were driven out by the Apache, and it was still Apache country when it first was seen by Cabeza de Voca, a lost Spanish soldier in 1535. Supposedly the Spanish discovered a silver mine and used Indian slaves for labor, its known as the "Lost Silver Mine" and is supposed to be somewhere on "Lost Mine Peak."

The Spaniards named the Rio Grande River, of which 107 miles flows along the southern border of the park. It was more or less rediscovered by white man in 1852. In 1881 a party of surveyers managed to navigate all three canyons. It took them 52 days. A second party of six men from the United States Geological Survey led by Robert T. Hill repeated the accomplishment in 1899.

The name Chisos means "ghost" and these are the mountains that make up the bulk of Big Bend. They received their name in 1882. White man had either driven out most of the Indians that lived in

the area or had deported them to Reservations elsewhere. All except Baja del Sol ("Under the Sun") a Comanche Chief. He knew what would happen so he made a bargain with the governor of the State of Chihauhau in Mexico. In return for allowing their tribe to live in Chihauhau his tribe would make war on the Mescalero Apache and wouldn't raid Chihauhau (but was free to raid other Mexican States). However, Baja del Sol couldn't stand being away from his beloved mountains and fled with some of his tribe back to the caves where they used to live. They managed to evade the American soldiers who were looking for them.

It was during this time that Baja del Sol along with his wife and her younger brother encountered a group of 30 Apaches with a captive Mexican boy (Domingo Parras). Baja del Sol prepared for battle in spite of the pleading of his wife because he felt if he didn't, the Apache would think he was afraid to keep his treaty with the Mexicans. He sent his wife and her brother on their way and then rode to the Apaches demanding the release of the boy. They refused and a battle began that lasted several hours in which Baja del Sol was killed, but he wounded many of his enemy before he was killed by a bullet. It made him a hero among the Indians for being a brave man.

Shortly after his murder people began to hear his footsteps and occasionally see his ghost wandering around his beloved mountain and began to refer to them as the "ghost mountains" or Chisos Mountains.

The Chisos was part of the Comanche Trail that the Indians used to ride into Mexico for raids. It became a hideout for rustlers and murderers, train robbers and men that became outlaws during the War Between the States. The notorious Black Jack Ketchum, a bank and trail robber, was among the most famous that was captured and hanged in Big Bend Country.

Many of the names have interesting stories behind them, for example, Dead Horse Canyon was a

box canyon where some smugglers had stashed horses from Mexico. However, they heard the Texas Rangers knew of their plans so they let the horses starve to death, rather than risk being caught taking them out. Dog Canyon in Dead Horse Mountain was named because of a pack of wild dogs that were left by the Indians who roamed there.

Emory Peak was named after the leader of the 1852 survey party Major William H. Emory. Boraches Spring is a corruption of Spanish for sandles (huaraches). It was an Indian encampment with many artifacts including a pair of sandles made out of rawhide from a nearby cave. Bone Springs acquired its name from the cattle bones around it and from animals that died when they became trapped in the marsh surrounding it.

In the early 1900s Hot Springs was homesteaded. The owner charged a nominal fee for bathing in the springs and later he built a bath house. By 1912 Mexican bandits called the Banderos Colorados or Red Flaggers began raiding the ranches. The ranchers requested protection and the Cavalry was sent down to patrol the area. Several outposts were established, one of them is Castolon

("Warden of a Castle"). It is within the park boundaries and preserved by the National Park Service. It has several buildings and the grave of an unknown soldier. This location also had the problem of Pancho Villa raiding over the border.

On May 5, 1916 the store at Glen Springs was raided under Lieutenant-Colonel Natividad Alvarez with some 200 men. Three United States soldiers plus a boy were killed, the ranch owner (who had an artificial leg) and his wife fled to the nearby hills to save their lives.

Just outside of the park boundaries can be found the ghost towns of Terlingua and Study Butte (named after a pioneer physician Dr. Bill Study). There are mining towns (for mercury) that existed from 1900 to 1946. A few people still live in Terlingua today.

When Everett Ewing was a Texas Ranger hunting for cattle rustlers, he saw Big Bend country for the first time and decided then it should become a state park. After he was elected to the state legislature he submitted a bill for that purpose. Texas Canyons State Park was created in 1933, but he changed his mind and decided it should become a National Park and campaigned along with friends to that end. Congress authorized the park in 1935. However, no progress was made so in 1941 Texas allotted $1,500,000 to acquire land. All but 25 sections were bought in one year. When the government was finally prepared to develop the park in 1944, Texas turned the land over to the National Park Service.

GEOLOGIC FEATURES

The three main features are the three steep-sided gorges, Boquillas, Mariscal, and Santa Elena.

Boquillas Canyon ("Little Mouth") is the longest (25 miles), the elevation at the rim is 3,500 feet and river level is 1,850 feet making the walls 1,650 feet high. The flood plain area contains cottonwood and willows. It is the widest of the three canyons and has a trail that leads up its side.

Near the mouth of the canyon is a *sandslide* where the weathered material has slid down. The wind has picked up some of this material and deposited it nearby as sand dunes. The canyon contains caves that have been formed by *wind-action* rather than solution. The wind picks up particles and through *abrasion* (wearing away by friction) has created a shallow cave. Fossils can be seen in the limestone sides of the canyon.

Boquillas Canyon is where the Mexican bandits crossed in 1916 and were chased back by the United States Cavalry.

Mariscal ("Marshall") Canyon is the most inaccessible of the three canyons. There is only a dirt road leading to the canyon, and no trails within it. It is cut through the southern end of the Mariscal Mountains and is 3,775 feet at the rim with river level at 1,925 feet. This makes the walls 1,850 feet high. It is considered the most spectacular of the three canyons.

Santa Elena Canyon is cut through a fault scarp in Mesa de Anguila. It's about 15 miles long with the rim at 3,661 feet and river level at 2,145 feet, so the walls are around 1,516 feet high. The walls in places seem to overhang the gorge, it is for this reason that sunlight strikes to the floor of the canyon for just a few hours a day. This canyon has a narrow flood plain area. Fossil shells can be seen in the walls.

All three canyons are carved in Cretaceous limestones. Within these limestones have been found giant pelecypods (clams) measuring 3 × 4 feet in size. Also found were bones of a crocodile *Phobosuchus* that was up to 50 feet long. For this region at that time was swampy. Remains of dino-

saurs including the horned dinosaurs, marine reptiles, primative mammals, plus ammonites oysters and other marine forms have been found.

At Tornillo Flats are brightly colored badlands with the colors being formed by iron oxide (FeO_2). These draws or canyons are dangerous because they can flood because of a sudden rainstorm instantaneously without any warning.

Some of the bedrock is capped with lava flows which erosion has carved into strange and weird shapes.

Tally Mountain used to be a flat topped mountain covered with good grass and it was called "cow heaven." But overgrazing has destroyed the grass and now after a sudden cloudburst there is nothing to hold or stop the water so the soil and rock is eroded away. Casa Granda ("big house") is also capped with volcanic rock. This is also true of the Tornillo Creek area.

In spite of the fact that there is usually less than 10 inches of rain per year there is a waterfall called Cat Tail Falls that is 200 feet high.

Near Persimmon Gap can be found all kinds of petrified wood. One log was found to be 10 feet in

diameter and 90 feet long. Just outside park boundaries along Terlingua Creek, fossil amber and also agate can be found.

Not all the bedrock in the park is sedimentary. The core of the Chisos Mountains are igneous intrusions called volcanic plugs and have formed dome-shaped mountains such as Vernon Bailey and Pulliam Ridge.

GEOLOGIC HISTORY

1. Deposition of Paleozoic beds in the Ouachita geosyncline some 300 million years ago.

Sedimentary rocks such as sandstones, shales and limestones were deposited. Animals such as trilobites, brachiopods and sponges lived in these shallow seas and left their remains.

2. During the Pennsylvanian period uplift occurred and the beds were buckled. Erosion.

The entire southwest was undergoing complex movement, the Ouachita Mountain, Wichita Mountains and Marathon uplifts raised the ocean floor into a series of folded and faulted mountains. Erosion almost leveled them. Southeast of Persimmon Gap on the hillside can be seen some remnants of these disturbed beds.

3. During the Mesozoic, submergence and deposition of Cretaceous beds.

The seas once again covered Texas and as the seas overflowed their troughs (to the west) and covered most of these Paleozoic mountains with sandstone and conglomerate. Deposition of the Glen Rose, Del Carmen and Santa Elena limestones during Lower Cretaceous time. These beds can be seen in Santo Elena Canyon and Sierra del Carmen cliffs.

4. During Upper Cretaceous some 100 million years ago, deposition of beds containing reptiles.

The formations deposited were the Boquillas and Pen formations. These limy muds created shales and flagstones (thin parallel layers of rock). The Boquillas formation can be seen by the Hot Springs area as resistant parallel layers.

5. Gradual uplift changing the area from a sea to a low swampy area. Formation of coal beds.

The climate was warm during the Upper Cretaceous and tropical junglelike vegetation covered the landscape. Dinosaurs roamed over the area cov-

ered with cycad and fern trees. This was where small rodentlike mammals could hide in the nooks and crannies. Crocodiles up to 50 feet long rested along the banks of rivers and swamps. This is the environment in which the Aguja and Javelina formations formed, and from which the bones for the bone exhibit came. The vegetation of the swamps was gradually turned into coal that exists today.

6. Laramide Revolution, formation Santiago Mountains, Mariscal and San Vicente cliffs. Downdropping between the Terlingua and Sierra del Carmen faults. Carving of canyons.

With the advent of the Laramide Revolution the Solitario Dome was created. This produced a large rift (100 miles long and 40 miles wide) valley bounded on the west by the Mesa de Anguila and on the east by the Sierra del Carmen. The Terlingua Creek area dropped forming a valley 100 miles long and some 35 miles wide. Many minor vertical faults occurred in between. The movement created a magma chamber and possible mineralization occurred at this time of mercury, lead, gold, silver, and copper.

As the area gradually uplifted the Rio Grande River was able to cut through the soft limestone sediments creating the three canyons.

7. Deposition of Cenozoic deposits the Big Bend Park beds.

The newly created highlands underwent erosion about 60 million years ago depositing gravel, sand and clay as alluvium. The dinosaurs were gone but mammals had evolved and left their bones behind to testify to their existence although they can no longer be seen today. Some of their bones may also be seen at the bone exhibit.

8. Volcanic activity, formation of Chisos Mountains.

All this unrest permitted the magma chamber to move towards the surface. Where it reached the surface the volcanic activity buried the new sediments under a layer of ash and lava. Igneous plugs formed the Chisos Mountains and folded the sediments and volcanic deposits, examples of these are the plugs such as Nine Point Mesa and Pummel Peak. The hot springs are testimony that the magma has not completely cooled.

9. Erosion of the softer beds creating the present landscape. Erosion created such features as
Erosion created such features as Mules Ears, and Lone Peak (small igneous plug surrounded by

sedimentary beds). Running water and weathering have created The Basin, a natural depression, that is not a volcanic crater as once thought. The water after a heavy rain leaves the basin through a V-shaped notch called The Window.

Overgrazing of the land by goats and cattle (before it became a park) has destroyed much of the vegetation that held down the soil, so erosion is rampant. But gradually the land is making a slow recovery to create that wonderful unique area called Big Bend country.

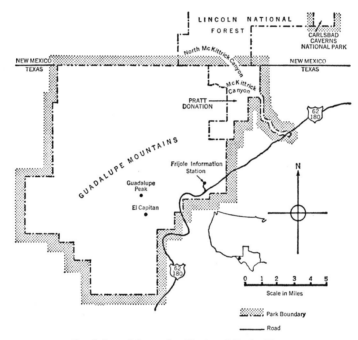

Guadalupe Mountains National Park, Texas.

Geologic Column of Guadalupe National Park

System	Series	Group	Shelf	Basin-Margin	Basin	Basin Member
Permian	Ochoan				Castile Formation	
	Guadalupe	Artesia	Tansill Formation	Capitan Limestone	Bell Canyon Formation	Lamar Limestone
						Unnamed Sandstone
			Yates Formation			McCombs Limestone
						Unnamed Sandstone
						Radar Limestone
			Seven Rivers Formation			Unnamed Sandstone
						Pinery Limestone
						Unnamed Sandstone
						Hegler Limestone
			Shattuck Sandstone Member		Cherry Canyon Formation	Unnamed Sandstone
			Queen Formation	Goat Seep Limestone		Manzanita Limestone
						Unnamed Sandstone
						South Wells Limestone
						Unnamed Sandstone
			Grayberg Formation	Getaway Limestone		Upper Getaway Limestone
						Unnamed Sandstone
						Lower Getaway Limestone
			Cherry Canyon Sandstone Tongue			Unnamed Sandstone
			San Andres Limestone	Cutoff Shale	Brushy Canyon Formation	
	Leonard		Yeso Formation	Victoria Peak Ls.	Bone Spring Limestone	
	Wolfcamp		Hueco Limestone		Wolfcamp (?)	

(Delaware Mountain Group spans the Basin column from the Capitan Limestone down through the Brushy Canyon Formation.)

CHAPTER 30
GUADALUPE NATIONAL PARK

Location: West Texas
Area: 81,077 acres
Established: October 15, 1961 (First opened to public 1970)

LOCAL HISTORY

There is evidence of Indians as far back as 12,000 years according to Carbon-14 dating. They lived in caves and shelters in the rocks and etched pictographs in them. They did not farm but followed the ripening of the plants from the valley floor (spring) to the high rims (fall). The roasting pits for mescal are found at every elevation for this reason. According to some of the pictographs, they also hunted animals that are now extinct.

The first references were from Spanish conquistadores who were exploring the land in the 1700s. Since they stayed near the rivers it is doubtful they ever came into the park. This is Mescalero Apache country and this also had a lot to do in discouraging visitors in the early days.

The term Guadalupe was first used on a map in 1828. Not much attention was paid to the region until the 1840s when the Mexican War occurred. Reconnaissance work was done in 1846 by the army which never got any closer than 100 miles (to the northeast); then in 1849 a series of surveys, all looking for wagon road routes.

First came the Ford-Neighbor Survey. John S. Ford was a Texas Ranger and Major Robert Neighbor the Indian Agent for Texas (Federal). In the Spring of 1849 they went around the edge of the Guadalupe Peaks exploring a southern and northern route between Waco and El Paso, Texas. That same year Lieutenant Francis Bryan went over that same route and confirmed the opinion of Ford and Neighbor, that the northern route was best.

That same year a wagon train was escorted from Fort Smith, Arkansas, to Santa Fe by the United States Army 5th Infantry led by Captain Randolph Marcy. From Santa Fe he traveled south to near Las Cruces, New Mexico and then decided to blaze a trail east across Texas. The expedition traveled around the southern end of Guadalupe Peak and camped in Guadalupe Canyon in September 1849.

John R. Bartlett (Commissioner of Mexican Boundary Survey) was in the area in 1850 and wrote up one of the best descriptions of the park region. Bartlett's Peak (fourth highest in the park) is named after him.

A Captain John Pope tried to establish a post in 1855, but gave up after a year because of lack of good water.

The Butterfield Stage Coach Line had a station south of the park (Pinery Station) at Pine Springs (east flank of Guadalupe Peak) that was used in 1859 only. Problems with lack of help and lack of water plus Indian trouble was responsible for the closing up of the station.

When the ranchers moved in the Indians retreated into their strongholds. It wasn't until 1869 that a Lieutenant Cushing dared to move against the Indians and destroyed their camp. For the next 10 years it was a hide and seek game, Army vs. Indians. The Indians lost.

After the War Between the States, ranchers came in and had to contend with rustlers, outlaws and Indians. By 1878-79 the army came in once again and moved the remaining Indians they could find out to a reservation.

Eventually the Mescalero Apache Indian was forced out by civilization. Battles between Mexico and the United States over many things, especially the salt flats, court litigations, congressional investigation and intervention by the United States Cavalry finally brought an uneasy peace in 1878.

In 1961 Wallace Pratt, a geologist, donated the North McKittrick Canyon area (perfect cross section of the reef) to the Federal Government as a National Park (total of 5,632 acres). For the sum of $1.5 million dollars (approximately $21.00 an acre) 72,000 acres were purchased from J.C. Hunter, Jr., of Abilene, Texas who had inherited the land from his father.

GEOLOGIC FEATURES

The Guadalupe Mountains are a portion of the largest fossil reef in the world, the Capitan Reef. The Guadalupe Mountains (and Carlsbad Caverns) are just the northwest tip of this huge horseshoe shaped reef that is 350 miles long. It is only exposed in Guadalupe and Carlsbad National Park, the Glass Mountains, and Apache Mountains. The rest is buried and can only be located by drilling.

The best exposure of the reef showing all three facies is in McKittrick Canyon with its 1,900 foot north wall. Within the canyon is an excellent study of *facies* (different rock types deposited in different geographic areas at the same time). Progressing from the reef to the backreef (or lagoon) are four separate facies:

(1) *coquina* (shell hash) deposited directly behind the reef in normal marine water; (2) pisolite (spherical accretionary particles) deposited by algae (?) as lenses shelfwards (most life could not exist under these conditions); (3) very fine-grained (microcrystalline) dolomite precipitated as calcite in very salty water, formed some distance away from the reef; (4) red siltstone (fine-grained) and sandstone

mixed with anhydrite. This change is typical in the lagoon or backreef behind the main reef that acted as a barrier.

The *reef talus* is a mixture of breccia, calcarenite (coquina) that is composed of pink or gray dolomite. The beds dip 20-35 degrees, and are 50 to 100 feet thick wedging out to only five feet.

Dolomite is usually a *secondary mineral,* that is, it is formed by replacement. In the case of dolomite the calcium in the limestone is replaced by magnesium. This is not an even process so frequently it occurs in patches.

Submarine slides can be seen in such areas as Bell Canyon in the Radar formation. These landslide deposits are a limestone breccia that is a sheetlike deposit of nonsorted nonstratified fragments of rock that are angular to subangular in shape. The fragments obviously came from the reef itself and the forereef facies. These deposits have blocks in them that are several feet in diameter and have a tabular form, they are underlain and overlain by the fine-grained basin deposits. As in modern slides that we can observe today, the large boulders (some of which weigh several tons) float on the fine muds in the slide and do not greatly affect the beds beneath.

The Capitan reef which now forms the Guadalupe Mountains was a *barrier reef* that had a lagoon between it and the land. It grew over its own talus

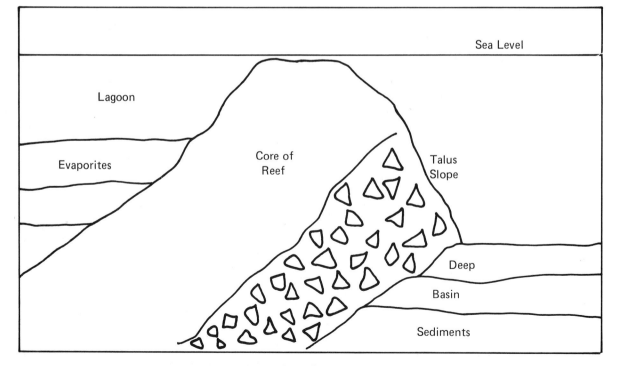

Reef Facies

towards the basin very rapidly. As a matter of fact, it migrated into the basin several miles and only built upwards some 1,300 feet indicating subsidence was very slow. The chief builders of the reef was not coral but stromatolites (fossil calcareous algae) and other types of algae, massive bryozoans (moss animals), individual vaselike (sycon) sponges and animals that might be corals (hydrocorallines ?). Other animals lived in the reef but were not frame builders such as brachiopods (bivalved marine animals), fusulines (single celled animals with calcite shell), bryozoans (not the massive variety) crinoids (sea lilies) and other animals such as gastropods (snails) and pelecypods (clams). The Capitan formation consists of both the talus upon which the reef sat and the reef proper.

Sandstone dikes can be seen in McKittrick Canyon (e.g., Yates formation), they are less than an inch in width to ten inches wide usually branching and ignoring the bedding planes. Most are not exposed for any great distance. They were formed in all probability by the filling up of cracks or fissures in the reef. Most of this sand was brought from another location by the wind (aeolian deposit). Some sandstone dikes are formed by sand being injected from below when it is saturated with water and can flow.

The Castile formation consists of alternate light and dark bands of calcite and gypsum. The calcite bands are stained with bituminous material so they are dark brown in color and are usually thinner by 1/3 than the gypsum layers.

This is probably the result of cyclic deposition or *varves*. The basin in which the Castile was deposited was isolated basin or *barred basin* which would permit water to enter periodically, such as once a year. The water would evaporate in a warm climate causing a concentration of the material in it. While the water was relatively fresh, calcite would precipitate out. As it grew more and more concentrated, the floating forms of life that were washed in when the water breached the barrier would die, and the decomposing bodies would mix with the calcereous ooze, coloring it brown. After salinity reached three times normal, either anhydrite or gypsum would precipitate out. Before salinity had a chance to reach ten times normal, a new surge of water would come in starting the cycle over again before halite (salt) had a chance to crystallize out. If for some reason it was not breached, gypsum would continue to form and thus form thicker layers.

When it finally was breached, the water level would be lower than usual, and would take on more than the usual amount of the less saline water. Hence it would take longer before it would become concentrated enough to precipitate anhydrite or gypsum and the calcite layer would become quite thick.

A pair of calcite-gypsum (anhydrite) bands would represent one year in time and six to nine feet of evaporation. A thick layer of thin bedded limestone would simply mean the water never became very saline (as can be seen at the top of some cuts). The inlet area was probably Hovey Channel (south of basin).

The beds are not undisturbed, an anticline and syncline can be seen in the Lamar limestone. There is also a cave in the limestone (Bear Canyon) nearby.

Penecontemporaneous deformation is obvious in many places where the debris that makes up the talus slope deformed the soft beds (semiconsolidated) as it slid over it. Thus deformation occurred at the time of deposition. Structures formed at this time are called *primary structures.* Other examples of primary structures are the *crossbedding*, found in some of the clastic beds, stratification or layering (e.g., Castile formation).

Besides the reef forming as a major feature, minor features also developed, such as *bank deposits.* These are shoal deposits or local mounds that formed at the rim of the basin. The main fossils are not those that are frame builders, but rather gregarious animals such as crinoids. These beds, unlike reefs, are stratified.

Turbidity currents which are distinguished in a body of water by the amounts of sediments they transport along the bottom, are responsible for transporting much of the material into the basins.

The *shelf deposits* are divided into five different phases: (1) reef derived material (coquina and limestone sand or calcarenite); (2) pisolites (rounded accretionary bodies); (3) fine-grained dolomite (replaced calcite); (4) evaporites (gypsum, calcite, anhydrite, salt); and (5) a land or *terrigenous* phase made up of quartz sandstone. All five phases grade one into the other as a *reef facies.*

The Guadalupe Mountains are a V-shaped range with two 55 mile arms, one extending NW and the other NE and are 20-30 miles wide. The apex of the V is El Capitan.

The most striking feature of the park is the 1,000 feet high El Capitan Cliff (8,078 feet), which is an ancient Permian reef of whitish colored limestone and the world's largest known fossil reef. Due north one mile is Guadalupe Peak which is 8,751 feet and the highest point in Texas. Range in

elevation of the park is from 3,000 feet to over 8,700 feet.

Some 4,000 feet of Permian rocks are exposed and were formed 230 to 280 million years ago in the Delaware subbasin of the Guadalupe Basin, a large saltwater inland sea.

The coquina is made up of fragments of clams, brachiopods, snails, bryozoans and sponges and algal particles. The sand is simply sand-sized grains of these shell fragments.

The pisolites are dolomitic in composition ranging in size from 1.5 to 6 mm in diameter. Most were flattened slightly. They accumulated layer by layer in a medium where the water would be agitated. They are most commonly seen in the Carlsbad Group.

A structure referred to as *tepee structures* can be seen in some of the backreef dolomite facies. These are simply beds deformed by folding into a chevron pattern that looks like the letter V upside down. These structures probably formed when the anhydrite in the joint system that criss-cross this formation took on water and converted into gypsum. When this happens, there is expansion of the

mineral which exerted enough pressure on the pisolitic dolomite to arch the beds upwards.

The reef area underwent faulting, fracturing the bedrock in many places. They have been filled in with veins of dolomite (especially in Tepee area and McKittrick Canyon). In some instances they have been deformed along with the beds that they developed in. They also have fibrous structure perpendicular to the bedrock walls.

GEOLOGIC HISTORY

1. Formation of Guadalupe Basin in West Texas area during the Permian.

The Guadalupe Basin was subdivided into five structural units, from southwest to northeast: The Marfa Basin in Mexico, the Diablo Platform in Mexico, the Delaware Basin in Mexico, New Mexico and Texas, the Central Basin in Mexico, New Mexico and Texas, the Central Basin Platform in Texas and Midland Basin in Texas.

During the Pennsylvanian it had been "deep" sea but with the uplifting to the north the basin formed. The deep Delaware basin was the only persistant marine area. It was more or less oval in shape and covered an area of around 10,000 square miles. Southwest of the basin was a channel to the open sea which periodically brought in fresh water to replace what was lost by evaporation.

2. Formation of the Capitan and Goat Seep reefs which are barrier reefs.

These reefs rimmed the Delaware Basin and were barrier reefs which were separated from the land by lagoons—similar to the Great Australian Barrier reef of today.

The Capitan reef was about 400 miles long and completely encircled the Delaware Basin. It was barely under water and yet a few miles from the reef the water was 1,800 feet deep and stagnant.

The main reef builders were calcareous algae, sponges, bryozoans, brachiopods and a variety of starfish. Corals in this area were not abundant and therefore not reef formers.

3. Development of the three typical reef facies and therefore a complex of interrelated formations that are all the same age.

a. Back reef facies—Gypsiferous shales, thin dolostones and sandstones of the shallow lagoons on the platform.
b. Reef and reef talus facies—vuggy limestone and dolostone with brecciated portions at the margin of the basins.
c. Normal marine deposits (forereef)—Carbonaceous limestones and sandstones on the floor of the basins.

4. During Wolfcamp time formation of Hueco limestone in subsiding Guadalupe Basin.

These are actually a mixture of limestones and interbedded dark shale accumulating some 1,600-2,200 feet of sediments. The shales are located in the center of the basin.

5. During Leonard time accumulation of the Yeso formation (shelf), Victoria Peak limestone (shelf basin margin), Bone Spring limestone, and Bushy Canyon formation (basin).

The Yeso formation are alternating beds of gypsum, sandstone (pale red), and silty dolomitic limestone that has salt layers. It was deposited on the shelf. Above the Yeso is the San Andres formation, which is a dark gray limestone and dolomite that contains oil, water and caves. It grades into the Cutoff shale.

The Victoria Peak limestone facies is a white limestone which accumulated on the seaward margins of the platforms as broad, shallow, limy banks. The thin bedded black Bone Spring limestone facies accumulated in the stagnant deeper portion of the basin. This section of the basin was so deficient in oxygen that the bottom dwelling organisms died and organic matter from above settled out and mixed with the sediments creating a black carbonaceous limestone. The Bushy Canyon was also deposited in the basin, in part above the Bone Springs. It is mostly sandstone (gray or yellow) with some fine sand and silt.

6. During Early Guadalupe time formation of the Goat Seep reef with the forereef and back reef facies.

The Grayburg and Queen formations were formed on the shelf. Cherry Canyon and Bell Canyon formations are members of the Delaware Mountain Group. The Getaway, a thin bedded limestone and the Goat Seep make up the reef. The Cherry Canyon sandstone formed in the main basin as the forereef facies where the water was deep, and well aerated. The Bushy Canyon had normal marine fauna, sandstone, and siltstone. The Cherry Canyon is divided into Lower Getaway limestone, Upper Getaway limestone, South Wells limestone and Manzanita limestone. All are separated by unnamed sandstones.

Behind the reef in the lagoon the water became concentrated to a brine, thus anhydrite and gypsum and gray shales and limestones formed. The water was so salty that organisms could not live, so no fossils are found.

There was a short period of erosion where the area was briefly above sea level, and then underwater again.

7. By Upper Guadalupe time the formation of the Capitan Reef made up of the Capitan limestone.

A much larger reef developed as the seas once again transgressed on the land and then slowly retreated.

The Bell Canyon sandstone interfingers with the Hegler limestone, Pinery limestone, Radar limestone, McCombes limestone and Lamnar limestone, which formed at the base of the reef. They are also part of the forereef facies. All are separated by unnamed sandstones. It represents the basin facies or forereef.

8. **Formation of the back reef or lagoon facies, the Carlsbad Group in which the adjacent Carlsbad caverns have formed.**

The Carlsbad Group consists of the Seven River formation, Yates formation and Tansill formation, a mixture of limestones, dolomite. Carlsbad Caverns were dissolved out of the Tansill formation and Capitan limestone.

9. **Regional warping caused water level to fall below the rim of the basins, exposing the reefs and platforms and thus land-locking Delaware Basin.**

The Marfa Basin was drained while the Delaware and Midland basins still contained water. But they became land-locked seas similar to the Caspian Sea of today. They had channels that opened periodically southward to the ocean.

10. **Formation of the Ochoan series in the land-locked sea as it evaporated and filled up.**

First the formation of the Castile anhydrite which is laminated with alternate light gray gypsum anhydrite bands deposited during the winter and dark brown calcite formed in the summer. On the surface it is gypsum, but below the level of the ground water table it has been converted to anhydrite (600 feet deep). These beds are about 1,800 feet thick. This was formed when water was three times normal salinity.

When the water became ten times normal salinity, the Salado Halite formation precipitated out. The Salado is about 2,000 feet thick with interbeds of potassium rich minerals—red sylvite, gray langbeinite, brownish bitter-tasting carnallite and pale-red polyhalite. It does not outcrop as it is extremely soluble, but is mined underground and supplies 90 percent of the United States potash fertilizers.

It took about 306,000 years to deposit the Castile and Salado formations (in a climate similar to Death Valley of today) and fill the deep depression that was the Delaware Basin.

A brief influx of normal sea water formed the Rustler dolostone with limited plant and animal life. Some anhydrite formed in the Rustler dolostone.

As the seas retreated to the south (never again to cover this region), a fine-grained red sandstone and siltstone was deposited as a thin blanket over the lowlands. These were the Dewey Lake Redbeds, they and the Rustler dolomite in places rest irregularly upon the Capitan limestone since they lapped over the edges of the basin.

11. **During Late Mesozoic Early Cenozoic the area was uplifted, and tilted to the northeast creating a fault on the abrupt west side of the park.**

This has created a fault line valley bringing Cretaceous beds in the valley in contact with the Permian beds that make up the hills. A graben has created the Salt Flat Bolson to the west of Guadalupe National Park, for the region now lacks an outlet. This uplift was in a series of movements, and not all at once.

12. **Erosion has dissected the region to its present state and has honeycombed the limestone with caves.**

The softer sediments have been removed by the erosional processes. In spite of the relatively small amount of rainfall there has been enough to form a number of caves in the limestone formations. Landslides still occur today, so the region is still undergoing change.

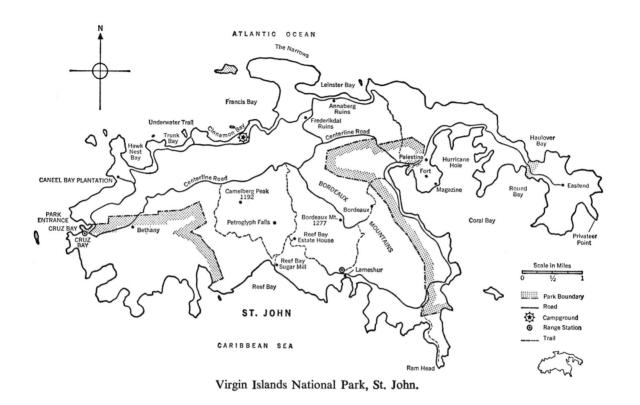

Virgin Islands National Park, St. John.

Geologic Column of Virgin Island National Park

Erathem	System	Group	Formation
Cenozoic	Neogene		Alluvium
	Paleogene		——— Dikes and Plugs ———
Mesozoic	Lower Cretaceous	Virgin Island Group	Diorite Rocks
			Tutu Formation
			Outer Brass Limestone
			Louisenhoj Formation
			Water Island Formation

Modified after Donnelly, Mathews, Cook et. al.

CHAPTER 31
VIRGIN ISLAND NATIONAL PARK

Location: St. Johns Island (Virgin Islands)
Area: 14,418.92 acres
Established: August 2, 1956

LOCAL HISTORY

The first settlers were a fairly peaceful group of Indians called the Arawak and Ciboneys. These were farmers who raised crops. They also spent a great deal of time smoking hemp (heroin) and getting high while lying in their hammocks. Potshards are found scattered over the island, especially near the spring at Cinnamon Bay. Lack of water probably was responsible for a relatively small Indian population.

Petroglyphs scattered all over St. Johns especially at Reef Bay suggesting that this region may have been an ancient shrine.

From the Amazon River Basin came the fierce Carib Indians. They either absorbed or exterminated the gentle Arawak and Ciboney Indians. These South American aborigines were cannibals.

Christopher Columbus is the official discoveror of the Islands in 1493. He called this archipelago "Las Virgenes" or The Virgins. This was in honor of St. Ursula (of Cologne) who was a martyr and her 11,000 virgins.

After Columbus, came the other countries, the Dutch, English, Spanish, French and Danish explorers. Following them were the pirates for this was part of the so-called "Spanish Main." Captain Kidd, Bluebeard and other pirates called this area the Sugar Islands. The pirate activity lasted around 100 years.

The Danes claimed the Virgin Islands in 1687, but didn't take over until 1717 when the planters and soldiers moved in. The planters brought in slaves for labor and divided St. Johns Island into estates. These were the first permanent settlements. A fort was also built (the ruins still exist today) in defiance of British claims to these islands.

With the slave labor imported from Africa (as many as 5,000) the plantations were able to raise sugar cane and cotton. By the year 1726 all land that could possibly be farmed was covered with these crops. This is why there are no virgin forests on St. John. All is second growth that has taken over the islands since the 1800s. Planting was accomplished by terracing the island, some of the walls still can be seen today.

A drought triggered off a slave revolt, which was a long bitter fight of several months. There was a heavy loss of life on both sides. The only planters that managed to survive were the ones that took refuge at Durlieu's Caneel Bay Plantation. It was located on a neck of land that they were able to defend even though they were besieged for months.

The revolt was finally put down by a soldier imported from Martinque (with France's permission) who was merciless. The few remaining slaves that revolted and were still alive built rafts and went over to the British Tortola's.

After the revolt the planters prospered more than ever. By 1789 the population was 167 whites, 2,200 slaves and 16 free Negroes. After 1802 importation of slaves was banned by Denmark. It was finally abolished in 1848. The plantations could not operate without the cheap slave labor so most of the planters freed their slaves and left. The population was reduced to less than 1,000. The freed slaves let the land go back to the bush.

During the 1920s the population was down to five whites and 500 natives. The so-called Capitol had two frame cottages, a barracks on stilts (United States Marines) and 10-15 board shacks. By 1965 the population rose to 1,300.

During World War I the United States bought the Islands from the Danes to keep the Germans away from its back door for 25 million dollars.

Lawrence Rockefeller by means of the Jackson Hole Preserve Corporation bought up land over the years. In 1956 he presented 5,000 acres to the United States as a National Park. The government

acquired more land in 1956 and in 1962 some 5,000 acres of offshore submerged land was added. It has the only underwater trail in a National Park.

GEOLOGIC FEATURES

The island is five by nine miles and has a total area of nineteen acres. The Bordeaux Mountains occupies about two-thirds of the island. Its elevation is 1,277 feet above sea level. The bedrock consists of lava flows, flow breccias, breccias, water deposited tuffs, and a limestone formation (interbedded). The drainage basins are located along fault zones with the smaller valleys along the joint systems and minor faults.

The *aquifers* or formations that transmit water are found in three types of material, they are: (1) valleys filled with alluvium; (2) beach sands; (3) volcanic tuffs and fractured volcanic rock.

There are no permanent streams on the island, only a few pools that are spring fed (usually in stream channels). These move along the fault zones or joint systems. The highly mineralized water acts as a laxative at first, but most people adjust to it very quickly.

Saprolite is highly weathered rock, it can be seen in the fractured rock zones only. Elsewhere it is missing and the soil rests directly upon the weathered bedrock that is anywhere from a few feet thick to 50 feet. One unusual exception to this is in a highly jointed region where weathering occurs down to 180 feet.

In the geologic past sea level used to be lower by at least 200 feet. This can be seen where the drop off occurs between the islands. Two additional erosional surfaces can be found at 280 feet and 900 feet above sea level, indicating that the island has not been stable during the Holocene.

An important feature of the islands are the *coral reefs.* The most common forms of coral found today are the antler, elkhorn, staghorn, brain and star corals. These form the main reef. Growing upon the reef are sponges, sea whips, sea fans, and sea plumes (gorgonians). Brightly colored tropical fish live among the crevices in the reef.

Corals are very sensitive animals and prefer to grow in water that ranges from 23° to 25°C (73°-77°F). The lowest limit is 18°C (64°F). The water has to be free of mud and silt and it has to be saltwater. The coral grows best on the side that faces the sea because the currents can bring an adequate food supply to them plus oxygen. The coral polyp must have sunlight and are very rarely found any deeper than 100 feet. The average rate of growth is approximately 14 mm per year (one foot in 20 years). If sea level rises faster than the reef can grow, the reef can "drown." The coral animal or *polyp* lives inside a cup-shaped depression called a *theca* which is the framework of the whole colony. The animals grow up and out in search of food, for they are dependent upon food coming to them. Different species have different shapes. The spaces between the structures are cemented together by an algae (coralline) called *nullipores.* Calcite forms inside the animal and over surfaces all to help bind the reef together.

Reefs are divided into three groups: (1) fringing reef; (2) barrier reef; (3) atolls. The fringing reef is a platform of coral that grows on the fringes of the island; the barrier reef has a lagoon between it and the land; while the atoll is a circular reef that encloses a lagoon in the center. St. John's Island has a fringing reef around it.

Contact metamorphism has changed the country rock to a hornblende hornfel. An igneous quartz-andesine dike can be seen at Caneel Bay.

Two very prominent strike-slip faults (movement is along the strike or trend of the beds) occur. The one on the west trends to the northwest; the one on the east to the northeast. There is also a series of normal faults (footwall moves up in relation to the hanging wall) that trend north-south and have offset the strike slip faults.

Hydrothermal alteration is the alteration of bedrock by hot water solutions. They form in this area red and white cliffs. The summit of Bordeaux Mountain is an example, also in sections along the southern shore located in the Water Island formation (from west to east), Chocolate Hole, Hart Bay, White Cliffs, John's Folly Bay and Red Point. The pyrite (FeS_2) has oxidized to produce the red stains, the white is caused by the formation *kaolin,* a clay mineral (formed by the decomposition of the feldspar minerals).

Contemporaneous slump structure occurs in the Outer Brass limestone and Tutu formation. While the sediments were semiconsolidated they slumped slightly producing elongate, miniature recumbent (lying on side) folds.

Replacement pegmatites are pegmatites formed by hydrothermal solutions (normal way) in metamorphic rocks that have been formed by replacement. They can be seen at Leinster Point.

Foliation and fracture *cleavage* form in metamorphic rocks. The foliation is the alignment of the minerals in the rock, while rock cleavage is the tendency of a rock to split along parallel planes which may or may not be parallel to the bedding planes. In the Virgin Islands, it occurs in the Tutu and Outer Brass beds striking to the east-west and dipping north.

GEOLOGIC HISTORY

1. Extrusion of Water Island formation (lava flows). Uplift.

The Virgin Islands formed by volcanic eruptions and deposition of limestones during the Cretaceous period or earlier on a fairly level seafloor. Some intrusion, mild deformation and metamorphism occurred during late Cretaceous or early Paleogene time. Then approximately 108 million years ago (during late Cretaceous time) an extrusion of almost two miles in thickness of volcanic rock poured out onto the ocean floor. These beds are called the Water Island formation. The east-west fault system was probably the source area. Movement may have occurred at the same time, causing the sharp north boundary. The formation consists of completely underwater lava flows with a few scattered layers of tuff. The lava flows are very basic in composition and are called *keratophyre* flows. Also present are *spilite* flows, another basic flow with chlorite and albite. These flows may be metamorphosed. Radiolarian tuffs are rocks consisting of pyroclastic debris with 5-10 percent of the rock being the SiO_2 radiolarian skeletons (single-celled animals). Flow breccia and tuffs also occur. About 15,000 feet are exposed in the south slopes of the Virgin Islands and can be seen near the intersection of Centerline Road and Bordeaux Mountain Road or near Chocolate Point Hole and along Centerline to Reef Bay trail. Uplift is responsible for the tilting to the north of these volcanic beds, forming a homocline with a dip 15°-90° (average 40°).

2. Uplift and erosion. Deposition of Louisenhoj (Loo-e-zna-hoi) formation.

The uplifting was very rapid, so the character of the rocks changed from submarine intrusions of the Water Island formation to the extrusion of pyroclastic debris which was deposited on land. Elevation occurred along the east-west fault system with the north block dropping down and being covered with sediments. Interbedded with this debris are the coarse beach deposits of weathered Water Island debris and layers of stream-deposited conglomerates. This all indicates that the Islands

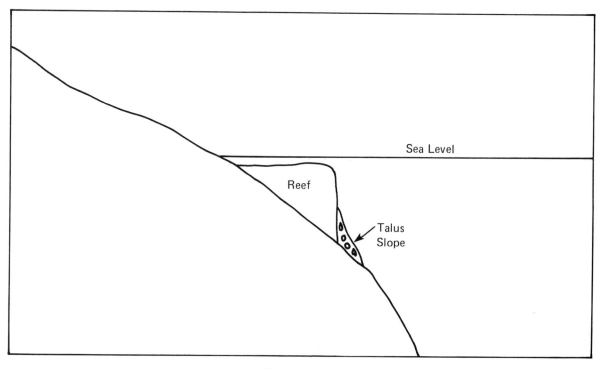

Fringing Reef

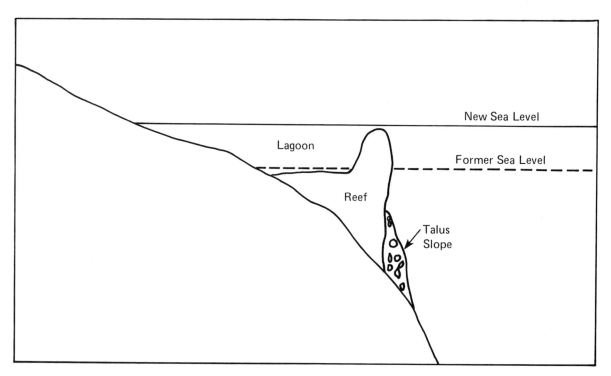

Barrier Reef

were above sea level. Composition of the rock is an augite andesite breccia and tuff that contains minor zones of conglomerate. These land deposited volcanic debris occurred on the flanks of volcanoes. They are with the interbedded stream and beach deposits and are called the Louisenhoj formation. It is late Cretaceous in age and can be seen near Cruz Bay and Rendezvous Bay. The vent is probably located in Pillsbury Sound between St. John and St. Thomas islands. The Louisenhoj covers the western half of the Island and is mainly cone debris (coarse). It has a bluish cast and is called by the natives "Blue Bit." Erosion leveled the top of the volcano and formed a smooth plain. Four foot blocks are found mixed with the ash deposits, thus silently testifying to the violence of the eruption.

3. Submergence, deposition of Outer Brass limestone.

Volcanic activity ceased for a while and the Virgin Islands once again became part of the sea floor with slow subsidence. Marine sediments were deposited in fairly deep water forming the Outer Brass limestone. This formation is thin-bedded with a high silica content even though it is primarily limestone. The reason is two-fold, first the presence of *Radiolaria* with SiO_2 tests (shells) and secondly the thin layers of tuff interbedded with the limestone (it also has a high silica content). The limestone was deposited below wave base since none of the material has been reworked. The tuff indicates volcanic activity nearby above sea level and makes up about ten percent of the Outer Brass limestone, the remaining 90 percent being a siliceous limestone. The late Cretaceous-early Paleogene intrusion metamorphosed (by contact metamorphism) part of the limestone. It can be seen at Annaberg Point on trails from Cruz Bay to Cancel, around Cinnamon Bay and the base of Leinster Hill (west side).

4. Uplift and tilting. Deposition of Tutu formation.

It was still during the Cretaceous (it lasted about 70 million years), when the land was elevated once again, but this time sharply tilted to the north, exposing the Louisenhoj and Outer Brass formations to erosion. From this debris of limestone and volcanic material was deposited the Tutu formation, which can be seen from Maho Bay to Leinster Point. The Louisenhoj, Outer Brass and Tutu formations are all members of the Virgin Island Group.

The Tutu formation is primarily a *volcanic wacke* (poorly mixed sediment consisting of volcanic particles). It also contains a *mega breccia* (large angular blocks cemented together) of limestone at Mary Point and Leinster Point. Some blocks have been found up to 100 feet long. They contain fossils but the metamorphism makes identification difficult. Outcroppings elsewhere indicate Upper Cretaceous age. It was deposited in an environment of uplifting and shedding of rocks from a rise that was steep walled.

5. Faulting, intrusion and metamorphism. Forming mountains and erosion.

Uplift and deformation continued which resulted in folding and faulting of the beds and considerable volcanic activity by the end of the Cretaceous period. Intrusion of diorite stocks (?) in The Narrows and diabase dikes by Caneel Bay occurred. Contact metamorphism was produced at a shallow depth. This region was probably attached to Puerto Rico then was uplifted and underwent erosion from youthful mountains to rolling plains, then the area was uplifted again (rejuvenation). This happened at least three times in the time span from late Cretaceous to early Paleogene.

6. Submergence and formation of coral reefs.

Near the end of Pleistocene, the water level was high and the currents had shifted. The land drifted to its present position so there was warm water that permitted fringing reefs to form made up of many varieties of coral, thus the present features of today developed.

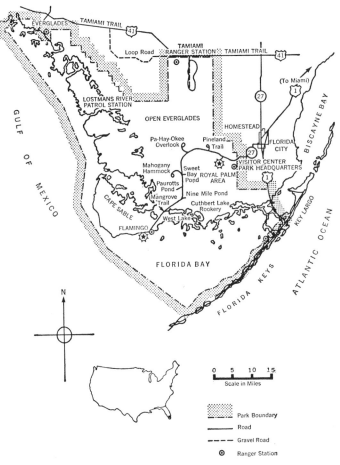

Everglades National Park, Florida.

Geologic Column of Everglades National Park

Erathem	System	Series	Formation	
Cenozoic	Neogene	Pleistocene (Sangamon)	Key Largo Limestone	Miami Oolite
		Pliocene	Caloosahatchee Formation	
			Tamiami Limy Sandstone Facies	
			Buckingham Marl Member	

Modified after Popenal, Simpson et. al.

PART V
FORMED BY
GROUND WATER

CHAPTER 32
EVERGLADES NATIONAL PARK

Location: S. Florida
Area: 1,400,533.00 acres
Established: June 20, 1947 (authorized May 30, 1934)

LOCAL HISTORY

Spain owned Florida for 306 years until the United States, in 1819, acquired it as a territory. In 1845 it became the 26th State of the Union. There were many explorers such as Ponce de Leon, Panfilio de Narvaez, Cabeza de Vaca, Hernando De Soto, but not one of these apparently entered the Everglades.

However, it was not uninhabited by man, the Calusa and Tequesta Indians lived there before the time of Christ. The Calusa even outlasted the 16th century Spanish explorers by some 250 years. The Seminole ("runaway or separtist" in Creek) Indian tribe is actually a collection of three tribes from Alabama and Georgia—the Creeks, Hitchiti and Yuchi. They banded together after the American Revolution. The women adapted their hairdos and dress after the Gibson Girls. In a war that began in 1835 and ended (with no treaty signed) in 1842 a few hundred managed to escape the United States Army and hide in the Everglades. The rest were captured and sent to Oklahoma. The descendants of those who escaped, form the now familiar Seminoles of today. They call the Everglades "Pahay-okee" or "river of grass."

A landscape artist in the mid 1920s used to wander around the Everglades and decided it should be preserved as a National Park. His name was Ernest F. Coe. The commercial hunters were against him and even threatened his life. But this did not deter him. In 1929 at his own expense, he flew to Washington D.C. and convinced the Park Service of the need to have this area preserved as a National Park. A bill creating the park was passed by Congress in 1935, as soon as the land was donated. For the next 20 years, Coe worked to establish the park. Finally the Florida Legislature voted the money to purchase the land and Coe's dream became a reality on June 10, 1947.

GEOLOGIC FEATURES

The main geologic feature is not even located in the park boundaries, it is Lake Okeechobee, or "Big Water." It is only 22 feet deep but still manages to cover 725 square miles while draining some 4,000 square miles. It is only 15.5 feet above sea level (Everglades is between 10 and 12 feet). Since it does not have a regular outlet, it simply overflows its south rim and a sheet of water gradually flowed to the glades.

Unfortunately, man has interferred with this natural cycle and has built canals and dikes to drain water off and depriving the Everglades of its lifeblood. Prior to the building of the canals, the land was gradually becoming higher with the deposition of silt. The canals now stop this process and silt up instead. Removal of water is causing the area to become lower because of *compaction* and *dessiccation* (loss of water from pore spaces).

The entire Everglades is a unique environment for its maximum elevation is only 12 feet above sea level and it averages only 10 feet above. To illustrate this point—the road to Flamingo has a sign which reads "Rock Reef Pass, elevation 3.1 feet." The road passes over a small ridge of limestone.

The main bedrock is the Miami Oolite (Pleistocene), a variety of limestone that looks like fish eggs cemented together. Oolitic limestone is presently forming today around the Bahamas. Requirements are (1) turbulent, agitating water; (2) warm supersaturated water; (3) tidal current action; (4) shallow banks.

The shallow banks keep the sediments above wave base so they are constantly being moved around. The shallowness permits the water to stay warm, hence is able to hold the calcium carbonate in solution until it becomes supersaturated. The tidal current brings fresh material in constantly. The calcite precipitates out and because of the agitation the crystals move around, and acts as a nuclei for additional material to coat. These round spheres are called oolites, they are from 1/16 to 2 mm in size.

Over this limestone layer is *muck, peat* and sandstone. The muck is a dark-colored soil with large amounts of decomposed organic material. The peat is the first stage in the formation of coal and is the material that remains after partial decomposition of plant material. The sandstone, in part, is brought down by longshore currents from the Carolinas.

Because of the continuous flow of water, *hammocks* have a tendency to develop. The trees are forced to grow with tangled, twisted roots that are designed to keep the trunks above water. They act as traps for sediments so gradually a land area builds up. In the saw-grass marshes, the alligator builds a nest two or three feet above water level. When they abandon them plants take over and add to them and trees develop upon them creating additional hammocks.

Potholes can also be found, they are depressions or solution pits formed in the soluble limestone. They range in size from small pits to large holes that can be measured in feet or occasionally tens of feet and may be several feet deep.

GEOLOGIC HISTORY

1. **During the Precambrian deposition of beds and pyroclastic debris in the southern part of Florida.**

This area was under water and not part of the continental area of today. The history is very sketchy and is only known from drill cores. It consists of granite, diorite and metamorphic rocks.

2. **Deposition of Paleozoic (Ordovician-Devonian) sandstone and shales.**

These beds are exposed in the northern half of the Florida Peninsula. They were probably stripped by erosion from the southern portion, instead Mesozoic beds are found resting upon the basement complex. Possible metamorphism of these beds occurred at this time.

3. **Erosion during Paleozoic and Mesozoic.**

During this time it was alternately exposed as low lying dry land or covered by shallow seas. Most of the time the peninsula of Florida was sinking slowly with horizontal Paleozoic beds being deposited on its surface. Thus today the Paleozoic "basement" rock are covered by more than 4,000' of sedimentary strata.

4. **During Eocene time Florida was a submarine bank of Ocala (?) limestone.**

This submarine bank separates the deeper waters of the Atlantic Ocean from the shallow water of the Gulf of Mexico geosyncline. The Floridian Plateau is a broad, flat region. It is a little more than 325' above sea level at its highest point. Most of it lies (more than half to the west) submerged beneath the water of both the Atlantic Ocean and the Gulf of Mexico.

5. **During Miocene time, arching of submarine bank during Ocala uplift.**

This created the peninsula of Florida as we know it today. Up to that point it was an offshore submarine bank undergoing deposition during Oligocene and Miocene time.

6. **Deposition of Pliocene beds.**

The uplift was gradual and most of the marls

formed in fresh or brackish water deposits. Marl is lime mixed with clay, it is probably the Buckingham Marl Member of Lower Pliocene Age.

The Upper Pliocene Caloosahatchee may underlie this area also, as these beds are found down around the Florida Keys.

7. Pleistocene glaciation, Florida Penninsula undergoing erosion because of lowering of sea level.

During the Pleistocene, sea level was as much as 300 or 400 feet lower than it is today, because of the water being locked up on land in the form of glaciers. In a region only a few feet above sea level this is a considerable drop and it exposed land that is now under water. Stream channels formed and produced a karst topography in the soft limestone and caliche. A series of terraces developed that appear as a series of "islands" through which the water of the Everglades drain. Outside the Everglades around the city of Miami is a ridge (remnant of the earlier beaches) through which springs issue. This ridge was the route that animals used from the tropics to migrate north and some forms of life to migrate south.

8. Deposition of Miami oolite during Sangamon interglacial stage.

During the interglacial stages sea level resumed its original level and the limestone deposited is the one that you see in the park today, it is called the Miami oolite. It is characterized by tiny calcareous spheres which resemble fish eggs. In places where it outcrops it has a rough pitted surface produced by the solvent action of rain water and subsurface water.

9. Erosion during Wisconsin glacial stage because of lowering of sea level.

Development of karst topography once again, although it is more obvious outside the park boundaries. Some homes are being lost due to sinking down in sinkholes (because of the weight of the house, the hidden cave collapses and the house collapses with it). The Miami Oolite is less than 5,000 years old.

10. Formation of Lake Okeechobee and Florida Everglades, deposition of sand marl and peat of today.

In the lower generally flooded parts of the Ever-

glades, the limestone is covered by peat or muck. When the hurricanes sweep mud inland from the coastal flats, the plants growing in the thin layer of soil over the oolite are smothered and die forming peat. Yet new homes for such plants are constantly being created. The diversion of much of the water is causing the marl to dry out and form mudcracks causing much of the land to subside. Because of this dessiccation (loss of water from pore space) there is a great loss of plant and animal life. Also the high organic content of this material allows it to catch on fire creating another hazard to life forms. Nature can repair the damage done by fires, hurricanes, droughts and flooding, but not the damage by man such as diverting the water away or building jet ports next to the park. Life is in a delicate balance from microscopic form of life to the larger forms. The chain begins with the fresh water plankton which the crayfish larve consume. They in turn are eaten by tadpoles which the little fish eat and are eaten in turn by the big fish, they feed the birds, crocodiles and alligators. Hence remove the water and the entire park dies.

The water in the park is both fresh and salt, so both environments are represented here. This is the reason why the worlds largest Mangroove Grove can exist here. The seeds germinate while still attached to the parent. When it detaches itself, it grows breathing roots to which coral and other marine-shelled animals attach. Eventually this becomes soil and adds to the land and hammock.

The sawgrass (belongs to sedge or bullrush family) is among the oldest growing plants that still exist today, it grew in these swamps at least 4,000 years ago. The three sided blades (up to seven feet long) have sharp teeth which when the grass decomposes leaves the spicules in the soil. They will even cut up the hands of persons handling the soil if they are not wearing gloves.

The ecology has already been upset by the accidental addition of water hyacinth (lavender blossoms) that are choking the canals and not allowing enough room for other forms of life. The National Park Service is importing mantees or sea cows, an animal that is related to an elephant (more than a whale) which weighs around 1,000 pounds. Its body is cigar-shaped, it has paddlelike fore limbs and a broad flat tail. They live in South America and their favorite food is water hyacinth, so they help to keep the canals open. Unfortunately poachers have found that sea-cow steak tastes good, so they are killing them off very rapidly.

Also endangered because of poachers are the crocodiles and alligators. Both live in the Everglades, the difference between them can be seen in the chart below.

Characteristics	Alligator	Crocodile
Shape of head	Wide	Slim
Teeth	No exposed side	Fourth side tooth exposed
Type of Water	Prefers fresh	Prefers salt
Diet	Turtles and Garfish	Prefers carion
Shape of snout	Blunt-nosed	Long-snouted
Color	Darker green	Lighter green

At one time 19 footers could be found but no longer. A very important function of these animals is the digging of "gator holes" for it is the only place that the fish and other aquatic forms of life can survive in times of low water. This in turn gives the birds a place to find food. Each form of life is very dependent upon one another and they all depend upon the flow of water through the Everglades to survive. So special care must be taken to preserve this unusual area.

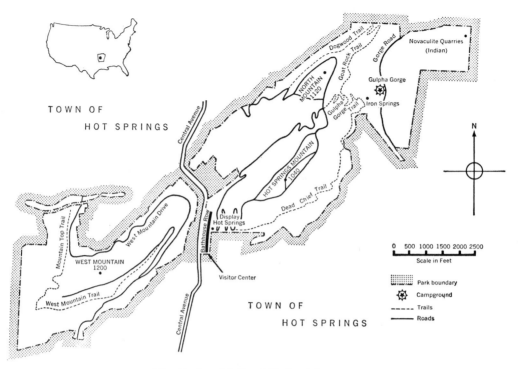

Hot Springs National Park, Arkansas.

Geologic Column of Hot Springs National Park

Erathem	System	Series	Formation	
Paleozoic	Pennsylvanian	Lampasos	Atoka Formation	
		Morrow	John's Valley Shale	
	Mississippian	Chester	Jackfork Sandstone	
			Stanley Shale	
			Hot Springs Sandstone	
		Kinderhook	Arkansas Novaculite	Upper Member
				Middle Member
	Devonian	Lower		Lower Member

Modified after Weller, Kummel et. al.

CHAPTER 33

HOT SPRINGS NATIONAL PARK

Location: S.W. Central Arkansas
Size: 3,535.24 acres
Established: March 4, 1921

LOCAL HISTORY

According to the Indians the 47 hot springs was the dwelling place of "The Great Spirit." Many battles were fought between different tribes for many years, to decide who would possess the springs. Finally the various tribes got together and decided the Great Spirit had meant this to belong to all who were sick, so these springs became neutral ground. They named this area "the land of the peace." Braves from enemy tribes would lie side by side in the hot mud to ease the rheumatism and other ailments acquired while riding the war path. A winding narrow cave which leads to the springs was discovered when digging the foundations for the Quapaw Bathhouse. It was probably used by the Indians.

In 1541 (a few months before his death) Hernando de Soto passed through the area and may have bathed in the waters.

Early settlers and trappers used the hot springs because in 1804 William Dunbar and Dr. George Hunter found wood buildings that had been erected as temporary quarters while people soaked at the hot springs. Dunbar and Hunter spent a month in this area (they were ordered by President Jefferson to explore the Ouachita River area) and analyzed the waters.

The first permanent house was built by Jean Emanuel Prudhomme (planter from Louisiana) after the Natchitoches Indians guided him to this area in 1807. Next came Isaac Cates who made a wooden trough so he could lie down in it and have the water flow over him. With him (the year was 1807) was John Periful (both were from Alabama) who acquired cabins from various sources and rented them out to visitors. This was to augment his income from hunting and trapping.

By 1820 the settlement had expanded and a log cabin inn had been opened by John Millard (it did not last long). Another hotel was erected in 1828 by Ludovicus Belding (he came from Massachussetts with his family). The first bathhouse (1830) was built by Asa Thompson, they were log cabins with tubs made out of planks. At the same time another building was built as a "vapor bath," it was erected over a water filled niche.

Because these baths were so popular Congress (to prevent exploitation by commercial business) set aside a total of four sections of land around the hot springs as a preserve in 1832. In 1836 Arkansas became a state and in 1851 the town that grew around the preserve was sold for private ownership. Once a rail line was built in 1875 the town grew. In 1882 the Army and Navy General Hospital was built for servicemen.

Today there are 17 concessioneer bathhouses (eight in city, nine in park) that are operated under Federal regulations. Only two of the springs are left natural (the Display Springs) so the tourist can see what they looked like. The remaining 45 are seated (to prevent contamination and loss of gas) and piped to a central reservoir (insulated so as not to lose heat). The water is piped to the various bathhouses, drinking fountains and jug fountains.

GEOLOGIC FEATURES

The park is ringed by five wooded mountains (West 1,200 feet; Hot Springs 1,040 feet; North 1,120 feet; Indian 1,120+ feet; and Sugarloaf 1,160 feet) which are a series of plunging folds trending northeast-southwest, they have been thrust faulted.

The *hot springs* issue from a fault zone in gray volcanic tuff at the base of Hot Springs Mountain. They have a combined flow of over a million gallons per day. The average temperature is 143°F and the composition of the water is almost the same for all of the springs. Unlike most hot springs the water does not have a bad smell. The chemicals found in the water (in order of their abundance) in

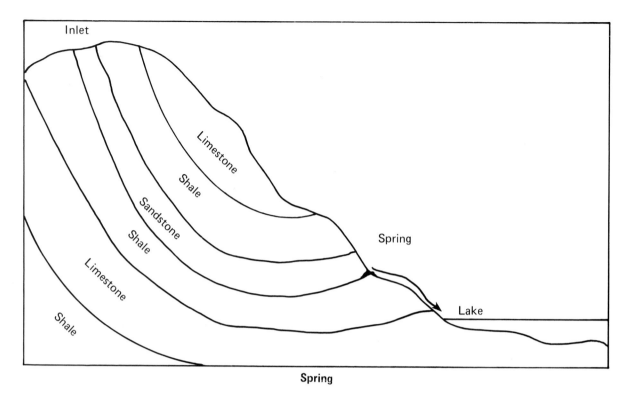

Spring

parts per million are: bicarbonate, silica and calcium, sulfate, magnesium, sodium, chloride, potassium, nitrate, flouride and phosphate. The gases are carbon dioxide, nitrogen, oxygen and radon. The age of the water has been determined by tritium analyses (a hydrogen isotope) it shows that some of the water is less than 20 years old (small amount), and most is considerably older.

The source of heat for these hot springs has never been solved. These are several theories, they are as follows:

a. Meteoric (rain) water travels down the fissures in the highly fractured rock and passes by a hot igneous intrusion.
b. Magmatic or juvenile (newly created water) comes from a magma chamber.
c. Very deep ground water is heated by the geothermal gradient.
d. Chemical reactions near the bottom of the springs.
e. Heat is produced by friction from faulting rocks.
f. Heat is produced by compression from the weight of overlying beds.
g. Heat is given off by radioactive minerals buried so deep that instruments can't detect them.

Novaculite occurs in the park (an Indian quarry is located in the northeast corner of the park). It is a flintlike rock that actually consists of minute quartz crystals cemented with chalcedony (kal-sid-knee) a cryptocrystalline form of quartz. If it is compact and stained (with iron oxide or maganese oxide), it is highly sought after because it occurs in all shades of reds, oranges, yellows, also green and gray. Sometimes it is sold as Arkansas Stone. It is used in lapidary work, and as whetstones. The Indians used it to make arrowheads and other stone tools.

The hot springs area is noted worldwide for its *quartz crystals* (cannot be collected inside park boundaries) that are found inside and outside the park. The best collecting locations are along the ridges of the mountains. The crystals are noted for their size, some are two and three feet long, but they occur in all sizes. The pincushion variety are small crystals in clusters, or thin needle crystals can be found also the variety called candle quartz (six times as long as they are thick).

Tufa deposits are formed by hot springs, they can be either of calcium carbonate or silica composition. The *blue-green algae* (single-celled plants) which live in some of the hot springs are responsible for the formation of this insoluble rock. They have to adapt to live in the hot water and one of the modifications is the ability to precipitate minerals out of the water. The 10-15 feet thick tufa deposits can be seen at the Display Springs beneath the vegetation.

Conglomerate and quartz veins (in the sandstone) also occur at Hot Springs National Park.

GEOLOGIC HISTORY

1. **Deposition of sedimentary beds in the Ouachita geosyncline, mostly lime, sand and muds.**

Some 26,000 feet of Paleozoic sediments were deposited in this geosyncline, with only 1/5 of them (6,000 feet) forming by the end of Devonian. The remaining 20,000 feet were deposited from Early Mississippian to mid-Pennsylvanian time. The early sediments are shales, sandstones, conglomerates, novaculite and a landslide formation (John's Valley shale).

2. **During Late Devonian-Early Mississippian time deposition of the Arkansas novaculite.**

Up to 950 feet of the novaculite was deposited. Ordinarily it is a white chertlike sedimentary rock (SiO_2 composition) that contains bedding planes, its texture is coarser than chert. Its hardness (seven on Moh's scale) and even texture makes it a desirable rock to use as an abrasive. In the hot springs area it has been stained with mineral matter (Fe_2O_3 and MnO_2) and is very colorful. Weathering (decomposition) on the ocean floor of the vol-canic ash that was blown and deposited in this region (source area probably to the south) is the most probable origin of this sediment.

3. **Erosion, nondeposition or removal of beds during mid-Mississippian time.**

The land areas were very low as shown by the relatively thin deposits formed. Either they did not supply sediments to the region or the seas temporarily retreated from the hot springs region allowing it to become land once again. Perhaps the sequence was there and has been lost in the thrust fault sequence that occurs in the region. Another possibility for the lack of deposits are the turbidity currents which may have carried these sands to a deeper portion of the basin. The contact between those beds are molds of an irregular surface that makes up the top of the shale layers, these are called *sole markings.* The orientation indicates first the sediments flowed down the slope, then parallel to the axis. At any rate some of the Mississippian beds are missing.

4. **Deposition of the Hot Springs Sandstone forming a disconformity between the sandstone and novaculite.**

The springs occur in a relatively small area along a fault located between Hot Springs Mountain and West Mountain that is located in a valley. The water issues mainly from the Hot Springs Sandstone that is Mississippian (Chester) in age being deposited some 300 million years ago. The porous sandstones also outcrop in the valley floor between West Mountain and Sugarloaf Mountain (because of the plunging folds). When it rains the water soaks into these beds, seeps down through the bedrock that is also highly shattered thus facilitating the transmission of the water. The water becomes heated by some method and finally emerges at the base of Hot Springs Mountain as a series of 47 springs.

5. Deposition of Stanley and Jackfork formations during Upper Mississippian time.

With the beginning deposition of the Stanley and Jackfort formation accumulation was very rapid. These beds were part of a delta on the south side of the Ouachita geosyncline. They make up a 1,200 foot wedge. The Stanley formation contains volcanic ash layers (at least five) which range from six to 85 feet thick and contain radiolarians and conodonts. This indicates the uneasy state the area was in. The Jackfort Sandstone indicates the source area was rising and streams with their increased gradients were able to transport coarser material. It contains poorly preserved pieces of plant fragments. The sand is light to dark gray in color and ranges from a fine to a coarse sand.

6. Ouachita orogeny during the Pennsylvanian which uplifted and formed the Ouachita Mountains. Erosion during Mesozoic.

The uplifting, first of all, folded the beds into a series of folds and also fractured them permitting secondary quartz and other materials to fill in these fissures. Next these intensely deformed, fractured beds, were thrust faulted northward in a series a huge thrust plates, probably during Permian time.

Uplifting means erosion and during much of the Mesozoic the streams began their stripping process removing the weaker, less resistant beds producing valleys and leaving the more resistant beds to stand as hills. Streams flowing over the surface became superimposed or antecedent depending upon the situation forming the wind and water gaps through the hills.

7. Final uplift during the Pleistocene.

Uplift produced additional folding and produced the series of fissures through which the springs presently flow. The erosional process is still dominant with some deposition occurring in the form of tufa deposits by the springs and modern day alluvium.

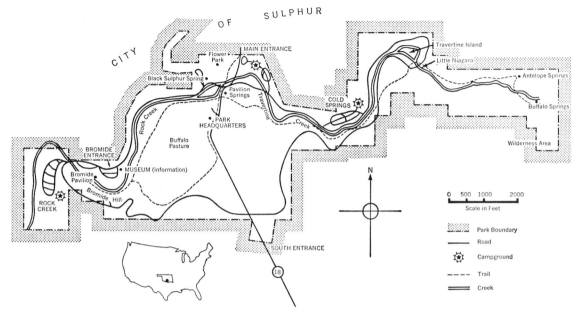

Platt National Park, Oklahoma.

Generalized Geologic Column of Platte National Park

Erathem	System	Series	Group	Formation
Paleozoic	Pennsylvanian	Virgilian		Vanoss Conglomerate
		Missourian	Hoxbar	
		Desmoinesian	Deese	
		Atokan	Dornick Hills	
		Morrowan	Springer	
	Mississippian	Chesterian		Caney Shale
		Kinderhookian		Woodford Shales
	Devonian	Ulsterian	Hunton	
	Silurian	Niagaran		
		Albion		
	Ordovician	Cincinnatian		Sylvan Limestone
		Champlainian		Viola Limestone

Modified after Retitti, Kummel, Clark and Stearn et. al.

CHAPTER 34

PLATT NATIONAL PARK

Location: S. Oklahoma
Area: 911.97 acres
Established: June 29, 1906

LOCAL HISTORY

Indians came to these cold water springs to drink the water (and not to bathe in it) long before white man ever appeared. One of the early white men to come to this park was Thomas Nuttal, a botanist, in 1902. Undoubtedly there were others there before him, but they left no written records.

Platt National Park is the smallest of the National Park System, a total of 1.5 square miles. It originally was part of the Old Indian Territory and belonged to the Chickasaw (and Choctaws) Nations.

They gave their "Peaceful Valley of Rippling Waters" to the United States in 1902 on the condition that it was to be preserved for the benefit of everyone and that the waters would always be free. This is the first (and only) property that the Indians have of their own accord given to the United States for conservation purposes. The United States promptly set it aside as Sulphur Springs Reservation. It was established as a National Park in 1906 and named after Senator Orville H. Platt, the United States Senator from Connecticut for 26 years, who was a good friend of the Indians.

GEOLOGIC FEATURES

The main geologic feature is the 32 large (and a few smaller) springs, each having a different chemical composition. Eighteen of them are classified as sulfur springs, four of them iron, three bromide, and the rest are simply freshwater springs (six large, several small).

These springs are located at the foot of the Arbuckle Mountains, which have been formed by folding and deep vertical faults. The beds are Precambrian and Paleozoic in age. They are a mixture of metasediments, and igneous rocks such as granite, gabbro, rhyolite and basalt. While the sedimentary beds are the usual conglomerate sandstone, shale and limestone, they also include dolomite, oolite, glauconite and chert. Each of these beds has a different chemical makeup that is responsible for the variation in the composition of the springs. The vertical faults have made all of these beds available to the ground water table.

Hence when the meteoric water falls upon the land and onto the vertical beds with their faults, it enters the various layers. As the water slowly percolates through these beds, it picks up different chemicals dependent upon where it flows. When the ground water issues out as a spring, it has its own unique composition.

Most of the springs issue from the base of Bromide Hill, a 140 foot high bluff. The two biggest springs in the park are at the eastern end, called Buffalo and Antelope springs. They are the source of Travertine Creek. Combined they issue five million gallons per day, which averages about 2,000 gallons per minute. The two springs were named after the animals which used to use them as a watering hole. Antelope Spring flows out of a group of rocks and Buffalo Spring comes up through a bed of sand. Both are freshwater springs which flow through bedrock that has very little soluble material. The water, however, contains calcium carbonate ($CaCO_3$), which is deposited as *travertine* or *calcareous tufa.*

Travertine Creek is named after this material, and it has deposited large amounts of the spongy substance around Travertine Island. Thick enough accumulations have occurred in the past to create waterfalls and rapids (e.g., Travertine Falls).

Black Sulfur Spring contains noticeable amounts of sulfur. Hillside and Pavilion springs are located nearby. Bromide Spring and Medicine Spring are the two main springs that issue from the base of Bromide Hill.

The connection between the amount of pre-

cipitation and flow of the springs is very close. If the weather has been dry the springs start drying up. This indicates the inlet is fairly close by.

GEOLOGIC HISTORY

1. Formation of igneous rocks during the Cambrian upon the Precambrian metasediments.

About 535 million years ago, gabbro and basalt formed in the vicinity of the Arbuckle Mountains. These were in turn covered by granite and rhyolite some 525 million years ago during the Cambrian period.

2. Transgression of Sauk Sea depositing. The Arbuckle Group.

During Middle Cambrian time this region underwent erosion. The Arbuckle geosyncline formed, depositing sandstone, glauconitic limestone (a complex hydrous potassium iron silicate), limestones and dolomites and chert. There was temporary retreat of the sea, which was known as the Sauk Sea. These beds were a facies change from the source area which was the Canadian Shield.

The sea regressed during early Ordovician time, producing an unconformity.

3. Transgression of the Tippecanoe Sea during the Ordovician.

The Tippecanoe Sea slowly transgressed over the exposed sediments of the Sauk Sea. Fossils indicate that the water was warm. The deposition of the Simpson Group (sandstone, shale and limestone) is probably a facies change of the St. Peter Sandstone deposited further north). These are the beds that provide the bromide and sulfur minerals for the springs in the park. They also contain gas and oil deposits elsewhere.

The Viola limestone was deposited conformably upon the Simpson Group. These beds contain some chert. There was a temporary transgression of the Tippecanoe Sea. When the seas returned the Ferndale limestone and Sylvan shale were deposited. Then once again erosion removed many of these beds.

4. Beginning of Arbuckle uplift during Pennsylvanian period.

In adjacent areas, the Silurian-Devonian Hunton Group was formed. There was deposition of the Mississippian Woodford and Caney shales and limestones, but they were later removed by erosion.

By late Paleozoic the *Wichita* and *Arbuckle* orogenies occurred, which resulted in the stripping of up to 15,000 feet of pre-Pennsylvanian beds covering the granitic-rhyolite basement of the Arbuckle Mountains. These sediments were deposited in the adjacent *Ardmore* and *Andarka* basins at the same time the area was being uplifted between 750 to 1,300 feet, then faulted (vertical faults) and finally folded.

The beds being formed were Pennsylvanian in age producing the Springer, Dornick Hills, Deese and Hoxbar groups. Finally during Virgilian (late Pennsylvanian) time the *Vanoss conglomerate* was deposited, which is the bedrock seen outcropping in Platt National Park.

5. **Criner Hills uplift formed, and was thrust faulted north towards Arbuckle Mountains.**

The uplift created more faulting, and additional folding. These processes created oil and gas traps and also the springs. Some of the rock layers have been tilted upwards where they are either exposed at the surface or are under permeable layers so that they may receive water. These rocks occur a short distance both north and east of the park. When the rain falls upon the land, it infiltrates the soil, enters the bedrock, and picks up the soluble minerals. It then passes through the tilted permeable layers (confined by the nonpermeable ones) and returns to the surface through fractures and fissures as the springs that created Platt National Park.

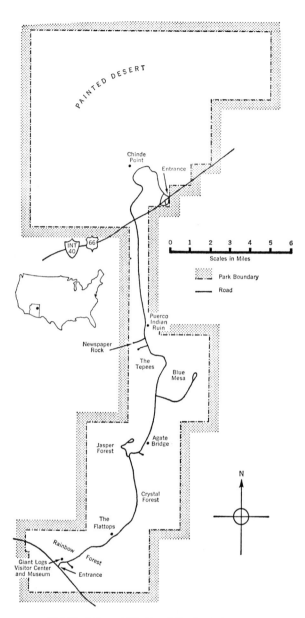

Petrified Forest National Park, Arizona.

Geologic Column of Petrified Forest National Park

Erathem	System	Series	Formation
Mesozoic	Jurassic	Lower	Glen Canyon Group
	Triassic	Upper	Chinle Formation
			Shinarump Conglomerate
		Middle	Moenkopi
		Lower	

Modified after Dunbar.

CHAPTER 35
PETRIFIED FOREST NATIONAL PARK

Location: East Central Arizona
Area: 94,189 acres
Established: 1962

LOCAL HISTORY

According to an Indian legend a goddess who was cold, tired and hungry came to rest in an area where there were hundreds of logs. She found a club and killed a rabbit with anticipation of having it for dinner. When she tried to build a fire with the logs they refused to burn because they were too wet. Angered she cursed the logs and turned them all into stone and that is how the Petrified Forest came into existence.

The Paiute Indians thought the logs were the shafts of arrows of the Thunder God Shinuau. The Navajo on the other hand declared they were the bones of the Great Giant named Yietso.

The prehistoric Indians lived in the region, from 500 A.D. to 1400 A.D. They range from the Basket Makers III culture of the 6th and 7th century to the Pueblo IV culture of the 14th century.

First the Indians lived in round pit houses, or shallow holes that were lined with slabs and had a dome roof consisting of poles, some brush and dirt. Then they advanced to rectangular houses made out of stone (or adobe) and finally to living in pueblos such as Agate House or the Puerco Ruins. Agate House within the park is built entirely out of the logs, the wood is held together by an adobe plaster. The Puerco Indian Ruins are pueblos arranged in a square around a central plaza, they were abandoned about 600 years ago. The entire complex probably has 125-150 rooms, but only three have been excavated. It was probably a two story dwelling which could hold about 100 families. Newspaper Rock is a large rock (sandstone) upon which petroglyphs have been carved. Additional petroglyphs can be seen near Rainbow Forest. The Indians used the petrified wood for a variety of things. Arrowheads, scrapers, hammers and other tools.

The first recorded visitor was Lieutenant Stitgreaves. He was an army officer who in 1851 explored many parts of northern Arizona, right after it was acquired from Mexico.

In 1853 an army expedition led by Lieutenant A.W. Whipple camped near the Black Forest. He described the place as trees being turned into Jaspar. He named the creek that drains the Painted Desert Lithodendron or "stone tree."

Lieutenant Beale led a caravan of camels through the area on his way to California in 1857. However, it stayed fairly unknown until 1878. This is when people began to settle in the region. When the Atlantic and Pacific (today the Santa Fe) railroad was built in 1883 in northern Arizona, the pillage began. Entire carloads of material was shipped to shops and jewelers. Mineral collectors blasted logs apart with dynamite to get the amethyst crystals growing inside. Many thousands of logs were crushed into a powder to be used as a commercial abrasive. It became a Forest Preserve in 1896, so between 1898 and 1900 the area was studied by a United States Geological Survey geologist Lester E. Ward. In 1906 John Muir discovered the Blue Forest (which he named). In 1921 Phytosaurs were discovered in the Blue Forest and were excavated between 1921 and 1929. Fossil leaves were found in the same area in 1933 and were removed and identified in the period from 1933 until 1940.

A crushing mill was to be erected near the logs for the purpose of crushing them into an abrasive for sandpaper. To prevent this Theodore Roosevelt set aside some 40 square miles in 1906 as a National Monument after the territorial legislature petitioned Congress to save the area for future generations. In 1931 it was enlarged to include the Black Forest and in 1962 enlarged again to include the Petrified Forest and elevated to the status of a National Park.

GEOLOGIC FEATURES

The main geologic feature is the petrified wood. The wood has been *silicified* or preserved by silica (SiO_2). The name for this material is *quartz*. Quartz is a very abundant mineral in the crust of the earth, especially on the continents. It occurs in two forms: The crystalline variety in which the atoms had time to arrange themselves in a near orderly pattern, it can form six-sided crystals if given the opportunity. Also the microcrystalline variety where the crystals are so small it is difficult to see them even with a microscope.

The crystalline varieties that occur in the park are the clear quartz usually referred to as *rock crystal* or the purple variety (because of the presence of manganese) called *amethyst.* The crypto-crystalline varieties are chalcedony (kal-sid-knee) a gray or blue translucent material where the crystals are arranged as narrow fibers in parallel bands or the more opaque varieties that are not banded such as flint (dark colored), chert (light colored) and jasper (red, yellow, brown, sometimes green or blue). The variation in colors is the result of impurities in the quartz, manganese oxide will produce purple, carbon a black rock, iron oxide black, brown or

various shades of red. If the quartz is pure it is either colorless or white, but with the impurities a variety of shades.

The wood became petrified when the logs were washed downstream and deposited on the sandbars formed on the floodplain along the sides of the river and were then buried. Nearby volcanoes blanketed the landscape with volcanic ash which has a high silica content. As the rainwater gradually percolated down through this ash ladened sediment, it dissolved some of the silica. The buried trees became saturated by the water carrying the silica and other minerals in solution. It gradually became more and more concentrated and finally began to precipitate out. First it just filled in the empty spaces but eventually it replaced everything including the cell walls. This was ion-for-ion replacement. Jasper, chert, flint and chalcedony were the most common replacement materials. In special cases where hollow spaces were available amethyst crystals lines the interiors. Many of the trees are so well preserved that the individual cells can be made out. The tree rings are clearly visible including all the detail on the exterior such as the bark, insect boring and even charred bark if the tree had been in a forest fire. In other instances, no internal fea-

tures were preserved only the external features from volume for volume replacement.

Fossils are some trace of former plant and animal life, they are found here. They are preserved in several ways. Fossil Ferns (19 species) have been found near the Second Forest in shale beds. They have been preserved as *impressions* that only show the external structure of the plant or animal or its outlines. Some are preserved as *carbonaceous residues* or *compressions,* where the actual plant is still there and is preserved by the process called *distillation.* Distillation occurs when all of the volatile materials such as hydrogen, oxygen and nitrogen evaporate leaving just a film of carbon which frequently preserves all the minute detail. If the rock with the carbonaceous film is soaked in hydrofluoric acid, the rock portion is dissolved and the film floats free which then can be transferred to a glass slide or piece of paper for further study. This process is known as *maceration.*

There are over 40 species of plants found as fossils in this region. The main ones are as follows:

a. *Calamites*—the ancestor of scouring rushes. They grew to heights of 30 feet and to a diameter of more than a foot.

b. *Araucarloxylon arizonicum*—a cone bearing conifer, related to trees still living today (Araucarias). This is the main tree found in the Petrified Forest.

c. *Woodworthina arizonica*—an extinct conifer or cone bearing tree.

d. *Schilderina adamanica*—a small swamp loving tree, the diameter of the trunk is very seldom greater than four feet, the interior has a series of odd radiating rays.

e. *Ferns*—a total of 19 separate species has been found.

f. *Flowering plants*—very primitive, the flowering plants did not become abundant until the Cretaceous period.

g. *Cycads*—a group of plants that have a base that resembles a pineapple and palmlike leaves. Still exists today.

h. *Coniferous Plants*—a large variety of cone bearing plants.

i. *Plant Fragments*—a variety of pine cones, pine needles, leaves, pieces of bark and other fragments.

Badlands have created the Painted Desert and Blue Mesa areas. The volcanic ash accumulated to great thickness in some areas. When this material weathers chemically (decomposition) the silica, manganese and iron are changed into a clay mineral called *bentonite*. When bentonite is dry, it is almost as hard as a rock, but when it is near water it acts like a sponge and absorbs it, thus increasing in size. Eventually it becomes unstable and will begin to flow down the slope. Physical weathering (disintegration) during the colder months (e.g., frost action) breaks the bentonite clay into smaller fragments. These are either picked up by the wind and blown away (aeolian process) or the sudden thunderstorms will remove the loose debris at the bottom of the slopes besides eroding the hills. This creates a landscape of rounded hills and pinnacles. If resistant material remains (e.g., petrified log or sandstone layer) on top, the soft material is protected below, forming a platform for the log or flat topped hills called *mesas* and *buttes* (depending upon size). Eventually the protective material is removed forming pinnacles (rounded miniature hills) and *tepees* (sometimes called haystacks) with their characteristic inverted cone shape.

The iron oxide and manganese oxides mixed in with the bentonite, shale, sandstone and conglomerate produce the marvelous array of colors thus creating a "Painted Desert." This along with the combination of the following factors:

a. Rainfall that usually falls in the form of cloudbursts (only nine inches per year).
b. Plants are unable to gain a foothold because of rapid erosion.
c. No organic matter to feed plants is found in bentonite.
d. Extremely long dry periods between cloudbursts.

No badlands will develop in the bentonite if there is sufficient rainfall that falls gently and will permit vegetation to grow.

Differences in the lithology and composition of the percolating ground water have created six separate forests.

a. Black Forest—dark colored fossil wood.
b. Blue Mesa—logs on top of pedestals of blue clay.
c. Jasper Forest—most of the logs have been replaced with Jasper.
d. Crystal Forest—Interior of the logs are lined with amethyst or rock crystal.
e. Long Logs—this area did not undergo as much uneven uplift so the logs with diameters three to seven feet and up to 150 feet long. Some of the trees have branches and insect borings.
f. Rainbow Forest—trees replaced with jasper, chalcedony chert, flint, agate, dendrites, onyx and other materials.

Phytosaur bones were found in Blue Mesa, these were large reptiles that resembled crocodiles in appearance but were not related to them. They are found up to 25 feet long and weighed as much as 1,000 pounds. They lived only during the Triassic period.

GEOLOGIC HISTORY

1. Deposition of the Moenkopi formation about 225 million years ago during the Triassic.

On a low coastal plain adjacent to the Cordilleran geosyncline streams meandered back and forth depositing red to brown and white sandstone and shales (with gypsum layers), and yellowish limestones. These beds contained fossils of reptiles (phytosaurs), amphibians (*labyrinthodont* with highly infolded teeth and *stegocephalians* more advance and thick-skulled), and fish fossils. These beds are from 300 to 600 feet thick. They grade from the thicker coastal deposits to thinner alluvial fan deposits with sand filled channels, mudcracks and animal tracks.

2. Uplift and erosion, formation of Shinarump conglomerate.

By Upper Triassic time the Moenkopi had been elevated and had undergone erosion. Fragments of the Moenkopi were scattered all over the surface as a gravel. When the large alluvial fans started developing, the finer material cemented the gravel together and it became the Shinarump conglomerate, the basal member of the Chinle formation (a basal conglomerate).

3. Deposition of Chinle formation during the Upper Triassic.

The Shinarump conglomerate marked the beginning of a broad flat alluvial plain that was to become the Chinle formation. It consists of brightly colored (various shades of yellow, pinks, orange, reds, white and gray) shales and siltstones. The colors are produced by such minerals as limonite (reds and yellows), hematite (reds, orange and pink), gypsum and volcanic ash (white and gray). The beds today above and below the surface of this ancient alluvial fan were lakes and swamps as well as slowly meandering rivers and streams. Along the sides of the rivers lived the Phytosaurs, Labyrinthodont and Stegocephalian amphibians. Various types of plants such as conifers, cycads, ferns and calamites grew along the banks and over the river plain.

4. Accumulation of logs on the sand bars, burial and preservation.

The trees died by various methods, some were killed by disease or insects (borings can still be seen), others were destroyed by forest fires (replaced charcoal found on some logs), some probably died of old age, others were washed out by floods. The trees toppled over and were transported downstream. On the trip downstream the smaller branches were knocked off, many times the bark was stripped and roots worn away, leaving only the logs.

When a stream meanders back and forth the velocity of the water changes. On the outside of a curve the velocity of the water increases (centrifical force) and permits erosion. On the inside the velocity is reduced and the stream is forced to deposit sediments so sandbars build up.

It is upon these sandbars that many of the trees became trapped. Rapid deposition buried them very quickly before decomposition had a chance to start. This rapid burial is very important and the reason why most plants and animals never become fossils, for they decompose before burial occurs. The deposits were up to 400 feet thick.

5. Deposition of volcanic ash from nearby volcanoes.

In California and Nevada crustal unrest was occurring and volcanic activity was rampant. The winds were able to pick up this ejected debris and carry it all the way to Arizona where it was deposited in layers on the alluvial fan. This material is necessary as it is the source of silica. Volcanoes were abundant in the west during the geologic past and it is very easy to find petrified wood out there. In the east the volcanoes were not as abundant, hence petrified wood is almost nonexistent.

6. Formation of the Glen Canyon Group (Triassic-Jurassic) and Cretaceous beds.

Gradually the seas transgressed upon this alluvial fan depositing the Glen Canyon Group and some 3,000 feet of marine beds (Cretaceous). This sea occupied the region for 100 million years as the Rocky Mountain geosyncline.

7. Occurrence of the Laramide Revolution creating the Rocky Mountains, uplift. Petrification of trees completed.

When the Rocky Mountains formed some 65 million years ago, the sediments that were 3,000 feet below sea level were elevated about 5,000 feet above sea level. This process of uplifting was uneven, so many of the logs are broken into sections. The groundwater passing through the bentonite beds leached out the silica and finished petrifying the logs. Differences in composition produced the different forests. Some of the trees are up to 160 feet long. Agate Bridge is 111 feet long and has partially been unburied leaving a portion of it over a 40 foot wide arroyo. In 1917 the National Park Service placed a concrete support beneath it to prevent its collapse since people insist upon walking over it.

As a matter of fact, there is a story about a cowboy (during the early days of the park) which bet he could ride his cow pony over the log. He took off the horseshoes of his horse (for better footing) and collected his bet of $10.00.

8. Erosion of Mesozoic sediments exposing the Chinle formation creating the Petrified Forest and Painted Desert.

With the uplift erosion took over and has stripped away the 3,000 foot blanket of sandstone, siltstone and limestone, exposing the Chinle formation. The individual layers react differently to weathering thus creating the Painted Desert and Petrified Forest. The resistant layers protect the less resistant. The bentonite zones form the badlands. It has been estimated that only 1/4 of the logs have been exposed with the remaining 3/4 waiting to have the 300 feet of Chinle removed from them. This is fortunate that there is so much of the wood left, for the tourists, in spite of the park rules, walk off with an average of 12 tons of petrified wood each year.

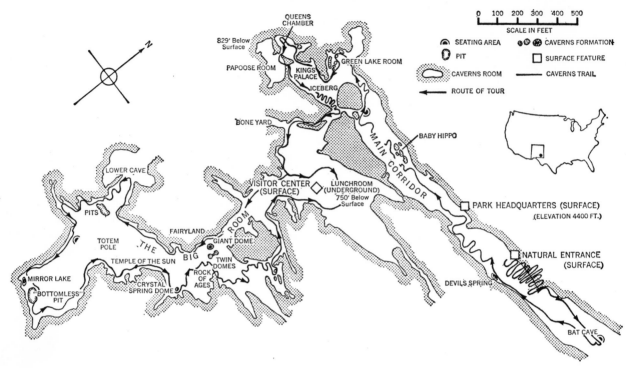

Carlsbad Caverns National Park, New Mexico.

Geologic Column of Carlsbad Caverns National Park

System	Series	Northwest Shelf		Marginal Zone	Delaware Basin	
Permian	Ochan				Rustler Formation	Delaware Mountain Group
					Salado Formation	
					Castile Formation	
	Guadalupe	Artesia Group	Tansill Formation	Capitan Limestone	Bell Canyon Formation	
			Yates Formation			
			Seven Rivers Formation			
			Queen Formation	Goat Seep Dolomite	Cherry Canyon Formation	
			Grayberg Formation			
		San Andres Limestone		Cherry Canyon Sandstone Tongue		
				Brushy Canyon Formation		
	?			Cutoff Shale		
	Leonard	Yeso Formation		Victoria Peak Limestone	Bone Springs Limestone	

Modified after Hayes.

CHAPTER 36
CARLSBAD CAVERNS NATIONAL PARK

Location: S.E. New Mexico
Area: 46,753.07 acres
Established: May 14, 1930

LOCAL HISTORY

The Basket Maker Indians lived in the area 1,000 to 4,000 years ago. They probably went into the cave only where daylight provided enough light to see by, because there is a sudden dropoff near the entrance. An Indian sandle was found at the bottom of the dropoff. Nearby are mescal cooking pits, that are huge piles (30-40 feet) of rocks with a central depression in the center. Pictographs occur. The best pictographs known are located at "Painted Grotto" in Slaughter Canyon. After the Basket Makers left the Apache came in and were living in the region when white man moved in.

In the late 1800s, the settlers knew of the cave. The first official record of someone entering it was in 1883 when Rolth Sublett's father lowered him (he was 12 years old) into the entrance which he probably explored with the available sunlight. In 1885 Ned Shattuck and his father, while looking for cows, saw a dark cloud come out of the cave and to their amazement the cloud proved to be millions of bats, who today consume around one ton of insects per night. After that discovery, people would refer to the cave as "Bat Cave." With the discovery of the bats a few people realized this meant there were guano (bat excrement) deposits in the caves that are nitrate-rich. The first person who tried to mine the guano (1903) was Abijah Long. He filed a claim on 40 acres around the mouth of the cave for the guano and other minerals. The mine shafts were dug into the entrance and the guano hauled out in cars. Profits were marginal and in a 20 year period six separate companies tried their luck. More than 100 million pounds were mined and sold to citrus fruit growers in California at $25.00 to $75.00 per ton. It was used for fertilizer.

One of the miners who worked for five out of the six enterprises was a local young man called James Larken White. He was extremely curious about the cave and explored it (1901) whenever he had a chance, sometimes with a companion, most of the time by himself. He managed to explore 20 miles of the cave on his own, stringing out a white cord, so as not to get lost. He had many misadventures but managed to survive each time. Jim White's description of the cave in 1923 was finally brought to the attention of Commissioner William Spry of the General Land Office. Spry assigned Robert A. Holley (mineral examiner) to investigate, which he did over a 30 day period and was very enthusiastic about the cave.

Geologist Willis T. Lee of the United States Geological Survey was convinced by the mayor of El Paso, Texas (Richard Burgess) to investigate the cave. He was also impressed by the cave. On October 25, 1923, President Coolidge made it a National Monument. It was later changed to a National Park by President Hoover on May 14, 1930.

Dr. Lee (sponsored by National Geographic) spent six months (1924-25) exploring and mapping the cave with Jim White's help as a guide. Even today it is not totally explored, at least 23 miles are mapped. Besides Carlsbad there are 50 other caves within the park that are known to exist. One of the more spectacular of these is New Cave in Slaughter Canyon. Parts of it were used to film "King Solomon's Mine." Wild Cave will have primitive tours but plans are to keep it "wild" or undeveloped. This also holds true for some of the new chambers in Carlsbad Caverns, e.g., Guadalupe Room discovered in 1966. Some of the areas would have to be destroyed in trying to develop it while others are too difficult to reach.

GEOLOGIC FEATURES

An opening existed at least 10,000 years ago for the bones of one of the ancestors to a jaguar

and an extinct ground sloth (giant) have been found within the caverns.

Carbon-14 dating indicates that bats have occupied these caverns for at least 17,000 years. Presently 14 species occupy the cave, the main species is the Mexican Free-tailed bat. They are so named because part of the tail hangs free and they spend about half the year in Mexico. The other four important species are the Fringed Myotis, Lump-nose, Pallid and Western Pipistrel bats.

The formation of caves occurs in stages. However, first the proper conditions must occur; (1) a joint system in limestone that is below the level of the ground water table (phreatic zone); (2) water with dissolved oxygen and dissolved carbon dioxide. The oxygen and carbon dioxide combine to form carbonic acid. This, in turn, can seep easily through the joint system and dissolve the limestone enlarging the openings.

The Bone Yard portion of Carlsbad represents the second stage, if the joints are closely spaced together the solution produces fins. The limestone develops a spongy appearance as the more soluble portions are removed and less soluble remain behind. The openings become larger and larger.

The third stage occurs when the incipient cave is elevated above the level of the water table. This spongy rock no longer has the support of the water and collapses, forming an opening in the bedrock called a cave. Solution continues and the cave becomes larger and larger. Many times certain layers of bedrock are not soluble so they determine in part the various levels of the cave. In other instances such as in Carlsbad the three main levels represent stages in the uplift of the Guadalupe Mountains. The first level from a reference point (the floor of visitor center) is approximately -250 feet (Bat's Cave). The second level is around -750 feet (-754 to be precise) this is the level of the Big Room which occupies an area equivalent to 14 football fields (14 acres) and has a 225 foot ceiling. the third level is at -830 feet (King's Palace and Lower Cave). This does not represent the lowest known level of the cave, however, which is -1,126 feet (from the reference point). This is the bottom of the pit of "Lake of the Clouds." However, geologists believe the lake represents a *perched pool* or in other words water trapped above the level of the groundwater table in the *zone of aeration* or *vadose zone.*

The features that most people associate with caves cannot form until the chambers are exposed to the air. The reason for this is something must force the calcium carbonate to crystallize or precipitate out of the water. The groundwater responsible for the solution of the cave is a very weak solution of carbonic acid saturated with calcium bicarbonate (soluble form). As the groundwater seeps through the joints it may eventually reach the ceiling of one of the chambers. The drop of water would then hang onto the ceiling. Since it is in an open, unconfined space it will lose some of the carbon dioxide gas. When this occurs the drop of water is unable to hold all of the calcium bicarbonate in solution so some of it will precipitate out at the edge of the drop of water. When that drop completely evaporates, there is left on the ceiling a tiny ring of calcite (just like the marks upon a glass dish). Another drop of water forms in the same location, evaporates and contributes a little more calcite to the ring. This happens millions of times and gradually a *soda straw stalactite* forms hanging from the ceiling.

The various decorative features in a cave are given the general name of *speleothems.* The icicle shaped pendants hanging from the ceiling are called *stalactites,* their counterparts on the floor are *stalagmites.* The little diagram below will help you to remember which is which.

S T A L A C T I T E
S T A L A G ∧∧ I T E

The soda straw stalactite, is a thin hollow tube through which the water moves and gradually it increases in length. If the tube were to become plugged, the water would collect on the outside and evaporate, contributing its share of calcite and the stalactite would develop the familiar icicle shape. If the flow were too fast some of the water would fall on the floor of the cave building up an icicle shaped mount called the stalagmite. If the two features eventually join together connecting the floor to the ceiling it becomes a *column* (e.g., Giant Dome, Veiled Statue).

The flow of water will determine if just stalactite form, just stalagmites or both. If flow is extremely slow, only stalactites develop for no water drops to the floor (Teeth of Whales Mouth). If flow is fairly fast water immediately falls to the floor to form stalagmites (Rock of Ages). If the flow is in between, a stalactite on the ceiling with its counterpart stalagmite on the floor (Frustrated Lovers). The distribution of the stalactites is to a large degree determined by the fracture system, for they have a tendency to form along the joints.

Rate of growth is extremely slow at Crystal Spring Dome. The growth is equivalent to a coat of paint (thin) every 85 years. In the section where the bats live, stalactites can be found with the body of the bat that has fallen on them being incorporated as part of the stalactite—a fossil of the future. About 95 percent of the spelothems are no longer "alive" or forming. The "dead" structures have a dull chalky-white appearance to them. The "live" features are frequently colored by iron oxide or manganese oxide and appear shiny because they are coated with a layer of water.

The spelothems are made up of a material called *travertine,* it is a form of calcite and is deposited by ground water and surface water. When it is translucent, banded and capable of taking a polish it is known as cave onyx. Travertine is frequently opaque because of clay and other impurities mixed with it, such as iron oxide. When the travertine is deposited on floors or along slanted walls it is called *flowstone.* When it forms features by dripping (e.g., stalactites) the proper name is *dripstone.*

A very unusual spelothem are the *helicitites* which resembles stalactites but appear to defy gravity for they twist and turn in many directions. They have the same beginning as a stalactite but the diameter of the tiny central tube is so small that capillary action occurs and gravity is no longer the controlling force. The size of the tube will not allow the water to flow down to form a drop at the tip. Instead the water is pulled down against the force of gravity because of the attractive force between the molecules of water, so direction of deposition is by chance. Secondly, calcite crystals do not have square sides, instead, the faces meet at angles. As the next crystal forms on one of these faces it will add its angle to the angle of the face upon which it rests. Crystals have a tendency to grow faster along the long or "C" axis, so orientation of the crystal axis in part will determine the direction of growth. Hence, the spelothem will begin to twist and curve and a helectite develops.

Cave pearls also form in Carlsbad and are a rare spelothems. Dripping water creates a small depression on the cave floor which will trap material such as bits of rock, or sand grains or bone fragments from bats. If the flow of water into the depression is rapid enough the fragments are constantly agitated and become coated layer by layer with calcite. Most of the pearls are small, about 1/4 inch in diameter, but some have been found up to an inch. They are usually round, although cylindrical ones have been found; their nucleus is usually a bat bone (they can be found in "The Rookery").

Iceberg Rock is a good example of a section of the ceiling that has fallen to the floor to enlarge a

cavern. It is called a *collapse block.* The stalactites on it are inclined which indicated the amount of rotation that has occurred since it fell and the direction of rotation. The scar from Iceberg Rock can be seen in the ceiling of the Green Room.

Slow weathering is occurring in the dry places and in several areas exfoliation of the dripstone or flowstone can be seen (e.g., region near Totem Pole Stalagmite).

The *regional joint system* can be seen near the seating area by Iceberg Rock. The Main Corridor have developed along these joints. Carlsbad Caverns has developed in the core of the Capitan Reef, the fracturing probably developed in this core shortly after it had undergone lithification. The jointing was probably caused by the compaction and uneven loading of the sediments (a limy mass) that make up the core, rather than being produced during the uplifting process as occurs in most cases.

The floor of the cave or passageways in many areas have been covered or filled in by sediments (clay, sand, silt and gypsum) which were brought in by streams or deposited by pools in the cave. Deposits in the cave indicate that in the geologic past there was more water available in the cave system than there is today. A stream periodically used some of the old passageways as its streambed and has deposited terraced silt beds some 20 feet high. A later stream eroded parts of these terraces and deposited limestone (locally derived) cobbles in its stream bed (direction of flow was away from the Pecos River Valley located east of the caverns).

GEOLOGIC HISTORY

1. Formation of Delaware Basin and Capitan Reef during the Permian period.

Between 200 and 250 million years ago the Delaware Basin in Texas and New Mexico was flanked on either side by the Central Basin Platform (east) and Diablo Platform (west). A massive barrier reef called the Capitan Reef formed. Carlsbad National Park is located in the northeast-southwest trending portion of the northern half of the reef. This reef enclosed approximately 10,000 square miles and is the world's largest known reef of this type.

2. Formation of the Capitan limestone and Tansill formation.

The Capitan limestone represents the actual reef and the reef talus upon which the reef grew. The Capitan limestone can be described as a massive, thick bedded limestone, light gray to pinkish gray color that ranges in texture from aphanitic (cannot see crystals with the naked eye) to finely crystalline. It was formed by the accumulation of lime-secreting algae, other plants and animals that frequent reefs (sponges, horn corals, bryozoans, etc.). The reef gradually built up an outwards over the debris that was created by wave action breaking the actual reef into fragment called *talus*. This was on the seaward side of the reef located to the south and southeast.

Behind the reef formed a lagoon or the back reef facies. The Tansill formation formed here. It is a thin bedded dolomite or perhaps a dolomitic limestone, yellowish gray in color. The upper portion is primarily siltstone. The Huapache monocline developed perpendicular to reef escarpment. The reef became several hundred feet high and from one to four miles wide with its top existing just below the surface. The lagoon also varied in width from one to twelve miles wide.

3. Permian Reef buried by Mesozoic sediments.

Gradually as time passed the lagoon began to fill up and finally the Delaware Basin itself. Over this surface which was now relatively flat crept the Mesozoic oceans, possibly Triassic formations formed but definitely Early Cretaceous marine sediments buried the reef region completely. The seas then retreated from this region for the very last time. The deposition of sediment produced stress and strain on the reef beds forming minute hairline fractures in the reef.

4. Laramide Revolution formed the Rocky Mountains and uplifted the region.

About 60 million years ago faulting occurred with uplift and minor intrusions of trachyte in the Castile formation (lagoonal facies). The uplift occurred in several stages.

Two major fractures developed, a series of normal faults parallel to the reef axis, another perpendicular to it. One fault in the Capitan Limestone parallel to the reef has 30 feet of displacement, another has 50 feet and may be due in part to the tension produced when the Walnut Canyon syncline formed. This syncline forms the canyon north of the caverns which (are located on the reef escarpment of today).

5. Phreatic stage of cave formation.

With the development of the fractures in the limestone the formation of the cavern could begin. The caves developed before the present erosional surface began. This is shown by the geologic evidence that the cave used to extend to higher levels (on the ridge) but these have since been removed by erosion (closed sinkholes with dripstone and flowstone fragmites). The phreatic stage (zone of saturation—below the level of the groundwater table) is first. With the entire cave below the water table, every available surface is in contact with water and the process of solution of any soluble material occurs, causing the bedrock to develop a spongy appearance. The evidence is everywhere that it had its beginnings underwater. Solution chambers separated by insoluble partitions (they may have swiss-cheese appearance), the chambers follow the joint system, the levels are enlarged sidewise as the result of lateral solution because of the concentration of the acids near the top of the water table all these testify to origin. An intermediate stage occurred where streams flowed through these channels and partially filled them with sediments.

6. Vadose stage of cave formation.

The last final uplift occurred elevating the Guadalupe Mountains, during Pliocene-Pleistocene time. The total elevation of reef was now several thousand feet. The beds were tilted eastward slightly (dip about 1°-3°). This raised part of the level of the caves above the level of the water table. Thus part of the system remained in the phreatic stage, and part in the vadose. The caverns have developed mainly in the reef core (Capitan Limestone) but the upper level is in the lagoon facies (Tansill formation). A portion of the Big Room has formed in the reef talus. Therefore, a walk through Carlsbad is a trip from the forereef facies to the backreef facies. This occurred before the Guadalupe Ridge was exposed by erosion.

With the water removed, the spongy rock collapsed forming caverns. Flowstone and dripstone structures could now start to develop such as the stalactite, stalagmites, helectites and cave pearls. In some sections of the cave, crystals of calcite have formed called *dogtooth spar.*

Some of the other geologic features that formed are the gypsum beds in the Big Room and Devil's Den tunnel. They were far more extensive in the past but have been removed by erosion of streams through the caves. The gypsum comes from the leaching of the Castile formation of the backreef facies.

Clues to the former higher water levels are the *lilly pads* by Temple of the Sun and Cave Man plus the *travertine terrace* by Grape Arbor that surrounded a former pool.

Devil's Spring near Whales Mouth is a backreef facies aquifer. It occurs in all probability along the uppermost siltstone bed of the Yates formation. The Yates lies below the Tansill and is a backreef facies the same age as parts of the Capitan limestone. The Main Corridor is along the zone that separates the reef core from the backreef.

Some of "live" formations are Baby Hippo, Papoose Draperies, Crystal Spring Dome (most active) showing the cave development is not over.

When did all this occur? Fortunately the geologist has a clue. At the -750 foot level bones of an extinct ground sloth (Nothrotherium) that lived during the Pleistocene were washed in. Covering the silt in which the bones are buried is flowstone with large stalagmites growing upon it. This means they could have formed no later than Pleistocene and perhaps as early as a few thousands of years ago.

It is conceivably possible that there is another cave as magnificent as Carlsbad located nearby, inasmuch as other caves have been found in the park. New Cave discovered in 1966 is a good example, it has many features that rival Carlsbad.

All these factors are a mute testament to the geologic past and an indication of the future.

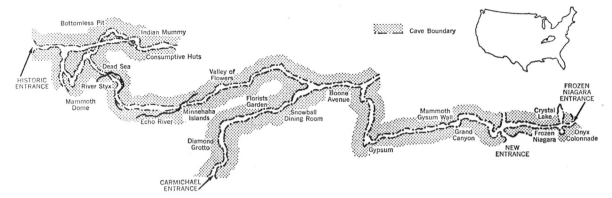

Mammoth Cave National Park, Kentucky.

Geologic Column of Mammoth Cave National Park

Erathem	System	Series	Formation
Cenozoic	Neogene	Holocene	Alluvium
Paleozoic	Pennsylvanian		Pottsville Group
	Mississippian		Glen Dean Limestone
			Hardinsburg Sandstone
			Golconda Limestone and Shale
			Big Clifty Sandstone and Shale
			Girkin Limestone
			St. Genevieve Limestone
			St. Louis Limestone
			Warsaw Limestone
			St. Payne Limestone
	Devonian		New Albany Shale

Modified after Linesay.

CHAPTER 37
MAMMOTH CAVE NATIONAL PARK

Location: Central Kentucky
Area: 51,354.4 acres
Established: July 1, 1941

LOCAL HISTORY

The Indians knew about the cave long before white man. The Woodland Indians lived in this part of Kentucky from 1,000 B.C. to about 900 A.D. Found inside the cave have been wooden bowls, woven sandles, bundles of reeds that were used for torches and remains of fireplaces. Chief City, a two acre cavern within Mammoth Cave was a gathering place (according to the relics) for both the ancient Indians and modern ones.

The Indians apparently went into the cave after gypsum (calcium sulfate), but for what purpose, is still under debate. Possible theories are as a medicine, a substitute for salt, a base for paint, possibly even for fertilizer.

In 1935 the mummified body of an Indian now known as "Lost John" was found underneath a five ton boulder. He was killed (according to Carbon-14 dating method) between 2,300-2,400 (2,370) years ago. He was 5'3" tall and about 45 years old. He was chipping some gypsum out along the ledge when the boulder fell upon him. When the body was first found it was very well preserved. Unfortunately it was taken out of the 54°F temperature and 87 percent humidity of the cave and placed in the sun. It was there that he developed a furry coat of fungus. So it was rushed back into the cave where it will not decay as long as it remains there. The body is now preserved in a glass case but is no longer on display.

According to legend the cave was discovered in 1799 when Robert Houchin chased a bear into the cave.

Potassium nitrate (saltpeter) occurs in the cave and was mined for use in black gunpowder during the 1812 war. Slave labor was used to shovel the saltpeter into wooden vats where the saltpeter was leached out with water. Hollowed logs were used as pipes to bring the water in. Some of the vats and pipes can still be seen today and are over 160 years old. At that time the cave was owned by a Mr. Wilkins and a Benjamin Gratz.

The cave and 200 acres originally sold for $40.00. During the 1812 war speculation pushed the price up from $116.67 in the morning to $3,000 by evening (after changing hands several times). The going price for part interest a month later was $10,000.

After the war it was sold to a Mr. James Moore, then to a Mr. Gatewood who owned it once before. He opened it to the public for tours. It was then sold to Frank Gorin who also purchased additional land around it (by the natural entrance) in 1837. After many discoveries of unusual nature were found it attracted many tourists. In 1837 Dr. John Croghan paid $10,000 for the cave and 2,000 acres (so as to control all entrances). He had 12 huts (brick, wood or stone) built for Tuberculosis patients (15). In 1842-43 they lived in the huts for several months, and even tried to grow plants around the huts to brighten things up. The theory was, constant temperature and humidity would be good for them. The results were, all got worse, in fact two died while living in the huts. The ones that lived in the huts the longest died within a three week period after they left the cave.

Stephen Bishop was 15 years old in 1837. He was a half Indian-half Negro slave, who loved to explore the cave. He placed a slender pole over the Bottomless Pit and became the first person to cross it. He discovered many of the features that have made the cave so famous today. He acted as a guide to many famous scientists and people. When he died in 1859 he was buried in what is now Old Guide Cemetary (within park boundaries).

Edwin Booth (the actor), Jenny Lind (the "Swedish Nightingale") and other famous people have visited this cave.

It was authorized on May 25, 1926 but didn't become established until 1941 and was later enlarged in 1948.

GEOLOGIC FEATURES

One of the most obvious features in the regions are the *sinkholes*. If a person were to fly over the Mammoth Cave area in a plane he would notice several things. A very definite boundary called the Dripping Springs escarpment marking the edge of the Mammoth Cave Plateau. North of this escarpment it is hilly, and streams are on the surface, with the Green River meandering back and forth. South of the escarpment is the Pennyroyal Plain that is low, relatively flat, and dotted with numerous circular depressions and round ponds, but practically no streams. The plateau and escarpment are the result of the sandstone and limestone beds that outcrop on the surface.

These circular depressions are called *sinkholes* or sinks. They formed when the ceilings of caves near the surface collapsed leaving a depression. Since they belong to a cave system, it is not very surprising to find that a stream will disappear into one sinkhole and emerge from another. An area that has many sinkholes is said to have a *karst topography.*

A rather unique feature seen here are the *domes* and *pits* that are apparently solution features along vertical joints. The abundant water source above (frequently a sinkhole) drips or more often flows down the sides and removes the limestone by both solution and abrasion (wearing away by friction) forming the high domes. The water continues to flow down the joint system that occurs in the floor and by the same process forms large circular pits. Sometimes a dome occurs over a pit such as Silo Pit (95 feet deep) and Roosevelt Dome (130 feet high), or may occur alone such as Wilson and Mammoth Domes, Bottomless Pit (105 feet deep).

Lily Pads form in pools because of the process of evaporation. The calcite is deposited around the edge of the pool as the water evaporates and forms a rim. If the pool exists a long time (formed in an irregularity on the floor) the rim may build up quite high—as much as a foot. If this irregular pool has scalloped edges it is then called a lily pad.

Draperies are a form of dripstone that often had their beginnings as a series of stalactites that joined together along a crack in the ceiling. The water that seeps down often picks up impurities so the draperies frequently have bands of color. On these occasions they may be referred to as "bacon."

Gypsum curls or rosettes can only form in the drier parts of the cave. Their composition is calcium sulfate and not calcium carbonate as most of the other cave features are. The oolitic limestone that contains marcasite has a tendency to form the gypsum growths. It forms where there is very slow seepage through the porous limestone. When the seepage reaches the surface it evaporates in the *open pores* of the rock and coats it with a layer of gypsum. Because the gypsum is forming in the pores from behind the coating it pushes it outwards into a *blister.* Eventually the blisters become large enough to burst and then they open into the flowerlike rosettes or long curly strands. The gypsum itself does not occur in the bedrock, to provide a source for the rosettes, such as can be found in Carlsbad. What probably happens is the marcasite (an orthorhombic form of iron sulfide, pyrite with the same composition is cubic) becomes oxidized and forms sulfuric acid (H_2SO_4). This acid reacts with the limestone to form gypsum $CaSO_4 \cdot 2H_2O$.

Because of the humid climate rivers flow through Mammoth Cave, they have enlarged many of the passageways through all five levels of Mammoth. This accounts for the horizontal fluting that can be seen along the sides of passageways. The depth of the concave surfaces formed depended upon the depth and volume of water passing through. Very frequently these streams have completely filled the passageways with sediments, other times filled them only partially producing a level floor.

Other spelothems that are seen in Carlsbad are also seen here such as stalactites, stalagmites, columns, helectites, dripstone and flowstone are found here also.

The atmospheric pressure determines the movement of the air in the caves. It "breaths out" when outside temperature is above 54° (temperature of air in cave) and "breaths in" if it is below.

GEOLOGIC HISTORY

1. **Deposition of Upper Paleozoic beds in the Appalachian geosyncline.**

   During the Devonian the New Albany shale was formed. This was followed by limestones Mississippian in age.

2. **Deposition of Mississippian limestones to a depth of 1,200 feet.**

   The source area at the beginning of the Mississippian was relatively low so mainly limestones were formed at first, although later sands and shales formed. These sediments were deposited into the geosyncline by streams and rivers.

Sequence of deposition was:

youngest	Glen Dean limestone
	Hardinburg sandstone
	Golconda limestone and shale
	Big Clifty sandstone and shale
	Girkin limestone
	St. Genevieve limestone
	St. Louis limestone
	Warsaw limestone
oldest	Ft. Payne limestone

3. Crustal uplift during the Pennsylvanian, formation of Pottsville Group.

Crustal unrest forced the seas to retreat and the region was gradually raised. The river flowed over this newly exposed sea bottom depositing sand and gravel which is known as the Pottsville Group. It is a basal conglomerate that grades upwards into a sandstone.

4. Appalachian Revolution forms Cincinnati Arch tilting beds to the northwest.

The entire State of Kentucky was arched into a huge arch that began in Ohio and did not die out until it reached Nashville, Tennessee. The angle of the slope is approximately 30 feet per mile.

5. Erosion, periodic uplift, rejuvenation of streams.

The uplift of the arch was gradual and in steps. Streams would have a low gradient which was increased with each additional uplift therefore would start carving valleys anew. Most of the later Pennsylvanian sediments were removed by erosion and deposited elsewhere.

The resistant sandstone beds formed two escarpments, the higher Pottsville escarpment and Lower Dripping Springs (Big Clifty Sandstone). The Mammoth Cave Plateau separated the two and Pennyroyal Plain and escarpment formed below the Dripping Springs escarpment.

6. Formation of cave system in the Girkin and St. Genevieve limestones.

As in most caves much of the formation of the caves was during the phreatic stage, where the ground water had easy access to the joint system. Since much of the solution occurred near the surface a karst topography formed. Uplift permitted the vadose stage to occur and its spelothem features develop.

7. Green River controlled the cave levels.

The tilting of the limestone beds force the Echo River which flows through Mammoth to drain into the Green River. The Green River, therefore, determines the base level of the Echo and other rivers that flow through Mammoth.

With each rejuvenation the Green River could increase the depth of its valley permitting the Echo River to enlarge and erode to lower and lower levels, producing the five separate levels.

As the upper levels were exposed to the air they began to develop the dripstone and flowstone features, while the lower levels, still under water were still in the process of forming.

8. Development of underground drainage system.

Originally all the streams flowed on the surface producing stream valleys, irregardless of whether the bedrock was sandstone or limestone. As the karst topography gradually developed, where the limestone outcropped the drainage began to go underground. It has reached the point today that the majority of streams flow through sinkholes and caves.

9. Formation of springs in the Big Clifty Sandstone.

Within the Big Clifty Sandstone is an impermeable shale layer that will not permit the meteoric water to reach the groundwater table, so it flows along the top of the shale layer until it reaches the escarpment and seeps out as a spring. This trapped water forms a *perched water table* because it occurs within the vadose zone.

10. Continued formation of Mammoth Cave.

The Big Clifty Sandstone protects the limestone layers underneath. This is important because of the control of the Green River over the development of the cave.

It allows the water to circulate and thus help to develop the caves and also determines the level of the caves and the time allotted to form these levels.

All these events put together managed to form some 150 miles of caves distributed among 325 known passageways.

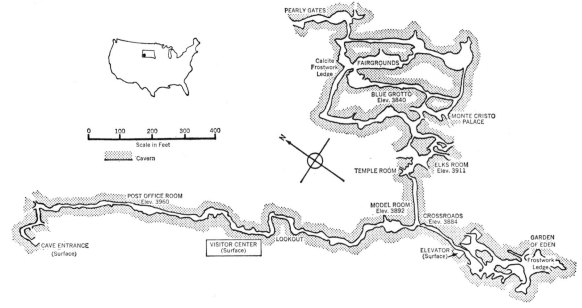

Wind Cave National Park, South Dakota.

Geologic Column of Wind Cave National Park

Erathem	System	Series	Formation	
Cenozoic	Neogene	Holocene	Alluvium	
	Paleogene	Oligocene	White River Group	Brule Clay
				Chadron Formation
Mesozoic	Cretaceous		Fall River Sandstone	
			Fuson Shale	
			Lakota Sandstone	
	Jurassic		Morrison Shale	
			Unkpapa Sandstone	
			Sundance Formation	
	Triassic		Spearfish Formation	
Paleozoic	Permian		Minnekahta Limestone	
			Opeche Formation	
	Pennsylvanian		Minnelusa Sandstone	
	Mississippian		Pahasapa Limestone	
			Englewood Formation	
	Ordovician		Whitewood Formation	
	Cambrian		Deadwood Formation	
Precambrian	Precambrian ?		Harney Granite and Metamorphics	

Modified after Deike, White et. al.

CHAPTER 38
WIND CAVE NATIONAL PARK

Location: S.W. South Dakota
Area: 28,059.26 acres
Established: January 9, 1903

LOCAL HISTORY

The Sioux Indians (or Dakotas) called the cave "Sacred Cave of the Winds." They had several legends about it, one stated that this was the cave through which their ancestors arrived at the Black Hills. Another says that Wokan Tanka, (the "Great Mystery") would send from this cave, the buffalo, out into their hunting grounds.

The area is filled with legends, even as to how the cave was discovered. One story claims it was discovered in 1877 by "Lame Johnnie" a notorious outlaw (and stage coach robber). Another story claims a cowpuncher was riding by when the wind blew his hat off and into a small hole. When he went after his hat he discovered the cave. Since the opening was too small he could not retrieve his hat and was forced to leave without it. The next day he rode by and the wind brought it back and was even nice enough to place it on his head. Probably neither are true.

The official discoverer is Tom Bingham, one of the pioneer settlers in the Black Hills. He was deer hunting and heard a strange whistling sound, so he began to search for the source. He found a small hole 8 by 12 inches in diameter through which the wind was rushing and making noise. Since the natural opening was too small another was dug near by (depth of six feet).

The region around the cave was open to claims so in 1890 the South Dakota Mining Company filed a claim on the cave. A Mr. Jesse D. McDonald and his sons (Alvin and Elmer) were hired to take care of the property. Alvin McDonald did most of the exploration, named many of the features in the cave, and kept a diary of all his mapping and exploration. He is now buried near the cave with a plaque marking the location.

In 1892, Jesse McDonald, Charles Stabler, John Stabler plus several other partners formed a company called "Wonder Wind Cave Improvement Company." They took the property over, opened up some of the passageways, put in stairs, etc., and opened the cave to the public. Magazine stories were printed about the cave in 1900 and 1901 and this brought it to the attention of the Government. President Theodore Roosevelt signed a bill to create the park in 1903.

GEOLOGIC FEATURES

Most important feature is the *boxwork* for which the cave is famous. The Mississippian limestone (Pahasapa) became highly fractured during the formation of the Black Hills. Calcite was deposited in these fractures as secondary filling by groundwater. Later the primary limestone was removed by solution leaving the secondary material behind as thin fins which crisscross one another. No other known cave in the entire world has the variety or amount of boxwork that Wind Cave has (only one cave in Europe in Czechoslovakia has some).

Cave Earth, of course, is found in all caves, in Wind Cave it is a thin veneer to several feet thick. The cave dirt comes from two main sources; first the insoluble residue from the limestone; secondly the material brought in from outside sources. Since the Pahasapa limestone is about 98 percent pure, it is unlikely that any of the soil is insoluble residue. More than likely it has been brought down from the surface during the infiltration process from the clays that cemented the breccias and material derived from weathering on the surface.

Break Down or collapse blocks are the portions of the ceiling and walls that have detached themselves from the surface of the cave and are now lying on the floor. It is thought that most of this occurred when the cave was in the phreatic zone because the boxwork on the underside is undamaged. This could easily occur if the cave were underwater, but not if it were dry.

Collapse Breccia is found in parts of the Pahasapa limestone where the ceiling of the cave collapsed and portions of the overlying Minnelusa fell along with Pahasapa as a plug into the cavern.

The above features are sometimes referred to as *primary features* inasmuch as they formed at the time the cave was developing.

The following features are called *secondary* for they developed when the cave was exposed to the air.

There are a few stalactites and stalagmites, but not very many. They will not be discussed here as they have been mentioned elsewhere. The same holds true for the various flowstone and dripstone features.

Of the secondary features, some were deposited below the level of the water table and others above. First is the *calcite lining,* it is primarily calcite, hematite and quartz. It was deposited when the cave was completely underwater along with or perhaps after the boxwork formed and coated all the walls. Several times the cave was drained and then resubmerged with the composition varying

from time to time. This occurred sometime between 50 million years ago (Eocene) or perhaps only 10 million years ago (Pliocene). Final draining occurred at the end of the Pliocene (two or three million years ago). Erosional and weathering processes have removed much of the lining in the upper levels.

Ice or calcite rafts will form in bodies of standing water in a cave. They are common in Wind Cave but are also found in Carlsbad. These are roughly rectangular pieces of calcite so thin that they are supported on top of the water by surface tension. They range in size from one to three inches across. Disturbance of any type, causes them to sink. New ones can form within a year so they are a feature that develops very rapidly. The area in the cave referred to as the Calcite Jungle has the floor covered with these rafts formed at a time when water level was higher.

Popcorn (globulites) is a series of layered nodules (1-10 mm) that is supported on a very small stalk (1 mm long). If it is covered with a bristlelike cluster of aragonite needles it is called *frostwork.* This is the second feature that the cave is very famous for. At first, popcorn was thought to have

formed beneath the level of water table, but at the present time it appears to be forming where dripping water splashes at the boundary between the top of the water and the air adjacent to a wall. Another possible origin is where water is seeping from the walls at the waterline. Frostwork may also occur as a coating of needles directly on the wall.

The needles have two separate compositions, aragonite and selenite. Aragonite is chemically the same as calcite but belongs to a different crystal system. Calcite is hexagonal-rhombohedral and aragonite is orthorhombic. Selenite is a clear, cleavable form of gypsum. The temperature seems to be a control over which forms. For example, aragonite has a tendency to form in areas where the mean annual temperature is between 56 and 63°F, but the cave is only 47°F, hence not much of it forms.

Other minor features that form are the calcite crystals called *dogtoothed spar*. Also, *dendrites* the crystals of iron or manganese oxides that form on the surface of a rock. Plus *stylolites* which are crenulated, striated, columns formed in limestone, which are the boundaries marked by the dark colored impurities left behind when the limestone went into solution because of pressure. They represent the points or zones of greatest pressure.

GEOLOGIC HISTORY

1. Deposition of Paleozoic and Mesozoic beds on Precambrian core.

The stratigraphic section is as follows:

Paleogene (Oligocene)	Brule and Charden formation	Plains and Badlands
Cretaceous	Fall River Sandstone	
	Fuson shale	
	Lakota Sandstone	Hogback
Jurassic	Morrison shale	
	Unkpapa Sandstone	
	Sundance formation	
Triassic	Spearfish formation	Red Valley
Permian	Minnekahta limestone	
	Opeche formation	
Pennsylvania	Minnelusa Sandstone	
Mississippian	Pahasapa limestone	Limestone Plateau
	Englewood formation	
Ordovician	Whitewood formation	
Cambrian	Deadwood formation	
Precambrian	Harney granite and metamorphics	Core

The differences in lithology have produced the division of the regions. These features did not develop until the Black Hills were raised. Wind Cave National Park encompasses all of the zones except the plains and badlands region.

2. Uplift of Black Hills about 70 million years ago.

During late Cretaceous-early Paleogene time (Laramide Revolution) the Black Hills dome was raised as an asymmetrical oval-shaped dome with a core of Precambrian granite (Harney granite). The dimensions are 150 × 74 miles. This core was overlain by Paleozoic and Mesozoic sediments. Prior to this event this region fluctuated between a large inland sea or a low plain area.

If the uplift occurred all at once with no erosion occurring the Black Hills would have been some 7,000 feet higher. Instead erosion has stripped these sediments from the core and deposited them elsewhere. Many of these redeposited beds form the nearby Badlands.

3. Formation of Wind Cave in the Mississippian Pahasapa formation.

The cave has developed in the upper 100-150 feet of the Pahasapa limestone. Back in the Paleozoic a karst topography developed in the Pahasapa prior to the deposition of the overlying Pennsylvanian Minnelusa formation. Some of these old sinkholes that are filled with the sandstone can be seen in Wind Cave.

As with any cave it developed along the regional joint system and went through the phreatic or vadose phases. The arid climate has not permitted a well developed karst topography to form.

Wind Cave began to develop possibly as early as middle Eocene (50 million years ago), but more likely as late as Pliocene (ten million years ago). At any rate, either date will make it one of the oldest caves in the world as most have formed only in the last one million years.

The uplift of the Black Hills created stress and strain, producing a fracture system along which the cave had developed.

Most development is lateral because the overlying sandstone does not permit cave development, nor does the lithology of the lower half of the Pasahapa limestone.

4. Development of cave zones.

Since development is mostly lateral no definite levels have formed. Total relief is only 450 feet. The differences in lithology has produced zones.

The upper zone forms in the breccia (angular debris formed in ancient sinkholes). No boxwork exists nor does the cave lining. It contains mainly crawlways that are small and mazelike.

Middle zone contains the biggest caverns, well developed boxwork and cave lining. The lowest has not had complete development of the passages, they are narrow and strongly controlled by the joint system and is called the "calcite jungle." It contains dogtooth spar crystals and cave rafts or "ice."

5. Development of boxwork and other features.

The fracture system that developed was filled in with secondary calcite, then, the more soluble limestone, was dissolved leaving the calcite "fins" or boxes. The colors are various shades of yellow, browns, pinks and blues. In some chambers (Fairground) the calcite is fluorescent and glows blue, green, red and violet under black light.

The boxwork has a preferred direction of 70-80° east of north and a second preference of N. 10-20° W. The boxwork even cuts through the geodes that occur in the limestone showing that the geodes occurred first and the fracture systems later.

6. Erosion and formation of topography.

The climate is presently arid today as shown by the intermittent stream (does not flow year round) that flows through Wind Cave Canyon and Red Valley. The Wind Cave Canyon creek may flow through the cave if it has sufficient water as evidence of a stream flowing through the cave can be found (off Garden of Eden).

Beaver Creek is the only permanent stream in the park and cuts across the ridges or hogbacks that are found in the region. The landscape is a combination of steep canyons and hills and flat upland areas. Origin of these flats (e.g., Bison Flats) is uncertain, they could be structural or erosional.

All in all Wind Cave is an extremely old cave, with at least 28 miles (as of late 1973) of passageways that have been mapped, and many other unmapped (but known) ones. Because of stratigraphy the cave can only enlarge along the strike (trend) of the beds, for up and down the dip it intersects either the surface or the watertable. It is conceivable that from 50 to 100 miles of passageways could exist.

It has derived its name because the process of "cave breathing." High air pressure on the surface (good weather) the wind enters the cave. Low air pressure (just before a storm) air moves out, if air pressure is equal no movement. The air flow may change several times a day, so Wind Cave is well named.

GLOSSARY

Ablation—The process of the wasting away of a glacier by melting and evaporation.

Abrasion—The wearing away of a rock by friction.

Aeolian—Transportation of particles by the wind.

Algae—A single celled plant, which may or may not be colonial.

Alluvial Fan—Fan-shaped deposit formed on the floor of a valley when a stream is forced to deposit because of the reduction in velocity.

Alluvial Plain—Formed by the joining together of several alluvial fans to form a continuous sheet.

Alteration—The changing of one material into another by the addition or subtraction of material.

Amethyst—A purple variety of crystalline quartz.

Amygdule—A vesicle in an igneous rock filled in with secondary material.

Angular Unconformity—An unconformity in which the erosional surface separates beds at an angle to one another.

Antecedent Stream—A stream that flows across a structure and developed before the structure formed.

Anticline—The upward folding of beds to form an arch.

Anticlinorium—A series of anticlines and synclines that are overall arched upwards.

Aquifer—A layer of bedrock that transmits water.

Arch—Beds which are curved upwards into an inverted U.

Archipelago—A group of islands located in a body of water.

Arete—A knife-shape ridge that separates two cirques.

Ash flows—A mixture of volcanic ash and gases that flows downhill in response to gravity.

Aureole—The metamorphosed bedrock around an igneous intrusion.

Autolith—An igneous inclusion containing fragments of the magma chamber (that crystallized earlier) which have been incorporated into the later material.

Avalanche Chute—An almost vertical, narrow trough carved by snow cascading down the mountain side.

Badlands—A region of clay formations that erode easily into a maze of low hills and valleys. Configuration is changed after each cloudburst.

Balanced Rock—A rock that rests upon a narrow pedestal of more easily eroded bedrock.

Balds—An open treeless area covered by a thick concentration of grass or shrubs on a mountain side.

Bank Deposits—Shoal deposits or local mounds that form at the rim of a basin in the ocean.

Barred Basin—An isolated basin that would permit the periodic entrance of water.

Barrier Reef—A reef that is separated from the land by a lagoon.

Basal Conglomerate—A conglomerate that was formed from the debris of an eroded surface.

Basement Complex—Highly deformed Precambrian rock upon which the younger sediments rest.

Batholith—A solidified magma chamber exposed by erosion and is more than 40 square miles in area.

Beach—The flat region along the shore line upon which the waves break and deposits material.

Bedding Planes—The surface of separation between two layers of rocks.

Bentonite—A type of clay that absorbs water and expands. It frequently forms badlands.

Bergschrund—The crevasse between the valley wall and head of a glacier. It is formed when the glacier pulls away from the wall.

Blanket Sand—A sandstone deposit that covers a large area but is relatively thin.

Blister—The first stage in the formation of a gypsum rosette on the ceiling of the cave. The gypsum is extruded from the pores of the bedrock and forms a bulge on the wall or ceiling.

Block Faulting—A type of fault that forms a mountain by the gradual uplift of a large block of bedrock.

Blue-green Algae—A very primitive form of algae, some species live in hot springs.

Boxwork—A solution feature formed when the secondary material that filled in fractures forms intersecting fins because of the removal of the original bedrock by solution.

Braided Stream—A type of stream pattern formed by the intertwining of many stream channels.

Breadcrust Bomb—A variety of volcanic bomb characterized by having a cracked crust that resembles the surface of a loaf of bread.

Buttes—A flat-topped hill with relatively vertical sides, capped by resistant material that protects the less resistant rock beneath. Buttes are smaller than mesas.

Calamites—The ancestor of the scouring rushes or horsetails.

Calcareous Tufa—Calcium carbonate material deposited by a hot spring.

Calcite Lining—A thin layer of calcite deposited on the walls of a cave during the underwater stage.

Calcium Carbonate—The chemical composition of the mineral calcite ($CaCO_3$). In large quantities it forms the rock limestone or marble.

Calving—The process of a section of ice breaking off of the end of a glacier and entering a body of water.

Carbonaceous Residue—The remains of fossilized plant material on a rock after all the volatile material has left.

Cave—A solution cavity of large size formed usually in limestone.

Cave Earth—The insoluble material that remains behind on the floor of a cave, plus the sediments brought in by groundwater.

Cave Pearls—Spherical accretionary bodies that form in pools on the floor of a cave. They usually have a nucleus of some type.

Cavitation—The process of wearing away bedrock by the collapse of microscopic vapor bubbles.

Chalcedony—A cryptocrystalline form of quartz (SiO_2), usually translucent.

Chlorastrolite—A green mineral and gemstone with an uncertain origin. It may be a variety of zeolite or prehnite, also called pumpellyite.

Cinder Cone—A type of volcanic cone made up of pyroclastic debris. They rarely exceed heights of 1,000 feet.

Cirque—A bowl-shaped depression formed at the head of a valley which contains a glacier.

Cleavers—A glacial ridge made of faceted angular volcanic rock.

Closed Drainage System—A drainage basin that has no outlet. The water flows into a lake that is not drained by a stream.

Cobble Beach—A beach that contains rocks that are cobble sized, that is between 64 and 4 mm.

Col—A saddle-shaped depression that forms a pass through a mountain. Formed when the headwalls of two adjacent cirques break down.

Collapse Block—A section of the ceiling of a cave that falls to the floor.

Collapse Breccia—A deposit of angular fragments of bedrock formed by the collapse of a structure.

Columnar Jointing—A variety of jointing formed by the shrinkage of the rock. The joints are in the pattern of polygonal columns. Usually forms in the basic igneous rock called basalt.

Column—Columns are formed in caves when a stalactite hanging from the ceiling joins together with a stalagmite growing upward from the floor.

Compaction—The reduction of pore space in a rock or sediment, usually occurs because of the weight of overlying beds.

Compressions—The remains of plant and animal preserved as a thin sheet of carbon upon a rock.

Conchoidal Fracture—A type of fracture in which the mineral or rock breaks with a curved surface.

Concretion—A secondary structural sedimentary feature formed when the cementing material in a rock precipitates around a nucleus (e.g., as a fossil).

Coniferous Plants—Plants that reproduce by cones.

Consequent Dip—The angle of slope produced by a formation along its initial tilt.

Contact Metamorphism—The type of metamorphism produced mainly by heat as the result of contact with a hot igneous intrusion.

Contemporaneous Slump Structure—Slump that occurs in semiconsolidated sediments that are still in the process of forming.

Continental Divide—A ridge along a mountain that acts as the dividing point for the drainage of the rivers as to which ocean their water will eventually flow.

Coquina—A sedimentary rock formed by the cementing together of shell fragments.

Coral Reef—A reef in the ocean formed by coral.

Cross Bedding—Layers of sediments deposited at angles to one another within a single formation. It is produced in an environment where the direction of deposition keeps shifting (e.g., delta or sand dune).

Cryptovolcanic Structure—A structure that appears to have formed by a volcano, however there is no evidence of any igneous activity nearby. Frequently formed by collapse of surface bed when the supporting structure beneath is removed by some agency.

Cuesta—An asymmetrical ridge formed by beds that have a dip of 30° or less.

Cycads—A type of tree that first appeared in the Mesozoic and still exists today. Resembles a huge pineapple with a large fern growing out of its top.

Debris Flow—A variety of mudflow made up of a mixture of different material of various sizes.

Decomposition—This is also called chemical weathering, it is the chemical breakdown of rock by the addition of other materials, such as water, oxygen, carbon dioxide, or by solution of the original rock.

Dendrites—Crystals of iron oxide or manganese oxide along some surface. The branching crystals resemble moss.

Desert Varnish—A coating of iron oxide or manganese oxide on a rock surface in an arid or semiarid climate.

Diastrophism—The process which forms mountains by folding, faulting or other forms of deformation.

Differential Weathering—The weathering of the same rock at different rates because of differences in intensity or composition.

Dikes—Igneous intrusions that cut across the bedding planes of the bedrock.

Diorite—An intermediate type of igneous rock that is made up of half light and half dark minerals.

Disconformity—A variety of unconformity in which an erosional surface separates two parallel beds.

Distillation—The process by which plants are reduced to carbon.

Dogtooth Spar—A variety of calcite crystals, that outwardly look like a dog's tooth.

Dome—A circular or oval anticline or uparching of beds.

Drainage Basin—The area that a stream drains.

Draperies—A travertine or onyx formation formed in a cave that hangs down as a sheet.

Drumlins—Asymmetrical inverted teaspoon shaped hills deposited by a glacier, they are made up of glacial till.

Dunite—An ultrabasic igneous rock that has a composition of 90 percent or more olivine.

Edifice—The actual volcanic cone that surrounds the vent from which the lava is ejected.

Entrenched—A stream, because of sudden change in base level, starts downcutting very rapidly without changing its course. It forms straight walled, steep sided canyons.

Euhedral Crystals—Crystals that are completely formed.

Eugeosyncline—The deeper portion of a geosyncline adjacent to the island arc system which contains mainly lava flows and graywackes.

Exfoliation—The process of the weathering and peeling off in layers of a rock (e.g., like the layers of an onion).

Exfoliation Dome—A dome formed by the process of exfoliation. They usually have a composition of quartzite or granite.

Facies—Deposition of different sediments in different geographic areas at the same time.

Fault—A break in the earth's crust in which there has been movement parallel to the break.

Fault Block Mountain—A mountain formed by the process of block faulting.

Fault Scarp—A cliff formed by faulting.

Fenster—A "window," a break or opening created by erosion through a thrust sheet or recumbent fold.

Fin—A thin narrow ridge formed by erosion.

Fiord—A glacial U-shaped valley that has been drowned by sea water.

Fire Fountain—Extrusion of lava into the air forced out by the action of effervescent gases.

Fissures—A crack or a break in the crust of the earth.

Flood Basalts—Basaltic lava flows that pour out of fissure eruptions in great volume.

Foliation—The alignment of minerals in a metamorphic rock.

Foraminifera—Single-celled animals that have a shell of calcite.

Fossils—Any trace of remains of a plant or animal that lived in the geologic past.

Fractionation—The separation of minerals in a melt into various layers because of specific gravity.

Frost Work—Needlelike crystals of aragonite or calcite on the walls of caves or other cave features.

Fumaroles—An eruption of steam from holes or fissures in a volcanic region.

Geanticline—A large upfolding of the crust of the earth that extends for hundreds of miles.

Geosyncline—A large linear trough that continuously sinks while receiving sediments.

Geothermal—The internal heat of the earth which increases with depth 1° for every 60 to 100 feet.

Geyserite—A deposit of silicon dioxide and water formed by geysers.

Geysers—Hot springs that eject water into the air periodically.

Glacial Erratics—Transported fragments of bedrock carried by glaciers and deposited elsewhere.

Glacial Grooves—Deep gouges in bedrock produced by rocks being dragged along bedrock by a glacier.

Glacial Pavement—Bedrock polished smooth as a glacier passed over it.

Glacial Plucking—The removal of sections of bedrock by freezing to the glacier and being plucked out as it moved.

Glacial Polish—A smooth shiny surface produced on bedrock by the abrasion of a glacier.

Glacial Rebound—The gradual return of compressed bedrock to its original elevation when the weight of glaciers has been removed.

Glacial Rock Basin—The formation of depressions in the bedrock of a glacial valley, produced by glacial erosion.

Glacial Steps—A series of steplike levels produced in a glacial valley by glacial erosion of beds of unequal resistance.

Glacial Striations—Scratches on the surface of the bedrock formed when rocks were dragged over its surface by glaciers.

Glacial Valley—A former V-shaped stream valley that has been modified to a U-shape because of glacial action.

Glacier—A body of recrystallized snow and ice that flows (or has flowed in the past) downslope in response to gravity.

Gneissic Texture—A texture of alternate bands of light and dark colored minerals produced in a rock.

Graben—A type of fault when one block drops down in relation to the two adjacent blocks.

Graded Bedding—A type of bedding in a sedimentary rock where the sediments are sorted according to size.

Graded Stream—A mature stream with a moderate gradient where erosion of sediments essentially equals deposition.

Gradient—The slope of a stream bed or hill, it is usually measured in feet per mile.

Granitization—The production of a rock that appears to be igneous, but formed by severe metamorphism and never crystallized from a liquid state.

Granular Disintegration—A form of physical weathering whereby the expansion and contraction of the various minerals causes them to separate from one another.

Graptolites—An extinct form of life that had a primitive backbone (notochord), lived in colonies and resembled the blade of a coping saw.

Graywacke—A sandstone that is mixed with clay derived from the decomposition of volcanic ash. Frequently called a "dirty sandstone" in contrast to a rock that is made up of only one mineral.

Greenstone—A metamorphosed basalt, usually a lava flow.

Gypsum Curls—A deposit of gypsum on the walls and ceilings of caves that has a rosette shape.

Halite—The mineralogical name for sodium chloride or rock salt.

Hanging Valley—A tributary valley whose floor is a considerable elevation above the floor of the main valley.

Hammocks—An accumulation of plants, trees and soil that form small hills or islands in a swamp.

Headland—A projection of resistant bedrock into the ocean forming a cliff.

Headward Erosion—The eroding back of a stream by the addition of tributaries to its headwaters.

Helectites—A formation in a cave made of travertine that resembles stalactites that defy gravity, and twist and turn in every direction.

Hogback—An almost symmetrical ridge formed by beds that have a dip of 45° or more.

Horn—A pyramid shaped peak of a mountain formed by the intersection of the head of three or more cirques.

Hornfel—A rock that has been baked by contact metamorphism. It is also the highest degree of a metamorphic facies.

Horst—A type of fault where the center block is elevated in relation to the two adjacent blocks.

Hot Spring—A thermal spring with a temperature of 15°F above the average annual temperature.

Hydraulic Action—The pulling and tugging, battering and shattering action of water, usually in the form of waves.

Hydrothermal Action—Reactions produced by the contact of a rock or mineral with hot water.

Hydrothermal Solution—A hot water solution that dissolves and transports minerals.

Ice—The crystal form of water, it will form below 32°F or 0°C.

Ice Cave—A cave produced in a glacier by meltwater streams flowing beneath a glacier or steam vents beneath a glacier.

Ice Margin Deposits—Glacial deposits formed at the edges of glaciers.

Impermeable—A material through which water cannot pass.

Impressions—The imprint of a fossil showing only the outline and external features.

Incompetent Beds—Bedrock that is weak and will deform easily under stress and strain.

Inliers—Segments of older bedrock completely surrounded by younger bedrock which has become isolated by erosion.

Intermittent Stream—A stream that does not contain water the year around.

Isoclinal fold—Any fold in which both limbs or sides are parallel.

Joints—A fracture in the earth's surface in which the movement has been perpendicular to the surface of the break.

Joint Sets—A series of parallel joints.

Joint Systems—A series of two or more intersecting joint sets.

Kaolin—A clay mineral formed by the decomposition of the feldspar minerals.

Keratophyre—An igneous rock usually a lava flow or igneous intrusion that contains albite and/or oligoclase, chlorite, epidote and calcite.

Kettle—A depression formed in glacial outwash when a block of ice became buried and then melted.

Kettle Lake—A lake that has formed in a kettle hole.

Klippe—An erosional remnant of an overthrust sheet that is isolated from the rest of the sheet.

Laccolith—An igneous intrusion with a known floor and a domed top.

Lagoon—The shallow protected area of water either between the land and a reef, or completely enclosed by a circular reef.

Lahars—A mudflow primarily of volcanic dust and ash.

Lath-shaped Crystals—Crystals that are much longer than wide, and have blunt ends.

Lava Cone—A volcanic cone or edifice made up of a series of lava flows. They are also called shield volcanoes.

Lava Stalactites—Stalactites of lava hanging from the ceilings of lava tubes or caves, formed by the splashing and dripping of the lava as the tube drained.

Lava Tube—A tunnel formed beneath the crust of a lava flow. It was created when the semiliquid flow drained out through a fissure in its side.

Law of Cross-cutting Relationships—A law used to determine the age relationship of igneous intrusions. It states that a rock is younger than the rock it cuts across.

Leucogranite—A granite made entirely of light colored minerals.

Lignite—A low grade form of coal, frequently brownish in color and soils the hands when it is touched. It may still have visible plant fragments.

Lily Pads—A pool found on the floor of a cave with scalloped edges.

Lithified—A sediment that has been turned into a sedimentary rock by compaction, crystallization, desiccation or cementation.

Loess—A deposit of buff-colored, silt-sized particles deposited by the wind.

Longshore Currents—A current of water that flows parallel to the shoreline.

Maceration—Process of soaking carbonized plant remains in a rock in hydrofluoric acid, which dissolves the rock and permits the plant to float to the surface.

Magma Chamber—A huge chamber of molten rock formed 35 to 40 miles beneath the crust of the earth.

Marble—Metamorphosed limestone.

Massive—Any rock that does not contain layering.

Mass Wasting—The removal of bedrock on a large scale by the force of gravity.

Mesa—A steep sided flat-topped hill capped with a resistant rock that protects the weaker rock beneath. It is of large areal extent.

Metamorphic Facies—A region of metamorphic rock that is formed within a particular range of temperature and pressure. It is a mappable unit, and contains specific metamorphic zones.

Metamorphic Zone—A metamorphic region that is characterized by the presence of specific metamorphic minerals. It is classified according to composition and is also a mappable unit.

Meteoric—Any form of precipitation that falls from the air such as rain and snow.

Micro-earthquakes—Earthquakes that are so slight that they can only be detected by instruments.

Migmatite—A rock that is alternate layers of thin metamorphosed rock and minute igneous intrusions.

Monadnocks—A resistant erosional remnant that stands above the surrounding erosional surface.

Monocline—A single fold or flexure with the beds above and below the fold parallel.

Moraine—Rock debris deposited in front of, along the sides, between or on top of a glacier. They form moundlike ridges after the glacier melts.

Mountain Glacier—Sometimes they are referred to as valley glaciers since they form within a single valley.

Muck—Soil that contains a very high percentage of decayed organic material and a high water and mud content.

Mud Cracks—Polygonal-shaped cracks formed on flood plain surfaces because of the process of dessication or loss of water from the mud, in other words—shrinkage cracks.

Mud Flow—A mixture of mud and water that flows down hill under the influence of gravity. The consistency ranges from muddy water to a very thick paste.

Mud Pots—Hot springs that have been unable to remove the mud by overflow, thus becoming choked with mud. Bubbling action produced by escaping gases.

Mud Volcanoes—Small cones of mud surrounding a very active mud pot. Formed because the mud is thrown out of the hole.

Native Copper—Copper that forms in bedrock as the element copper. It is not combined with any other element.

Natural Arch—An arch formed by some process of erosion.

Neve—Granular snow or ice formed by the recrystallization of snow.

Nip—A depression formed at the base of a sea cliff by wave action.

Normal Fault—A fault in which the footwall moves up in relation to the hanging wall.

Novaculite—A form of extremely small quartz crystals cemented together by chalcedony to form a flintlike rock.

Nuee Ardenteees—Fiery clouds of superheated steam and droplets of lava, that give the cloud weight so that it will rush down the hillside at speeds as high as 60 miles per hour.

Nullipores—A variety of seaweed that secretes calcium carbonate, sometimes called a coralline seaweed.

Nunataks—Mountain peaks that project above the glacier that has buried them.

Obsidian—A natural volcanic glass.

Outliers—Segments of younger bedrock completely surrounded by older bedrock which has become isolated by erosion.

Outwash Plain—The combination of the sand and gravel outwash deposited by several glacial meltwater streams that meander back and forth in front of a glacier in a braided stream pattern.

Oxbow Lakes—Abandoned stream meanders that are filled with water.

Paint pots—Mudpots in a volcanic area that contain iron oxide and other chemicals which color them.

Pater Noster Lakes—A series of glacial lakes in rock basins connected by a single stream.

Peat—The partially decomposed plant material that accumulates in layers in swamps.

Pediment—A sloping rock plain formed at the base of a mountain in an arid or semiarid climate covered with a thin veneer of gravel.

Pele's Hair—Thin hairlike strands of volcanic glass.

Pele's Tears—Solidified glassy droplets of lava.

Penecontemporaneous Deformation—Deformation of sediments at the time of deposition or shortly afterwards.

Peneplain—A surface of old age that slopes seaward and has a gently undulating surface. Concept not accepted by all geologists.

Perched Pool—A pool, usually in a cave, that exists above the level of the groundwater table.

Perched Water Table—A water table trapped in the zone of aeration (vadose zone) by impermeable beds above the level of the ground water table.

Peridotite—The same as dunite, an igneous rock that contains 90 percent or more olivine.

Phenocrysts—The larger of the two distinct sizes of crystals in an igneous rock.

Phreatic Explosion—The violent conversion of groundwater into steam. They usually have a low temperature and do not eject debris.

Phytosaur—An animal that lived during the Triassic period and in outward appearance resembled an alligator, but was not related.

Piedmont Glacier—Two or more valley glaciers that have joined together in the main glacial valley.

Piercement Structure—Plastic rock is injected into the surrounding bedrock.

Pillow Lava—Lava that is extruded underwater and forms a glassy skin with each surge. They resemble a series of pillows, one stacked upon another.

Pinnacle—A small turret or peak that may or may not support a resistant rock remnant.

Pit—A hole or opening in the ground.

Pit Crater—A volcanic crater that is a pit and has no cone erected around it.

Placer Deposit—An accumulation of weathered out minerals or elements (e.g., gold) with a high specific gravity in stream gravels.

Plug Dome Volcano—A volcano formed by a thick pasty mass of lava forcing its way to the surface and flowing out into a dome-shaped structure. The adjacent bedrock is shattered and accumulates on its sides.

Polyp—An animal that lives inside of cuplike structures such as coral or bryozoans.

Popcorn—A series of layered nodules that form in caves that are supported on a thin stalk.

Pothole—A hole in bedrock on the bottom of a stream formed by the abrasive action of smaller rocks being swirled around like a drill by the stream water.

Primary Structural Features—Features formed in sedimentary rock at the time of deposition or shortly afterwards, such as cross-bedding ripple marks and stratification.

Pumice—Solidified lava expanded by gases, it is so light that it will float on water. It resembles a sponge because of the numerous bubble holes or vesicles.

Pumpellyite—A green mineral sometimes of gemstone quality. See chlorastrolite.

Pyroclastic—Lava that has solidified in mid-air after being ejected from a volcano. It is classified according to size.

Quarrying–Also known as glacial plucking. It occurs when rock is frozen to the glacier and is yanked out when the glacier moves.

Quartz Crystals–Crystals of silicon dioxide that are six-sided and end in a point.

Radiolaria–Single-celled animals that live in marine water and have a SiO_2 skeleton.

Rapids–An area where a stream flows very rapidly over rocks or a series of small ledges with turbulent flow.

Rebound–When a material such as a rock resumes its original thickness after the force that compressed it is removed.

Reef Facies–The reef core that is made up of vuggy limestone or dolomite and lies over the breccia zone.

Reef Talus–The portion of the reef that is broken into angular fragments (breccia) by wave action. It is formed at the base of the reef core.

Regional Joint System–The trend of the joint system in a specific area.

Regional Metamorphism–Metamorphism on a large scale, it may be measured in 10s or 100s of miles. It is usually produced when an area undergoes deformation.

Regression–The retreat of the ocean from the land by the elevation of the land, the lowering of sea level or a combination of both.

Relief–The difference in elevation between the highest and lowest points in an area.

Replacement Pegmatites–Pegmatites formed by hydrothermal solution in metamorphic rocks that have been formed by replacement.

Resistant–Bedrock that does not weather as rapidly as the surrounding bedrock.

Reverse Fault–A fault in which the hanging wall moves up in relation to the foot wall.

Rhyolite–The fine-grained equivalent of the igneous rock granite. The main minerals are quartz, feldspar and mica.

Rhyolite Dome–A dome or upwarped structure with the composition of rhyolite.

Rift Zone–A zone where two sections of the crust of the earth have separated. Frequently a block of the crust will drop in between or lava will extrude out of the opening.

Ring Dikes–A series of igneous dikes that radiate out from a central point. They usually form when a portion of the overlying bedrock collapses into the magma chamber.

Ring Fractures–A series of fractures in bedrock that radiate from a central point. They are usually formed when unsupported rock collapses (e.g., over a magma chamber). They become filled with lava or magma to become ring dikes.

Ripple Marks–A series of miniature symmetrical or asymmetrical ridges formed on the floor of a body of water. They are produced by either wave action or a rapidly flowing current.

Roche Mountonnees–They are often referred to as sheepbacked rocks. They are asymmetrical knobs of bedrock with one smooth gentle side and a steep jagged side (formed by glacial plucking) developed when a glacier passed over the bedrock.

Rock Crystal–Another name for quartz crystals.

Rock Falls–The rapid detachment and falling of large segments of bedrock from a cliff.

Rock Flour–Bedrock that has been pulverized by glacial action into a fine powder.

Rock Glaciers–A glacierlike body of mostly rock with some ice that moves slowly down a valley in a lobate shape. They usually form at the higher elevation where the temperature is so low that water exists only in the form of ice.

Rock Slides–A variety of landslide in which bedrock detaches itself from some plane of separation and moves very rapidly downslope under the influence of gravity.

Sandbar–An accumulation of sand deposited on the inside of the curve of a stream because of the reduction of velocity of the water. Or sandbars may form off shore in an ocean, again because of the reduced velocity of the water.

Sandslide–The cascading of sand downslope.

Sandstone Dikes–Dikes of sandstone created either by sand filling up fractures or welling up from beneath into openings.

Saprolite–A rock that is completely decomposed but still in place.

Sapwood–The living part of the tree, usually a yellow band, as long as this portion of the tree survives it can grow over most injuries.

Scour and Fill–A portion of the stream bed that is first eroded and then filled with material.

Sea Arches–A portion of a headland in the ocean which has a tunnel drilled through it by wave action.

Sea Caves–The enlargement of a joint in a sea cliff by wave action to produce a cave.

Sea Mounts–An underwater or submarine mountain.

Sea Stack–The remnant of a sea arch that has collapsed and has left a pillar of rock standing.

Sea Walls–An accumulation of large blocks of bedrock that became detached during a storm and tossed up onto the beach.

Secondary Minerals–Minerals that form in a rock after the rock has formed.

Sheeting–The breaking up of granite into large sheets (e.g., the process of exfoliation).

Shelf Deposits–Sedimentary deposits formed in shelf areas of the sea. The depth of the water is less than 600 feet deep.

Shield Volcanoes–A volcanic edifice or cone made entirely of a series of lava flows. They have a very gentle slope of just a few degrees. Another name for them is lava cone.

Sial–The *si*licon, *al*uminum layer of the crust of the earth of which the continents are made. Main rock is granite.

Silicified–The replacement of organic material with quartz.

Sill–A tabular igneous intrusion that lies parallel to the bedding planes of the country rock.

Sima–The *si*licon, *ma*gnesium portion of the earth's crust. It underlies all the continents and makes up the ocean floor. The main rock is basalt.

Sinkholes–A collapsed cave that has formed a depression on the surface. They may or may not be filled with water.

Slaty Cleavage—Cleavage in a rock that resembles that of slate, it will split into thin sheets. The direction of cleavage may or may not be parallel to the original bedding planes.

Slicks—They are also called heath balds. They are a treeless area that contains a dense mass that is almost impenetrable of mountain laurel and rhododendron.

Slump—The downward and slightly outward movement of sections of sediments or bedrock as a unit (or may be a series of units). It is a type of fast mass wasting.

Snow—Crystals of ice, they belong to the hexagonal crystal system which explains the development of the six points (three axis in one plane at 60° angles to one another, the fourth axis is perpendicular to the first plane).

Soapstone—A metamorphic rock that used to be the mineral talc.

Soda Straw Stalactites—These are very thin strawlike stalactites that are simply a hollow tube.

Sole Markings—An uneven surface produced by the scouring of turbidity currents as they move downslope.

Solftaras—Fumaroles that emit sulphur dioxide along with the steam.

Solifluction—A type of slow mass wasting that occurs in permafrost regions. Since there is no place for water to escape because of the frozen ground beneath the sediments become saturated and will flow slowly down a slope that is 1 or 2°.

Spatter Cones—Small volcanic cones formed by thick blobs of lava collecting around the base of a fire fountain.

Spatter Rampart—A ridge formed by large blobs of lava that accumulate next to a fissure.

Spelothems—A general term for any feature formed in a cave.

Spheroidal Weathering—A variety of physical weathering in which a block of rock, especially granite and sandstone weathers gradually in a series of concentric spheres down to a central core.

Spindle Bomb—A variety of volcanic bomb that becomes twisted and tapered at both ends.

Spit—A sandbar that is attached at one end to the land.

Spring—An opening in bedrock from which groundwater flows out.

Stagnant—When a body of ice or water no longer moves downslope, but stays in one place.

Stalactites—A pendant-shaped formation that hangs from the ceiling of a cave. Its composition is travertine.

Stalagmites—An inverted pendant-shaped formation that has built up on the floor of a cave that is composed of travertine.

Stream Piracy—The process of one stream stealing the headwaters of a less favorably located stream.

Stream Valley—A V-shaped valley that is carved by a stream.

Strike—The trend of a formation, structure of fault on the surface.

Strike-Slip Fault—A type of fault in which the movement is in the horizontal direction and slips along the strike of the fault.

Stromatolites—A variety of calcareous algae.

Stylolites—Secondary structural features that are dark crenulated lines which represent points of contact where limestone or marble under pressure have gone into solution and have left the impurities behind.

Sublimination—The process of a gas changing directly into a solid or a solid into a gas. The liquid stage is by-passed.

Submarine Slides—Landslide that occur underwater.

Submergent—An area that used to be land but is presently covered with water.

Subsequent Stream—A stream which has had its course altered to flow along a structure, in other words a structurally adjusted stream.

Superheated Water—Water that is heated well above the boiling point but does not boil because of the pressure it is under.

Superimposed Stream—A stream that is gradually let down onto a buried structure that existed before the stream and now cuts across the structure.

Synclines—A trough or downfolding of beds.

Synclinorium—A series of anticlines and synclines that are overall arched downwards. It is a large scale feature that is measured in hundreds of miles.

Talus—The loose rock debris that accumulates at the base of a cliff.

Tarn—A lake that occupies the bottom of a glacial cirque that no longer contains a glacier.

Temporal Transgression—A formation that crosses time boundaries. Deposition of the lower beds began in one age and deposition of upper beds ceased during another age.

Tepee Structure—Chevron type folds produced in sedimentary beds.

Terrigenous—Land derived sediments.

Theca—The cuplike shell made of calcite in which the coral or bryozoan polyp live in.

Tholeiitic Basalt—A basalt that contains very little olivine but has other crystals present that are the result of separation and settling.

Thompsonite—A mineral frequently found as amygdules in basalts.

Thrust Fault—A low angle reverse fault, that may be thrust a distance of several miles.

Transgression—The advance of the sea onto an area that was once land. It is the result of lowering of the land, an increase of sea level or the combination of both.

Travertine—A calcium carbonate deposite formed in caves or by hot springs.

Tree Molds—Lava that is cool enough to become chilled around a tree and preserve its exterior outline.

Tufa—A calcium carbonate deposit or limestone with a spongy appearance deposited by either springs or streams.

Tuffs—Pyroclastic debris of volcanoes consolidated into a rock (mostly dust and ash in size).

Turbidites—Sediments deposited by a turbidity current, they are usually muds but may be coarse graded material.

Turbidity Currents—A current in a body of water that is distinguished by the sediments that are being transported.

Ultrabasic–A rock or material that is made up primarily of ferromagnesium minerals.

Unconformity–A break in the sequence of deposition either because of erosion or nondeposition in the area involved.

Vadose Zone–The zone of aeration in the crust between the groundwater table and the surface. Water is held in the pore spaces against the pull of gravity.

Valley Glacier–A glacier that forms within a single valley, also called a mountain glacier.

Varves–Seasonal deposition of sediments forming a series of parallel bands.

Vent–An opening of some type through which material may pass.

Vesicle–A gas or bubble hole in a volcanic rock.

Volcanic Bombs–Pyroclastic debris thrown from a volcano in a molten state which solidifies before striking the ground. They are larger than 32 millimeters in diameter.

Volcanic Dikes–See dikes.

Volcanic Mudflows–See Lahars.

Volcanic Neck–The congealed lava in a volcanic vent that is exposed after the edifice has been removed by erosion.

Volcanic Plug–A large bulbous igneous intrusion of a pasty mass of lava.

Volcanic Sink–See pit crater.

Water Gap–A valley that cuts across a ridge with a superimposed or antecedent stream flowing through it.

Waterfall–A stream that flows over a ledge or cliff and falls free for a distance.

Wave Cut Bench–A bench produced by wave action at the base of a sea cliff. It was the former floor of a nip before that portion of the sea cliff collapsed.

Wave Cut Cliff–A cliff at the end of a headland carved by wave action.

Weak Beds–Formations that are easily eroded.

Weathering–The breakdown of bedrock by a combination of physical and chemical processes.

Weather Pits–Pits in bedrock produced by weathering because the bedrock is not as resistant to weathering as other portions of the bedrock.

Welded Tuff–Pyroclastic debris that is fused or welded together by heat.

Wind Gaps–An abandoned water gap, usually develops because of stream piracy.

Window–See fenster.

Xenolith–A section of country rock (local bedrock) that was not completely assimilated by the magma chamber, so is completely surrounded by the igneous rock.

Zone of Aeration–See vadose.

Zone of Saturation–See phreatic.

BIBLIOGRAPHY

Albert, Lewis S., Acting Superintendent (July 20, 1974) Lassen Volcanic National Park, Mineral, California; U.S. Dept. of Interior, National Park Service, Personal Communication.

Alden, William C. (1932) . . . Physiography and Glacial Geology of Eastern . . , Washington: U.S. Gov. Print. Off.

———. (1932) Physiography and Glacial Geology of Eastern Montana and Adjacent areas, Washington: U.S. Geological Survey, Professional Paper 174.

Alt, David D. and Hyndman, Donald W. (1973) Rocks, Ice and Water: The Geology of Waterton-Glacier Park, Missoula, Montana: Mountain Press Publishing Co.

———. (1972) Roadside Geology of the Northern Rockies, Missoula, Montana: Mountain Press Publishing Co.

American Automobile Association (1956) Western Tour Book, 1956-57 edition, Washington, D.C.: American Automobile Association.

Anderson, C.A., Camp, C.L., Chaney, R.W., Louderback, G.D., Williams, Howel (1941-42) Bulletin of the Dept. of Geological Sciences, Volume XXVI, Ohio State University: University of California Pub.

Atwood, Wallace W. (1964) The Physiographic Provinces of North America, New York: Blaisdell Publishing Co., 12th Print. 1st Ed., 1940, Ginn and Co.

Baars, Donald L. (1972) Red Rock Country, Garden City, New York: Doubleday/Natural History Press.

Bailey, E.H., Irwin, Wm. R., Jones, D.L. (1964) Franciscan and Related Rocks and Their Significance in The Geology of Western California, California Division of Mines and Geology, Bulletin 183.

Bailey, Edgar H., Editor (1966) Geology of Northern California: California Division of Mines and Geology, Bulletin 190.

Bakkar, Elna and Lillard, Richard G. (1972) The Great Southwest, Palo Alto, California: American West Publishing Co.

Barnett, John (1969) Mount Rainier National Park, Fresno, California: Awani Press.

Bateman, Paul C., Clark, Lorin D., Huber, N. King, Moore, James G., and Rinehart, C. Dean (1963) The Sierra Nevada Batholith: A Synthesis of Recent Work Across The Central Part, Washington: U.S. Geological Survey Professional Paper 414-D.

Behrendt, John C., Tibbetts, Benton L., Bonini, Wm., Lavin, Peter M. (1968) A Geophysical Study in Grand Teton National Park and Vicinity, Teton County Wyoming, Washington: U.S. Geological Survey Professional Paper 516-E.

Bennison, A.P., Renfro, H.B., Feray, Dan E. (1974) Pacific Northwest Region—Washington, Oregon (Idaho in part), Tulsa, Oklahoma: American Asso. of Petroleum Geol. Map 6.

Bibliography of National Parks and Monuments (1941) Western Museum, U.S. National Park Service.

Bibliography of National Parks and Monuments West of Mississippi River (1941) Western Museum, U.S. National Park Service.

Blank, Richard H. Jr. and Christiansen, Robert L. (1972) Volcanic Stratigraphy of The Quaternary Rhyolite Plateau in Yellowstone National Park, Washington: U.S. Geological Survey Professional Paper 729-B.

Books, Kenneth G. (1972) Paleomagnetism of Some Lake Superior Keweenawan Rocks, Washington: U.S. Geological Survey Professional Paper 760.

Brooks (1970) Geography and Geology of Alaska, Cincinnati, Ohio: Geo. A. Flohr Co., U.S. Geological Survey Professional Paper 45.

Brooks, Maurice (1967) The Life of The Mountains, New York: McGraw-Hill Book Co.

Brooks, Myrl (1974) Voyageurs National Park, International Falls, Minnesota: U.S. Dept. of Interior, National Park Service, Personal Communication.

Cambrian Subcommittee, Howell, B.F., Chairman (1944) Correlation of The Cambrian Formations of North America, N.Y.: Bulletin of The Geol. Soc. of America, Vol. 53, pp. 993-1003, Reprinted 1955,57,64.

Carlsbad, Caverns National Park (1967): U.S. Gov. Print. Off., National Park Service, U.S. Dept. of Interior, Pamphlet.

Castolon, Big Bend Natural History Asso.: U.S. Dept. of Interior, National Park Service, Pamphlet, Map.

Cheney, Eric S. and Stewart, Richard J. (1974) Sodic Gneisses in Archean Greenstone Terranes: Subducted Graywacke Melange Belts?, University of Washington, Seattle: Unpublished Thesis.

Christiansen, Paige W. and Kottlowski, Frank E. (1967 2nd Ed.) Mosaic of New Mexico's Scenery, Rocks, and History; Scenic Trips to The Geologic Past No. 8, New Mexico: New Mexico Bureau of Mines and Mineral Resources, 1964, 67.

Christiansen, Robert L. and Blank, H. Richard Jr. (1972) Volcanic Stratigraphy of The Quaternary Rhyolite Plateau in Yellowstone National Park, Geology of Yellowstone National Park, Washington: U.S. Geological Survey Professional Paper 729-B.

Churkin, Michael Jr. (1973) Paleozoic and Precambrian Rocks of Alaska and Their Role in its Structural Evolution, Washington: U.S. Geological Survey Professional Paper 740.

Clark, Earl (1970) Mt. Rainier—The Live Time Bomb in Seattle's Back Yard: Science Digest.

Clark, Thomas H. and Stearn, Colin W. (1968 2nd Ed.) Geological Evolution of North America, New York: The Ronald Press Co.

Clary, David A. (1972) "The Place Where Hell Bubbled Up," Washington: U.S. Gov. Print. Off.

Climate and Seasons of Olympic National Park (1973): U.S. Dept. of Interior, National Park Service, Personal Communication, Mimeographed.

Cobb, Edward H. (1974) Synopsis of The Mineral Resources and Geology of Alaska, Washington: U.S. Geological Survey Bulletin 1307.

Cobban, Wm. A. and Reeside, John B. Jr. (1952) Correlation of The Cretaceous Formations of The Western Interior of The United States: Bulletin of The Geol. Society of America, Vol. 63, pp. 1011-1044, 2 Figs. 1 Pl.

Cohee, Geo. V. and Wright, Wilna B. (1972) Contributions To Stratigraphy Changes in Stratigraphic Nomenclature by The U.S. Geol. Survey, 1971, Washington: U.S. Geological Survey Bulletin 1372-A.

———. (1974) Changes in Stratigraphic Nomenclature of the U.S. Geol. Survey, 1972, Washington: U.S. Geological Survey Bulletin 1394-A.

Cook, Laurence F. (1961) The Giant Sequoias of California, Washington: U.S. Gov. Print. Off.

Cooke, Wythe C., Gardner, Julia, Woodring, Wendell P. (1943) Correlation of The Cenozoic Formations of The Atlantic and Gulf Coastal Plain and The Caribbean Region: Bulletin of The Geol. Society of America, Vol. 54, pp. 1713-1723.

Cooper, G. Arthur, Chairman (1942) Correlation of The Devonian Sedimentary Formation of North America, New York: Bulletin of The Geol. Society of America, Vol. 53, pp. 1729-1794, 1 Pl., 1 Fig., Repr. 1955, 57,58,64.

Cosner, Oliver J. (1972) Water in St. John. U.S. Virgin Islands: National Park Service and Gov. of Virgin Islands, Caribbean District—Open File Report.

Crandell, Dwight R. (1969) The Geologic Story of Mount Rainier, U.S. Geological Survey Bulletin 1292.

Darton, N.H. (1909) Geology and Water Resources of The Northern Portion of the Black Hills . . . , Washington: U.S. Gov. Print. Off.

Deike, Geo. H. III and White, Wm. B. (1961) Preliminary Report on The Geology and Minerology of Wind Cave: National Park Service, Mimeographed.

Denver Mountain Area (1948): U.S. Dept. of Interior, Geol. Survey, Map, Text on back of map.

DeVoto, Richard H. (1972) Paleozoic Stratigraphy and Structural Evolution of Colorado: Quarterly of The Colorado School of Mines, Vol. 67, No. 4.

Diller, J.S. (1889) Geology of Lassen Peak District: U.S. Geological Survey, Annual Report 8, pp. 395-432.

———. and Patton, Horace Bushnell (1902) The Geology and Petrography of Crater Lake Nat. Park, Washington: U.S. Gov. Print. Off., U.S. Geol. Survey.

———. (1895) Geologic Atlas of The United States, Lassen Peak Folio, California, No. 15.

Dole, Hollis M. (1968) Andesite Conference Guidebook, State of Oregon: Dept. of Geol. and Mineral Ind., Bulletin 62.

Dunbar, Carl O., Chairman, (1942) Correlation Charts Prepared By The Committee on Stratigraphy of The National Research Council, New York: Bulletin of The Geol. Soc. of America, Vol. 53, pp. 429-434, Repr. 1956, 58, 64.

———. and Waage, Karl M. (1969) Historical Geol. 3rd Ed., New York: John Wiley and Sons, Inc.

Dutton, Capt. Clarence Edward (1884) Hawaiian Volcanoes: U.S. Geol. Survey, Annual Report, 75-219 pp.

Dyson, James L. (1957) The Geologic Story of Glacier National Park, Kallispell, Montana: Glacier Natural History Asso., Special Bulletin No. 3.

Eardley, A.J. (1962) Structural Geology of North America, 2nd Ed., New York: Harper and Row.

Editors of American Heritage Magazine (1963) The American Heritage Book of Natural Wonders: American Heritage Publishing Co.

Editors of Sunset Books and Sunset Magazine (1965) National Parks of The West, Menlo Park, California: Lane Magazine and Book Co.

———. (1969) California National Parks, Menlo Park, California: Lane Magazines and Books Co.

Fagerlund, Gunnar O. (1954) Olympic National Park, Washington: Washington, D.C.: Natural History Handbook Series No. 1, Dept. of Interior, National Park Service, Rev. 1965.

Fenneman, Nevin M. (1931) Physiography of Western United States, 1st Ed., New York: McGraw-Hill Book Co.

Feray, Dan E., Renfro, H.B. and King, Philip B. (1970?) Texas, Tulsa, Oklahoma: American Assc. of Petroleum Geol., Map 7.

Fisher, Geo. W., Pettijohn, F.J., Reed, J.C. Jr., Weaver, Kenneth N. (1970) studies of Appalachian Geol.; Central and Southern: Interscience Pub., Division of John Wiley and Sons.

Fonda, R.W. and Bliss, L.C. (1969) Forest Vegetation of The Montane and Subalpine Zones, Olympic Mount., Washington: Repr. Ecological Monographs, 39, pp. 271-301, Pamphlet, Personal Communication.

Frebold, Hans (1953) Correlation of The Jurrasic Formations of Canada: Bulletin of The Geological Soc. of America, Vol. 64, pp. 1229-1246.

Frome, Michael (1967) National Park Guide, New York: Rand McNally and Co., Copyright 1968, 69, 70, 71, 72.

Fryxell, Fritiof M., Love, J.D., Reed, John C. Jr., Frison, Dr. Geo., McCurdy, C.H., Loope, Lloyd (1968) Grand Teton National Park: U.S. Geol. Survey, Dept. of Interior, Map.

Geology of The Capitan Reef Complex of The Guadalupe Mountains Culberson County, Texas and Eddy County, New Mexico Field Trip Guidebook, May 6, 7, 8, 9. 1964, Roswell, New Mexico: Roswell Geol. Society.

Geology of The Delaware Basin and Field Trip Guidebook, Sept. 29, 30; Oct. 1, 1960: West Texas Geol. Soc.

Getze, Geo. (1971) Coast Volcanoes: Los Angeles Times, Repr. Fresno State College Alumni Journal, Mimeographed Report.

Glacier National Park (1971): U.S. Gov. Print. Off., U.S. Dept of Interior, National Park Service, Pamp.

Glaciers of Olympic National Park (1973): Dept. of Interior, National Park Service, Mimeographed, Personal Communication.

Glimpses of Our National Monuments (1930) Washington, D.C.: U.S. Gov. Print. Off., U.S. Dept. of Interior, National Park Service.

Great Smoky Mountains National Park (1967): U.S. Gov. Print. Off., Dept. of Interior, National Park Service Pamphlet.

Gregory, Herbert E. (1957) Bryce Canyon National Park: U.S. Dept. of Interior, U.S. Geol. Survey, Text on back of map, Map.

———. and Evans, Richard T. (1957) Zion National Park: U.S. Dept. of Interior, U.S. Geol. Survey, Text on back of map, Map.

Guggisberg, C.A.W. (1970) Man and Wildlife, New York: Arco Publishing Co.

Hansen, Harry, Writer's Program (1970) Colorado: A Guide to The Highest State, New York: Hasting House.

Hartley (1974) Outline of The Geological History of Isle Royale: National Park Service, Mimeographed Report.

Hayes, P.T. (1957) Geologic Quadrangle Maps of U.S., Geology of Carlsbad Caverns E. Quadrangle, New Mexico: Washington: Dept. of Interior, National Park Service Map.

———. and Koogle, R.L. (1958) Geologic Quadrangle Maps of The U.S., Geology of Carlsbad Caverns W. Quadrangle—New Mexico, Texas, Washington: Dept. of Interior, Geol. Survey, Map. Personnel Comm.

Haynes, Jack E. (1912) Yellowstone National Park, Haynes Official Guide, 26th Ed., St. Paul, Minnesota: The Pioneer Co.

Hess, H.H., Bowin, Carl O., Donnelly, Thomas, Whetten, John T., Oxburgh, E.R. (1966) Geology of St. Thomas and St. John, U.S. Virgin Islands, New York: The Geol. Soc. of America, Memoir 98.

Hietanen, Anna (1968) Belt series in the region around Snow Peak and Mallard Peak, Idaho, Washington: U.S. Gov. Print. Off.

Howell, J.V. (1957) Glossary of Geology and Related Sciences, Washington, D.C.: The American Geol. Inst.

Huber, N. King (1973) Glacial and Postglacial Geological History of Isle Royale National Park, Michigan, Washington: Geological Survey Professional Paper 754-A.

———. (1973) The Portage Lake Volcanics (Middle Keweenawan) on Isle Royale, Michigan, Geol. of Isle Royale Nat. Park, Michigan, Washington: U.S. Geol. Survey Professional Paper 754-C.

Huntington, Ellsworth (1928) The Secret of The Big Trees Yosemite, Sequoia, General Grant Nat. Parks, Washington: U.S. Gov. Print. Off., U.S. Dept. of Int.

Hurlbut, Cornelius (1952) Dana's Manual of Minerology, 16th Ed., New York: John Wiley and Sons, 3rd Print. 1955.

Iddings, Joseph Paxson (1888) Obsidian Cliff, Yellowstone National Park: U.S. Geol. Survey, Annual Report 7, pp. 249-295.

———. (1891) The Eruptive Rocks of Electric Peak and Sepulchre Mountain, Yellowstone Nat. Park: U.S. Geol. Survey, Annual Rp. 12 pt. 1 pp. 569-664.

Imlay, Ralph W. (1944) Correlation of The Cretaceous Formations of The Greater Antilles, Central America and Mexico: Bulletin of The Geol. Soc. of America, Vol. 55, pp. 1005-1045.

———. (1952) Correlation of The Jurassic Formations of North America, Exclusive of Canada: Bulletin of The Geol. Soc. of America, Vol. 63, pp. 953-992.

———. and Reeside, John B. Jr. (1954) Correlation of The Cretaceous Formations of Greenland and Alaska: Bulletin of The Geological Soc. of America, Vol. 65, pp. 223-246.

Irwin, Wm. P. (1960) Geologic Reconnaissance of The Northern Coast Ranges and Klamath Mountains, California: California Division of Mines, Bulletin 179.

Jenny Lake Nature Trail (1971) Moose, Wyoming: Grand Teton Natural History Assc., Dept. of Interior, National Park Service, Pamphlet.

Jensen, Paul (1964) National Parks, A Guide to The National Parks and Monuments of The U.S., New York: Golden Press.

Kaune, Robert W. Jr., Park Naturalist (1973) Rain Forests of Olympic National Park: U.S. Dept. of Interior, National Park Service, Mimeo., Per. Comm.

Keefer, Wm. R. (1971) The Geologic Story of Yellowstone National Park: U.S. Geol. Survey Bulletin 1347, 2nd Ed. 1972.

Kenney, Nathaniel T. (1974) The Other Yosemite, Washington, D.C.: National Geographic Society, National Geographic Magazine, Vol. 145, No. 6, pp. 762-780.

Khudoley, K.M. and Meyerhoff, A.A. (1971) Paleogeography and Geological History of Greater Antilles, Memoir 129, Boulder, Colorado: The Geol. Soc. of America, Inc.

King, Philip B., Neuman, Robert B. and Hadley, Jarvis B. (1968) Geology of The Great Smoky Mountains Nat. Park, Tennessee and North Carolina, Washington: Geological Survey Professional Paper 587.

———. and Stupka, Arthur (1961) The Great Smoky Mountains: U.S. Dept. of Interior, U.S. Geological Survey Map, Text on back of map.

Kistler, Ronald (1974) Hetch Hetchy Reservoir Quadrangle, Yosemite National Park, California—Analytic Data, Washington: U.S. Geological Survey Professional Paper 774-B.

Kleinpell, Robert M. (1938) Miocene Stratigraphy of California. Tulsa, Oklahoma: The American Association of Petroleum Geologists.

Kummel, Berhart (1961) History of the Earth, An Intro-
duction to Historical Geology. San Francisco: Freeman,
W.H. & Co.

Langford, J.A., Gipson, Fred (1952) Big Bend A Home-
steader's Story. Austin, University of Texas Press.

Lee, Willis T. (1917) The Geologic Story of Rocky Mountain National Park Colorado. Washington: Gov. Print.
Off.

Livesay, Ann, Revised by Preston McGrain (1962) Geology
of the Mammoth Cave National Park Area. Kentucky
Geological Survey, Special Publication 7.

Loomis, B.F. (1966) Pictorial History of the Lassen Volcano. Published in cooperation with the National Park
Service and Loomis Museum Association.

Marshall, Kay, and Colbert, Edwin (1965) Stratigraphy and
Life History. New York: John Wiley & Sons, Inc.

Matthes, Francois E. (1937) The Geologic History of
Mount Whitney. Sierra Club Bulletin, v. 22, no. 1 p.
1-18.

———. (1955) The Mount Rainier National Park. U.S. Dept.
of Interior, U.S. Geological Survey, Text on back of
map, Map.

———. (1965) Glacial Reconnaissance of Sequoia National
Park, California, Washington: U.S.G.S. Professional
Paper 504-A.

———. (1950) Sequoia National Park A geological album.
Berkeley; University of California Press.

———. (1958) The Story of the Yosemite Valley. Dept. of
Interior, Geological Survey, Text on back of map, Map.

Matthews, William H. III (1968) A Guide to the National
Parks Volume I: The Western Parks. New York: The
Natural History Press.

———. (1973) A Guide to the National Parks. Garden City,
New York: Doubleday/Natural History Press.

McKee, Bates (1972) Cascadia The Geologic Evolution of
the Pacific Northwest. New York: McGraw-Hill Book
Co.

McLearn, F.H. (1953) Correlation of the Triassic Formations of Canada. Bulletin of the Geological Society of
America, v. 64, p. 1205-1228.

Mech, David L. (1966) The Wolves of Isle Royale. Washington: Gov. Print. Off., U.S. Fauna Series 7.

Melbo, Irving Robert (1950) Our Country's National Parks,
Volume 1. New York: The Bobbs-Merrill Company, Inc.

———. (1950) Our Country's National Parks, Volume II.
New York: The Bobbs-Merrill Company, Inc.

Melson, William G. (1970) America's Sleeping Volcanoes
Part I. Science Digest.

Mendenhall, Walter C. (1902) Geology of the Central Copper River Region, Alaska. U.S. Geological Survey, Geological Survey Professional Paper 41.

Meyer, Frederick A., and Boswell, Brenda. The Redwood
State Parks. Sacramento: Dept. of Parks and Recreation
State of California.

Mississippian Subcommittee, Weller, J. Marvin, Chairman
(1963) Correlation of the Mississippian Formations of
North America. Bulletin of the Geological Society of
America, v. 59, p. 91-196.

Murie, Adolph (1971) The Wolves of Mount McKinley.
Washington: Gov. Print. Off., Fauna Series No. 5.

National Geographic Society (1966) America's Wonder-
lands, New Enlarged Edition. Washington: National
Geographic Society.

National Park Service (1973) National Parks and Landmarks. Washington: U.S. Gov. Print. Off.

Newman, William L. (1972) Geologic Time. U.S. Gov.
Print. Off., Pamphlet.

Oakeshott, Gordon B. (1971) California's Changing Land-
scapes. New York: McGraw-Hill, Inc.

Oething, Philip, Feray, Dan E., and Renfro, H.B. (1966)
Geological Highway Map No 1—Mid-Continent Region.
Tulsa, Oklahoma: The American Association of Petro-
leum Geologists.

———. (1967) Geological Highway Map No. 2—Southern
Rocky Mountain Region. Tulsa, Oklahoma: The American Association of Petroleum Geologists.

———. (1968) Geological Highway Map No. 3—Pacific
Southwest Region. Tulsa, Oklahoma: The American
Association of Petroleum Geologists.

Oil and Gas Investigations. Dept. of Interior, U.S. Geological Survey Map OM-203.

Ordovician Subcommittee of the Committee on Stratigraphy of the National Research Council. Twehhofel,
W.H., Chairman (1954) Correlation of the Ordovician
Formations of North America, New York: Bulletin of
the Geological Society of America, v. 65, p. 247-298.

The Origin of Crater Lake, National Park Service, Mimeographed.

Pennsylvanian Subcommittee of the National Research
Council Committee on Stratigraphy, Moore, R.C., Chair-
man (1944) Correlation of Pennsylvanian Formations of
North America. New York: Bulletin of the Geological
Society of America, v. 55 p. 657-706.

Permian Subcommittee of National Research Council's
Committee on Stratigraphy, Dunbar, Carl O., Chairman
(1960) Correlation of the Permian Formations of North
America. Bulletin of the Geological Society of America,
v. 71, p. 1763-1806.

Petitti, Jamie S., Acting Chief Park Interpreter, Platt National Park (1974) Sulphur, Oklahoma: U.S. Dept. of
Interior, National Park Service, Personal Communication.

Plummer, Fred G. (1899) Mount Rainier Forest Reserve,
Washington: U.S. Geological Survey 21st Annual Report, Part V.

Popenal, W.P., Imlay, R.W., Murphy, M.A. (1960) Correlation of the Cretaceous Formations of the Pacific Coast
(United States and Northwestern Mexico). Bulletin of
the Geological Society of America, v. 71, p. 1491-1540.

Pough, Frederick H. (1953) A Field Guide to Rocks and
Minerals. Cambridge: Riverside Press.

Precambrian Symposium (1966) The Relationship of Mineralization to Precambrian Stratigraphy in certain Mining
Areas of Ontario and Quebec. Toronto: Business and
Economic Service Ltd., Special Paper No. 3, The Geological Association of Canada.

Reed, John C. Jr., Hack, John T. Resume of the Geology of
the Blue Ridge. U.S. Dept. of Interior National Park
Service, Mimeographed. (date unknown)

Renfro, H.B., and Feray, Dan E. Geological Highway Map No. 4–Mid-Atlantic Region. Tulsa, Oklahoma: American Association of Petroleum Geologists, 1970.

———. (1970) Geological Highway Map No. 5–Northern Rocky Mountain Region. Tulsa, Oklahoma: American Association of Petroleum Geologists.

A Report on the Everglades National Park in Florida–Geology and Soils. Dept. of Int. National Park Service, Mimeographed.

Rice, C.M. (1955) Dictionary of Geological Terms. New York: Edwards Brothers, Inc.

Richmond, Gerald M. (1974) Raising the Roof of the Rockies. Rocky Mountain Nature Association Inc.

Roberts, Albert E. (1972) Cretaceous and Early Tertiary Depositional and Tectonic History of the Livingston Area, Southwestern Montana. Washington: Geological Survey Professional Paper 526-C.

Rocky Mountain National Park (1969) U.S. Gov. Print. Off., Pamphlet.

Rodgers, John (1970) The Tectonics of the Appalchians. New York: John Wiley and Sons, Inc.

Rohn, Arthur H. (1971) Wetherill Mesa Excavations, Mesa Verde National Park. Washington: Gov. Print. Off.

Rowell, Galen (1974) Climbing Half Dome the Hard Way. National Geographic Magazine v. 145 No. 6 p. 782-791.

Ruhle, George (1972) Roads and Trails of Waterton-Glacier National Parks. Minneapolis: John W. Forney.

Ruppel, Edward T. (1972) Geology of Pre-Tertiary Rocks in the Northern Part of Yellowstone National Park, Wyoming. Geological Survey Professional Paper 729-A.

Russell, Richard J. (1955) Guides to Southeastern Geology. New York: Geological Society of America.

Ruth, Kent (1957) Oklahoma: A Guide to the Sooner State. Norman: University of Oklahoma Press.

Scenic Wonders of America (1973) Pleasantville, New York: Reader's Digest Association, Inc.

Schaffner, E. Ray, Chief Park Naturalist, Shenandoah National Park (1974) Luray, Virginia: U.S. Dept. of Interior, National Park Service, Personal Communication.

Schnable, Jon E. and Goodell, H. Grant (1968) Pleistocene–Recent Stratigraphy Evolution, and Development of the Apalachicola Coast, Florida, Boulder, Colorado: Geological Society of America, Inc., Special Paper No. 112.

Shaler, Nathanial Southgate (1889) The Geology of the Island of Mount Desert Maine. U.S. Geological Survey Annual Report 8, p. 987-1061.

Sigafoos, Robert S., and Hendricks, E.L. (1972) Recent Activity of Glaciers of Mount Rainier, Washington. Washington: Geological Survey Profession Paper 387-B. 387-B.

Simpson, Charles Torrey (1920) In Lower Florida Wilds. New York: The Knickerbocker Press.

Sims, Richard H., Superintendent, Crater Lake National Park (1974) National Park Service, Personal Communication.

Smedes, Harry W., and Prostka, Harold (1972) Stratigraphic Framework of the Absaroka Volcanic Supergroup in the Yellowstone National Park Region. Geological Survey Professional Paper 729-C.

Smith, George Otis (1903) . . . Contributions to the Geology of Washington. Washington: Gov. Print. Off.

———, and Russell, Israel Cook (1898) Glaciers of Mount Rainier. U.S. Geological Survey Annual Report 18 Part 2 p. 349-415.

Stearns, Harold T. (1966) Geology of the State of Hawaii. Palo Alto, California: Pacific Books.

Stephenson, Lloyd W., King, Phillip B., Monroe, Watson H., Imlay, Ralph W. (1942) Correlation of the outcropping Cretaceous Formations of the Atlantic and Gulf Coastal Plain and Trans-Pecos Texas. Bulletin of the Geological Society of America, v. 53, p. 435-448, 1 Pl.

Stewart, Richard J. (1970) Petrology, Metamorphism and Structural Relations of Graywackes in the Western Olympic Peninsula, Washington, Unpublished Dissertation.

———. (1971) Structural Framework on the Western Olympic Peninsula, Washington. Geological Society of America, Abs. with Programs 3(2):201.

———. Zeolite Facies Metamorphism of Sandstones in the Western Olympic Peninsula, Washington. Seattle: University of Washington, Unpublished thesis.

Stirling, Matthew W. (1955) Indians of the Americas. Washington: National Geographic Society.

Stupka, Arthur (1960) Great Smoky Mountains National Park. Washington: Gov. Print. Off., Natural History Handbook Series No. 5.

Subsurface Geology of Wind Cave. Dept. of Interior, National Park Service, Mimeographed.

Sunset Books Editorial Staff (1969) Redwood Country and the Big Trees of the Sierra. Menlo Park, California: Lane Books.

Surface Geology of Wind Cave National Park. Dept. of Interior, National Park Service, Mimeographed.

Swartz, Charles K. (1942) Correlation of the Silurian Formations of North America. New York: Geological Society of America, Bulletin, v. 53, p. 533-538 1 Pl.

Tabor, Roland W., Branch of Western Environmental Geology Geological Survey (1974) Menlo Park, California, Personal Communication.

———. (1972) Age of the Olympic Metamorphism Washington: K-Ar Dating of Low-Grade Metamorphic Rocks. Menlo Park, California Geological Society of America Bulletin, v. 83, p. 1805-1816.

———., Cody, W.M., and Yeats, Robert S. (1970) Broken Formations and Thrust Faulting in the Northeastern Olympic Mountains, Washington. Geological Society of America, Abs. of Programs.

———. Sample Locations and Regional Metamorphic Zonation Shown by Isochronic Lines Fig. 10, Olympic National Park, Mimeographed, Map.

———. (1974) Structure of the Core Rocks of the Olympic Mountains, Washington: A Case for Subduction? Expanded abstract from Northwest Scientific Association, Annual Meeting, Vancouver, May 10, 1974.

Tectonic History and Mineral Deposits of the Western Cordillera—Symposium (1966) Geology Division of Canadian Institute of Mining and Metallurgy, Special Volume 8.

Tilden, Freeman (1951) The National Parks. Revised and Enlarged Edition, 1970. New York: Alfred A. Knopf, Inc.

Triassic Subcommittee, Reeside, John B., Chairman (1957) Correlation of the Triassic Formations of North America exclusive of Canada. Bulletin of the Geological Society of America, v. 68, p. 1451-1514.

Udall, Stewart L. (1966) The National Parks of America. New York: G.P. Putnam & Sons.

University of Minnesota (1973) Resources Basic Inventory—Primary Development Areas, Voyageurs National Park, Minnesota Part I: Ecosystems Analysis. University of Minnesota.

University of Minnesota (1973) Resources Basic Inventory—Primary Development Areas, Voyageurs National Park, Minnesota Part II: Annotated Bibliography of Basic Literature. University of Minnesota.

Van Cleave, Philip F., Staff Interpretive and Environmental Specialist, Carlsbad Caverns & Guadalupe Mountains National Parks, (1974) National Park Service, Personal Communication.

Vitaliano, Dorothy B. (1973) Legends of the Earth. Bloomington: Indiana University Press.

Warren, Henry C. (1971) Lassen's Geology, National Park Service, Mimeographed Report.

Western Cenozoic Subcommittee, Weaver, Charles E., Chairman (1944) Correlation of the Marine Cenozoic Formations of Western North America. Bulletin of the Geological Society of America, v. 55, p. 569-598.

White, Walter S. (1972) The Base of the Keweenawan, Michigan and Wisconsin. U.S. Geological Survey Bulletin, 1354-5.

Williams, Howel (1972) Crater Lake—The Story of its Origin. Los Angeles: University of California Press.

——. (1956) Crater Lake National Park, Oregon. U.S. Geological Survey, Dept. of Interior, Text on Back of Map, Map.

Winter, T.C., Cotter, R.D., and Young, H.L. (1973) Petrography and Stratigraphy of Glacial Drift, Mesabi-Vermilion Iron Range Area, Northeastern Minnesota. Washington: U.S. Geological Survey Bulletin 1331-C.

Wolff, Roger G., and Huber, N. King (1973) The Copper Harbor Conglomerate (Middle Keweenawan) on Isle Royale, Michigan, and its Regional Implication. U.S. Geological Survey Professional Paper 754-B.

Wright, Thomas L. (1971) Chemistry of Kilauea and Mauna Loa Lava in Space and Time. U.S. Geological Survey Professional Paper 735.

Writers' Program, Arizona (1956) Arizona: The Grand Canyon State. New York: Hastings House.

——., Arkansas (1941) Arkansas. New York: Hastings House.

——., California (1939) California: A Guide to the Golden State. Wilmington, Delaware: Scholarly Publishing.

——., Florida (1947) Florida. New York: Oxford University Press.

——., Kentucky (1954 revised) Kentucky: A Guide to the Bluegrass State. New York: Hastings House.

——., Montana (1939) Montana: A State Guide Book. Teaneck, New Jersey: Somerset Publishing.

——., New Mexico (1962) New Mexico: A Guide to the Colorful State. New York: Hastings House. 1962.

——., Oregon (1941) Oregon: End of the Trail. Wilmington, Delaware: Somerset Publishing.

——., South Dakota (1952) South Dakota: A Guide to the State. New York: Hastings House.

——., Texas (1969 Revised) Texas: A Guide to the Lone Star State. New York: Hastings House.

——., Washington (1941) Washington: A Guide to the Evergreen State. Wilmington, Delaware: Somerset Publishing.

——., Wyoming (1940) Wyoming: Guide to Its History, Highways and People. Fair Lawn, New Jersey: Oxford University Press.

Wylie, W.W. (1882) Yellowstone National Park, Great American Wonderland. Kansas City, Mo.: Ramsey, Millet & Hudson.

You asked About the Geology of Big Bend. National Park Service, Mimeographed.

APPENDICES

APPENDIX A
TYPES OF FOLDS

1. Warp

Over several miles beds are very gently folded but appear to be horizontal if you are just standing and looking at them.

2. Monocline

A single flexure, beds above and below—folds are parallel.

3. Homocline

All the beds dip in the same direction.

4. Anticline

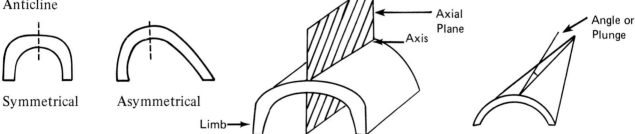

Symmetrical   Asymmetrical

Beds arch upwards, sides are called limbs, highest (or lowest) point *axis,* imaginary plane which bisects fold *axial plane.*

5. Syncline

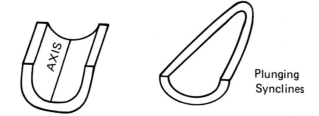

Plunging
Synclines

Beds arch downwards, axis is at *lowest* point of fold.

6. Isoclinal Fold

Can be in any position, but both limbs are *parallel*.

7. Overturned Fold

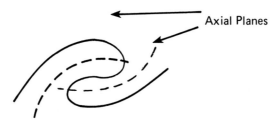

Axial Planes

Both limbs point in the same direction (usually) the axial plane is inclined from the horizontal.

8. Recumbent Fold

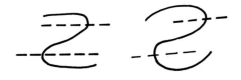

Axial plane is horizontal.

9. Dome

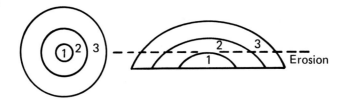

Erosion

Circular anticline, oldest beds found in middle of eroded dome.

10. Basin

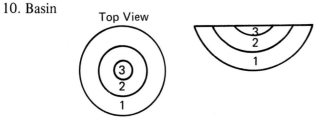

Top View

Circular syncline, oldest beds found on the edge, *youngest* in the middle.

11. Anticlinorium

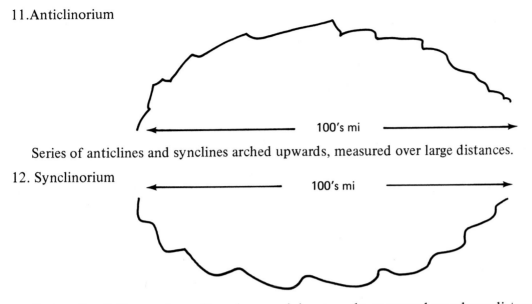

100's mi

Series of anticlines and synclines arched upwards, measured over large distances.

12. Synclinorium

100's mi

Series of anticlines and synclines depressed downwards, measured over large distances.

TYPES OF FAULTS

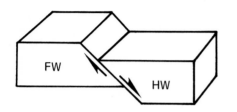

Normal Fault

Footwall moves up in relation to the hanging wall.

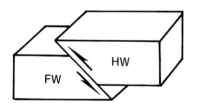

Reverse Fault

Footwall moves down in relation to the hanging wall.

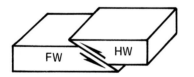

Thrust Fault (Low angle reverse fault)

Hanging wall moves up in relation to the footwall angle up to 10''.

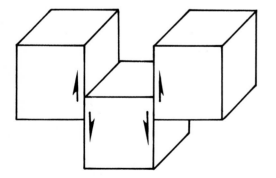

Graben Fault

Center block moves down in relation to two adjacent blocks.

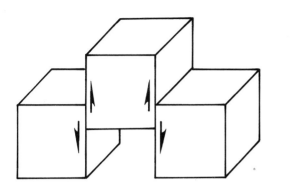

Horst Fault

Center block moves up in relation to two adjacent blocks.

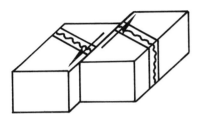

Strike-Slip Fault

Movement in a horizontal direction.

APPENDIX C

IGNEOUS ROCK IDENTIFICATION

Composition and Color	Acidic (light)		Intermediate (medium)				Basic (dark)	Ultra Basic (dark)
Feldspar present	Orthoclase		Orthoclase and Na-Plagioclase		Na-Ca to Na-Plagioclase	Na-Ca to Ca-Na-Plagioclase	Ca Plagioclase	Some Plagioclase
Accessory Mineral	Muscovite and Biotite		Biotite		Hornblende	Augite	Augite Olivine	Olivine
Quartz	More than 5%	Less than 5%	More than 5%	Less than 5%	More than 5%	Less than 5%	Less than 5%	none
Phaneritic Nonporphyritic	Granite	Syenite	Quartz Monzonite or Granodiorite	Monzonite	Quartz Diorite	Diorite	Gabbro	Peridotite, Pyroxenite, Hornblende Anorthesite
Phaneritic Porphyritic	Granite Porphyry	Syenite Porphyry	Quartz Monzonite Porphyry or Granodiorite Porphyry	Monzonite Porphyry	Quartz Diorite	Diorite Porphyry	Gabbro Porphyry	
Aphanitic Nonporphyritic	Rhyolite (Felsite)	Trachyte (Felsite)	Latite (Flesite)	Latite (Felsite)	Dacite (Felsite)	Andesite (Felsite)	Basalt	
Aphanitic Porphyritic	Rhyolite Porphyry	Trachyte Porphyry	Latite Porphyry	Latite Porphyry	Dacite Porphyry	Andesite Porphyry	Basalt Porphyry	
Glassy Nonporphyritic	Obsidian, Pitchstone, Perlite							
Glassy Porphyritic	Obsidian Porphyry, Pitchstone Porphyry,							
Vesicular	Pumice, Scoria							
Amygdaloidal	Amygdaloidal Basalt							
Fragmental	Tuff, Breccia, (Agglomerate)							
Pegmatite	Simple (same minerals as granite) including graphic granite / Complex (same minerals as granite plus rare minerals) / Ferro-Magnesium (made up of ferro-magnesium minerals)							

Note: Any shade of red, no matter how dark is considered a light colored mineral. Any shade of green no matter how light is considered a dark mineral.

APPENDIX D

CLASSIFICATION OF SEDIMENTARY ROCKS

Clastic Sedimentary Rocks		
2mm+	Rounded fragments cemented together	Conglomerate
	Angular fragments cemented together	Breccia
2mm to 1/16mm	Sand sized graims (mainly quartz)	Sandstone
	Sand sized quartz plus 25%+ feldspar	Arkose
	Sand sized quartz, plus clay and mica	Graywacke
1/16mm to 1/256mm	Silt-sized particles, has gritty feel	Siltstone
	Silt-sized, buff color, conchoidal fracture	Loess
Less than 1/256mm	Clay-sized particles with no bedding planes	Mudstone
	Clay-sized particles with thin bedding planes	Shale

Nonclastic Sedimentary Rocks		
$CaCo_3$	Soft crystalline material	Limestone
	Extremely fine-grained limestone	Lithographic Limestone
	Resembles fish eggs cemented together	Oolitic Limestone
	Shell fragments cemented together	Coquina
	Shells of microscopic plants and animals	Chalk
	Deposited in caves	Travertine
	Translucent banded travertine	Cave onyx
	Deposited by springs (hot or cold)	Tufa
	Limestone mixed with mud	Marl
SiO_2	Dark, dense, hard material, conchoidal fracture	Flint
	Light, dense, hard material, conchoidal fracture	Chert
	Minute quartz crystals cemented with chalcedony	Novaculite
	Skeletons of single celled plants (diatoms)	Diatomite
$(Mg, Ca) CO_3$	Massive accumulations of dolomite	Dolostone
$CaSO_4 \cdot nH_2O$	Massive accumulations of gypsum	Gyprock
$NaCl$	Massive accumulations of halite	Rocksalt
C	Varieties are lignite, sub-bituminous and bituminous	Coal

APPENDIX E

CLASSIFICATION OF METAMORPHIC ROCKS

Foliated	Very thin bedded. Will have a ring to it when tapped	Slate
	Wavy, satiny surface, minute mica flakes	Phyllite
	Scaly appearance with overlapping grains	Schist
	Alternate light and dark bands of minerals	Gneiss
Nonfoliated	Resembles conglomerate but breaks across grains	Conglomerite
	Resembles sandstone but breaks across grains	Quartzite
	Soft, coarsely crystalline material	Marble
	Contains mostly hornblende	Amphibolite
	Clay rock that has been baked	Hornfel
	May be fibrous, extremely soft, green	Serpentinite
	Slippery feel, gray or green, extremely soft	Soapstone
	Dense, conchoidal fracture, greenish cast	Greenstone
	Shiny black, conchoidal fracture, lightweight	Anthracite coal

INDEX